Physiological Acoustics

BY

ERNEST GLEN WEVER

AND

MERLE LAWRENCE

Physiological Acoustics

PRINCETON NEW JERSEY

PRINCETON UNIVERSITY PRESS

1954

L. C. Card 53-6389

ISBN 0-691-08018-6

Third Printing, 1971

Printed in the United States of America

by Princeton University Press, Princeton, New Jersey

TO JULIUS LEMPERT

Preface

HEARING consists of a complex series of events beginning with the application of physical vibrations to the ear and ending with a perception of sound. We are concerned here with the initial portion of this series, in which the ear's operations are simply acoustic: in which sound vibrations impinge upon the exterior part of the ear, are transmitted inward to the sensory cells of the cochlea, and produce a mechanical stimulation of these cells. The further processes in the sensory cells and in the elements of the auditory nervous system are electrophysiological in nature and lie beyond the boundaries of physiological acoustics. It must be said, however, that though these further processes are not the immediate matter of our discussion they are ever present in the background to stimulate and guide our thinking about the primary problems of sound reception.

As indicated here, our field embraces only auditory reception, and we have not included any discussion of the non-auditory effects of sonic or ultrasonic vibrations, such as the disruptive and thermal effects that may be observed in isolated cells and small organisms.

The treatment begins, in the first three chapters, with a consideration of the anatomy of the ear, the physical nature of sound, and the methods that have been developed for the study of auditory processes. The discussion in these early chapters is largely on an elementary level and serves to introduce many of the terms and principles that will be employed later on. Also the presentation here of the chief methods of investigation simplifies the later description of specific experiments. This introductory part of the book leads up to a consideration of the conditions underlying the ear's sensitivity, and thus brings into focus the problem of the basic function of the middle ear mechanism.

This functional problem and the manner of its solution then come in for searching study, and lead to a formulation of the principles governing the operation of the middle ear as a mechanical transformer. We thus confront the problem of how energy in the form of aerial waves is effectively utilized by the ear.

Following this examination of the general problem of the ear's handling of vibratory energy we take up more specifically the efficiency and fidelity of the process. The question of distortion in the ear is given detailed attention because the opinion has long prevailed that the middle ear is the seat of aural distortion and there has been much reluctance to accept the contrary view, to which our evidence leads, that this structure carries out its task with great fidelity and that it is the sensory cells of the inner ear that introduce distortion into the sounds that we hear.

A major problem, and one given the most particular consideration, is how sounds enter the cochlea and how they act upon its internal structures in producing their patterns of sensory stimulation. After a discussion of the trends of present thinking on this problem we present what we regard as the simplest and most satisfactory theory of this process.

Two chapters are devoted to derangements of the conductive mechanism and the forms of deafness that result. Special attention is given to otosclerosis, one of the most frequent causes of conductive deafness and of singular interest because its symptoms have finally yielded to otologic treatment.

In the final chapters we bring together many of the facts and principles developed earlier and seek to evaluate them further and to show the present status of our knowledge in this field.

Near the end of the book is a glossary of abbreviations and symbols, and thereafter are a number of appendixes containing physical, mathematical, and anatomical data that are commonly useful in connection with auditory problems. There is also a list of references containing the sources specifically mentioned in the text.

The discussion of the various topics is systematic: the existing facts and the more credible opinions are reviewed critically, the points of agreement and conflict are indicated, and our interpretations and evaluations are given. Many of the results that are presented here represent our own investigations carried out over a number of years. Some of the more recent experiments were designed expressly to answer questions raised in the writing of this book, and several of these experiments are reported here for the first time. Chief among these new experiments are the following: (1) an experimental test of the drum-lever hypothesis (page 87), (2) the location of the rotational axis of the cat's ossicular chain (page 101), (3) the lever action of the ossicles (page 105),

(4) further experiments on the locus of distortion in the ear (page 163), (5) the electrical conductivity of the cochlea (page 282), and (6) a further study of the patterns of response in the cochlea as shown by electrical potentials (page 310). In addition, new data are given on the practical limits of drum membrane displacement (pages 89, 149), the constancy of the ossicular lever ratio (page 153), the maximum tension of the cat's stapedius muscle (page 193), and the lengths of the fluid pathways of the cochlea (page 286). Also, further results are presented (page 271) on our experimental test of the traveling wave theories, an investigation also stimulated by the preparation of this book but which has already been reported in current journals. In connection with all these experiments we gratefully acknowledge the assistance given by a contract with the Office of Naval Research and also the assistance of Higgins funds allotted to Princeton University.

Our thanks are offered to many friends who have stimulated and aided us in this work. Because the actions of the ear are often made clear and meaningful only in a contemplation of its aberrations and failures, we feel fortunate in having received over many years the benefits of a close association with Dr. Julius Lempert, Dr. Philip E. Meltzer, and other members of the Lempert Institute of Otology, and through both discussion and experiment in their company of having come face to face with many of the clinical problems of hearing. We are indebted to Dr. Stacy R. Guild of the Johns Hopkins Otological Research Laboratory for supplying us with photographs of the drawing of the cochlea shown as Plate 3 and of the human stapedius muscle shown as Plate 7, and to both Dr. Lempert and Dr. Dorothy Wolff of the Lempert Institute of Otology for making available to us the sections of the tensor tympani muscle shown in Plates 5 and 6. Dr. Lempert also loaned us the original drawings for Plates 8 and 9. The frontispiece is from "Three Unpublished Drawings of the Anatomy of the Human Ear" by the late Max Brödel, reproduced with the permission of the publishers, the W. B. Saunders Company. And finally we express our appreciation to Suzanne Wever, who carried out the exacting task of typing the final copy of the manuscript.

E. G. W. and M. L.

February 1953

Contents

List of Plates

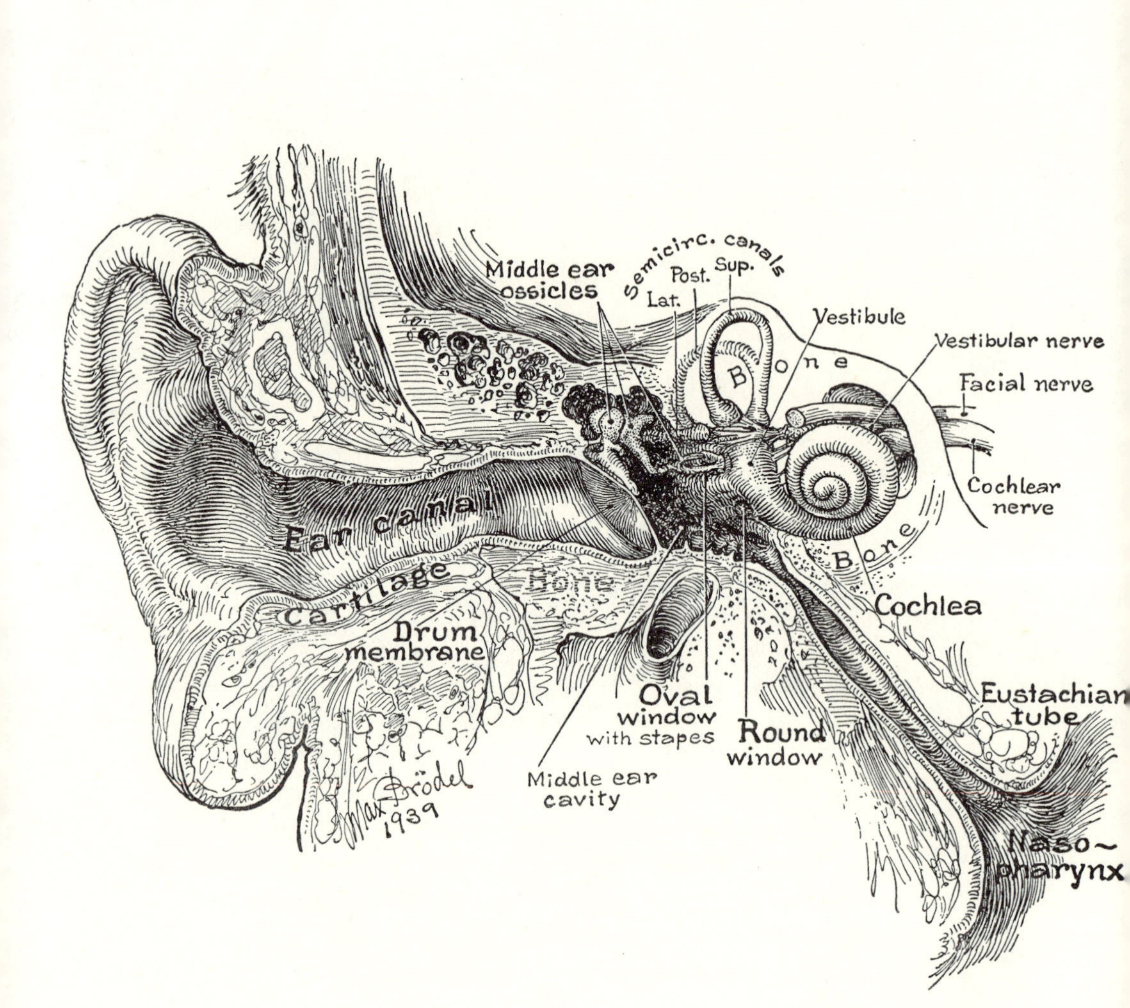

Middle ear ossicles
Semicirc. canals
Post.
Sup.
Lat.
Vestibule
Vestibular nerve
Facial nerve
Cochlear nerve
Bone
Ear canal
Cartilage
Drum membrane
Oval window with stapes
Round window
Middle ear cavity
Cochlea
Eustachian tube
Naso-pharynx
Max Brödel 1939

1

The Ear and Its Structure

HEARING is the result of two kinds of processes, one mechanical and the other physiological. Sounds outside the body strike the external ear and are conducted through the peripheral auditory apparatus to the deep-lying sensory cells, and so far their actions are purely mechanical. In these cells, however, an important transformation takes place. A new action arises that is not mechanical, but as far as we can ascertain is electrochemical. Electrochemical likewise are all the further processes by which the sensory changes, or rather other changes that are representative of them, are relayed onward through the complicated pathways of the nervous system.

In this book we are concerned with the first part of this chain of events, the mechanical part, which constitutes the principal field of physiological acoustics. In the chapters to follow we shall study the ear as a piece of acoustical apparatus and evaluate its service in bringing the vibratory energy of the outside world into play with the final receptor cells of the cochlea.

HISTORICAL ORIENTATION

To gain perspective on our field of study we begin with a historical survey. The early Greeks, like Empedocles in the fifth century B.C., conceived of sounds much as we do today as vibratory movements propagated through the air, and they were aware of the fact—as indeed anyone may discover for himself—that hearing is the result of the passage of these vibrations into the ear.

Beyond this elementary fact, however, Empedocles and his followers had little understanding of the hearing process. They suffered from two serious handicaps. Primarily, their knowledge of the ear's anatomy was woefully limited. They knew about the drum membrane and the tympanic cavity beyond it, but the other, deep-lying structures escaped their notice. Therefore it was natural for them to regard the tympanic cavity as the seat of hearing.

Secondly, in their consideration of the hearing process the ancient Greeks were further handicapped by an idea that then dominated all sensory theory and in fact pervaded all scientific thinking of the time. This idea, known as the principle of resemblances (by which all causes of phenomena are sought in similarities of substance and form) is found almost universally among primitive peoples as the basis of incantations and magic. It forms the basis of homeopathic medicine as the belief that a disease symptom may be made to vanish by doing something that by itself produces that same symptom. For example, a fever is treated by applying heat and frostbite by rubbing with snow. This magic principle as applied to sensation led to the assertion that "like is perceived by like," and as applied to hearing it meant that for anyone to perceive the vibrations of the air outside the body it was necessary for him to have within the ear another quantity of air of a special sort, the "implanted air." This air, supposedly introduced into the tympanic cavity at the birth of the individual and retained there throughout life, was considered as resounding in some way to the motions of the air outside and thereby giving an impression of sound.

Galen,* toward the end of the Greek period (about A.D. 175), took a step forward by recognizing the importance of nerve excitation in the sensory processes in general, and he knew about the auditory nerve, having seen its bundle of fibers passing out of the internal auditory meatus to the brain. Thus he shifted the seat of hearing from the tympanic cavity inward to what he conceived as an expansion of the auditory nerve and called the "neural membrane." However, he had no knowledge of the inner ear itself and, what is more remarkable, he seems to have had no acquaintance with the drum membrane either, but this membrane may well have been absent in the old, poorly preserved skulls that he must have used for study. The "neural membrane" he located at the end of an oblique and tortuous passageway that evidently comprised

* For Galen's works, in Latin translation, see the list of references at the end of the book.

In general, this list is to be consulted whenever an author's name is mentioned. When more than one title is listed after an author's name, each is given a number, and a particular title is referred to in the text by adding its number, italicized and within parentheses, after the name. Sometimes an author's name is mentioned without a number even though the listing will be found to contain more than one title; in such a case all the titles are pertinent to the point under discussion.

the external auditory meatus and its expansion into the middle ear cavity. He still retained the old belief in the implanted air and only transferred it to a more remote locus; lying deep in the passageway it now served as an intermediary agency, the means by which the "neural membrane" was stimulated.

DISCOVERY OF THE CONDUCTIVE MECHANISM

No progress was made in the theory of the action of sound on the ear until the sixteenth century, when the great anatomists of that age, in the course of their comprehensive scrutiny of the human body, brought to light most of the heretofore hidden parts of the ear. By the middle of that century the essential features of the conduction apparatus were well recognized. The two larger ossicles were discovered first, but by whom is not known. Berengario da Carpi in 1514 mentioned them briefly, and later, in 1543, Vesalius described them in detail and gave them their present names of malleus and incus. Then Ingrassia (1546) discovered the third ossicle, the stapes, and the two windows of the cochlea. Fallopius (1561) carefully described the ossicles and their articulations, and also distinguished the two principal divisions of the inner ear and gave them their present names of cochlea and labyrinth.

Eustachius (1564) described the tensor tympani muscle and the tube connecting the tympanic cavity with the pharynx, now known by his name. Both of these structures had been seen earlier without any clear comprehension of their functions. The second tympanic muscle, the stapedius, was first accurately described by Varolius (1591).

After the principal parts of the middle ear had been identified it became possible for Coiter in 1566 to present the first systematic account of the transmission of sound by the ear. In his book, *De auditus instrumento*, which has the distinction of being the first treatise dealing exclusively with the ear, he traced the path of vibrations from their entrance into the external auditory meatus through the drum membrane and the auditory ossicles to the cochlea and the labyrinth. He believed also in an alternative path, by way of the air of the tympanic cavity, which he thought was excited by the movements of the malleus. In this connection he argued against the implanted-air hypothesis, pointing out that the opening of the Eustachian tube to the pharynx made it impossible for the tympanic contents to have any distinctive characteristics. In the cochlea and labyrinth, which he considered to be filled with air, the sounds became amplified "as in a musical instrument" and their movements affected the terminal twigs of the auditory nerve.

During the century following Coiter's systematic review some slight progress continued in the discovery of details of the ear's anatomy and in the understanding of its functions, but the next event of importance was the publication of DuVerney's book, *Traité de l'organe de l'ouie*, which first appeared in 1683.

DUVERNEY'S OBSERVATIONS

Joseph Guichard DuVerney was a physician of prominence in Paris, a professor of anatomy and surgery, and a medical counselor to the king. Along with these attainments, his study of the ear stamps him as a scientist of uncommon skill and discernment. This work deserves our most careful consideration, for not only is it representative of the most advanced thought of its time but it is the point of reference for much that came afterward. The problems that he raised stimulated many of the discussions and investigations of the years to follow, and certain of them are still of active concern to the student of hearing.

DuVerney began his treatment with a careful description of the anatomy of the various parts of the ear and then went on to discuss the particular functions of these parts. He was a skilled dissector, and he displayed many of the ear's anatomical features with more exactness and clarity than anyone had done before. Also he was keenly interested in how the apparatus worked and tried to explain the operations in detail. For the most part his explanations are reasonable, and sometimes they are substantiated with well-conceived tests. The anatomical mistakes that he made are understandable in view of the limitations of his observational method, for he did not have the benefit of the microscope or of means for fixing and staining tissues; these valuable technical aids were still far in the future. His functional errors, which are sometimes serious from a modern point of view, are of course to be blamed upon the primitive state of acoustic science at that time.

DuVerney regarded the external ear as a "natural trumpet"

whose purpose was to collect and amplify sound waves and to convey them to the delicate parts within. He asserted that persons who have lost the auricle do not hear very well and have to resort to the use of the palm of the hand to remedy their defect. The obliquity of the external auditory meatus, he correctly pointed out, serves to protect the drum membrane against external injury and to keep out dirt and insects. This form also, he thought, serves in a fashion like the convoluted form of the concha to augment the intensity of sounds by repeated reflections.

The drum membrane is the proper receiver of sounds, he said, but is not absolutely necessary, for persons without this membrane are able to hear by applying the teeth to a vibrating instrument.

The drum membrane is tuned, DuVerney thought, by means of muscles. He mistook the external ligament of the malleus for a muscle, and so he described two muscles attached to this ossicle and serving to regulate the tension of the drum membrane. This tension is increased and relaxed according to the particular kinds of sounds to be heard; indeed, he asserted that it would be impossible for the drum membrane to take up vibrations without suitable tuning to them. He supported this position by some experiments with two lutes, and showed that on plucking a string of one lute a sympathetic vibration was set up in the other only if one of its strings was tuned to the same note or to a note in harmonic relation. He concluded that the drum membrane must be tensed for high tones and relaxed for low tones.

DuVerney sought to trace the vibrations inward to the ultimate receptors. The movements of the drum membrane are communicated directly to the malleus, then to the incus, and finally to the stapes, and thereby pass to the petrous bone and the labyrinth. Also, he admitted, it is possible to conceive that sounds pass inward by an alternative route, by way of the tympanic air, but it was his conviction that this aerotympanic pathway was much inferior to the other. Again he referred to his experiments with the two lutes, and in this connection he found that a communication of vibrations from one instrument to the other occurred readily when both were resting on a table top, but only feebly when this solid connection was broken by raising one into the air. He thus strongly favored the ossicular over the aerotympanic route of conduction. It is of further interest that his explanation of conduction from the teeth involved the ossicular route also: he traced the vibrations through the jaw bones to the temporal bones and thence through the ossicles to the inner ear.

DuVerney's further discussion is marred by the erroneous belief, which was general in his time, that the spaces of the inner ear were filled with air. This air he regarded as the implanted air of the ancients, which Perrault had located here when the old place in the tympanic cavity had been made to seem degraded by the presence of the Eustachian opening to the pharynx. DuVerney supposed that the movements of the stapes were communicated first to this air in the vestibule and finally to the air of the cochlea and semicircular canals.

DuVerney, like his immediate predecessors, regarded the bony spiral lamina as the final responsive structure of the cochlea, for, he said, it has the proper mechanical properties: it is hard, dry, thin, and brittle—the attributes well known in musical instruments as fitting them for vibration. Unlike the others, he knew about the basilar membrane, but he regarded it simply as a delicate means of supporting the bony spiral lamina on its outermost edge. Its connection with the cochlear wall he saw as dividing the spiral canal into two scalae, and as he did not know of the helicotrema he regarded this division as complete.

The stapedial movements he conceived as entering the upper or vestibular scala and acting upon the spiral lamina from above. At the same time, in line with his somewhat grudging admission of the aerotympanic route, he supposed that the vibrations along this path entered the round window and beat upon the spiral lamina from below. Such double action, he thought, ought to give particularly vigorous stimulation.

DuVerney mentioned that some students had doubted that the semicircular canals served as a primary receptor for sounds, and according to them had only the accessory function of reinforcing the entering sounds by resonance. However, he took the position that this part of the labyrinth was a true acoustic receptor. His reasons were plausible: the canals have a suitable structure, they are served by the same nerve, and certain animals, such as birds, have no cochlea yet they hear with these other organs.

Despite his inclusion of the semicircular canals as a part of the hearing organ, DuVerney placed the main emphasis upon the cochlear apparatus, and he went on to describe the tapered form of the spiral lamina and to develop his theory that by reason of its varying width this structure is differentially tuned to the range of audible frequencies. Because this bony lamina is wider at the basal end of the cochlea and narrower toward the apex, he located the low tones in the base and the high tones in the apex.

FURTHER DEVELOPMENTS OF AUDITORY THEORY

Two further developments were necessary to bring this conception of sound conduction in the ear close to its modern form. The first was the discovery that the cochlear spaces are filled with fluid. Several persons had noticed that fluid was present in these spaces, but Cotugno in 1760 was the first to maintain that only fluid is present: that it fills the whole space. Meckel provided the conclusive evidence for this view by an ingenious experiment. He exposed some fresh specimens of human temporal bones out of doors in freezing weather and observed that the labyrinth when opened was filled with ice. This proof of the pervasiveness of the labyrinthine fluid sounded the end of the implanted-air hypothesis by crowding it out of its last reposing place.

The next and most important development was the identification of the true auditory receptors. DuVerney had considered the spiral lamina as ideally suited to be a resonator and he thought that the auditory nerve fibers were expanded upon its surface. But two influences led away from this position and toward the acceptance of a soft structure. One was a classical bias in favor of the "neural membrane" that Galen had spoken of, with a little of the magic of the old principle of resemblances to reinforce it; thus Valsalva argued that a soft tissue ought to be served by a "soft" part of the auditory nerve. This argument becomes intelligible when we recall that classically the sensory and motor nerves had been designated as soft and hard respectively. Valsalva thus was referring to the sensory—i.e., auditory—division of the auditory-facial bundle that was then regarded as one nerve. He characterized these soft sensory tissues in the labyrinth as a whole as the "zona cochleae," which is his name for the membranous spiral lamina.

The second influence was derived from the growing interest in the process of resonance and the search for resonating structures in the ear. DuVerney had indicated only a broad kind of selectivity for the spiral lamina, but as time went on this conception grew more and more specific. Haller in 1751, in reporting what is evidently the interpretation of DuVerney's theory that was then current, spoke of the spiral lamina as made up of a series of strings of varying lengths.

Cotugno brought these two influences together. He accepted from Valsalva the idea of a membrane as the receptor and conceived this membrane as made up of a series of vibrating strings. This conception of the basilar membrane as a resonator and as the final receptor organ met with general acceptance during the latter part of the eighteenth century. Further progress had to await the discovery of finer details of cochlear anatomy.

The anatomists had achieved this much in the understanding of the ear by the methods of gross dissection and by viewing with the unaided eye or with simple lenses. Then about 1830 the compound microscope, which had long been known in principle, was made into a practical instrument through improvements in the art of lens-making. Thereby a new world of objects was brought into view, and soon the ear claimed its share of searching study.

With the aid of this new instrument, Huschke in 1835 made out several new features of the cochlear structure. He saw the limbus and its outward extension as the vestibular lip, and distinguished these parts from the bony spiral lamina, whose outer edge he now called the tympanic lip. Reissner discovered a new membrane, now known by his name, which runs all along the cochlea and divides the space on the vestibular side of the basilar membrane into two parts, the scala vestibuli and the cochlear duct. The cochlear duct contains endolymph, a fluid entirely separate from the perilymph of the vestibular and tympanic scalae.

The most important event of this period was Corti's discovery, in 1851, of the complex sensory structure lying on the basilar membrane. He saw the tectorial membrane, the hair cells, and the "rods" of Corti, and made out in considerable detail the relations of these new parts and others that had already been recognized.

Many investigators took part in the elaboration of further details, and by the end of the century the picture was well filled in. Retzius and Held made outstanding contributions, especially to our knowledge of the forms and positions of the hair cells and their innervation by the auditory nerve fibers. By this work the hair cells were finally identified as the ultimate receptor elements of the auditory apparatus.

THE ANATOMY OF THE EAR

Out of the many centuries of inquiry reviewed in the foregoing pages comes our modern conception of the anatomy of the ear, which will now be sketched as a foundation for the functional treatment to follow.

The three major divisions of the ear—outer ear, middle ear, and inner ear—are represented in the frontispiece.

The outer ear consists of the visible parts, an expanded flap called the auricle or pinna and a short, curved tube, the external auditory meatus. The meatus is about 2.5 cm in length and leads inward to the tympanic membrane or drum membrane, which forms the external boundary of the middle ear.

The middle ear apparatus lies in a cavity of complex form in the outer, mastoid portion of the temporal bone. This cavity is filled with air, conveyed to it through a tube—the Eustachian tube—that leads to the pharynx. The middle ear cavity has three divisions, the tympanum, the epitympanum, and the mastoid antrum. The tympanum is the main part of the cavity that lies immediately behind the drum membrane. It is a greatly flattened cavity, extending only a little way inward from that membrane, but is expanded in the other directions to give it a cross-sectional form approaching a square. Above it and extending backward and laterally is the epitympanum. Extending still farther backward and laterally is the mastoid antrum, from which are connections with the numerous and irregularly placed mastoid air cells.

The ossicular conductive mechanism lies partly in the tympanum and partly in the epitympanum. This mechanism consists of the tympanic membrane and a chain of three small bones, the auditory ossicles.

The tympanic membrane, roughly oval in outline, is placed obliquely in the meatus as the frontispiece shows. Its edges are firmly held in a little groove in a bony ring formed by the walls of the meatus, except in a region at the upper border where this ring is incomplete. This region is known as the notch of Rivinus. Attached to the membrane along a radius running from this notch to the center of the membrane is the manubrium or handle of the malleus, which draws the center (umbo) of the membrane inward and gives it the form of a flat cone. The main portion or pars tensa of the membrane is stiffly held, whereas a triangular portion or pars flaccida, bounded superiorly by the notch of Rivinus, is relatively lax.

According to measurements made by Tröltsch and Bezold and summarized by Siebenmann (2), the mean diameter of the tympanic membrane along the line of the manubrium is 9.2 mm and the diameter perpendicular to this line is 8.5 mm. The inward deflection of the umbo is 2 mm. A calculation from these dimensions gives a mean area of the membrane of 69 sq mm. Schwalbe gave a similar figure of 69.5 sq mm for this area.

A line extended through the manubrium of the malleus divides the tympanic membrane into anterior and posterior halves and another line at right angles to this one and through the umbo gives a division into quadrants.

The first ossicle, the malleus, is attached to the drum membrane, as just described. The second ossicle, the incus, extends from the malleus to the third ossicle, the stapes. The stapes has its footplate implanted in the oval window, and is firmly held there by an annular ligament. The individual forms and features of these ossicles are shown in Fig. 1.

The three ossicles are joined by means of articular ligaments and the system is suspended in the middle ear by eight other ligaments. Two of these ligaments lead to muscles, the tensor tympani and stapedius muscles, whose contractions alter the rigidity of the system and have other functions discussed in Chapter 10. The remaining ligaments are the superior, anterior, and lateral ligaments of the malleus, the posterior ligament of the incus, the ligament between malleus and drum membrane, and the annular ligament of the stapes.

The superior ligament of the malleus (as seen in Plate 1) extends from the head of this ossicle to the roof of the epitympanum. It is somewhat variable in different ears and often is described as only a strand of mucous membrane. The anterior ligament of the malleus runs from the anterolateral portion of the neck of the malleus and the anterior process of the malleus to the petrotympanic fissure, which lies in an anterior direction in the fossa that lodges the condyle of the mandible. The lateral ligament of the malleus runs from the lateral portion of the neck of the malleus to the edges of the notch of Rivinus. The posterior ligament of the incus anchors the short process of this ossicle in its fossa in the

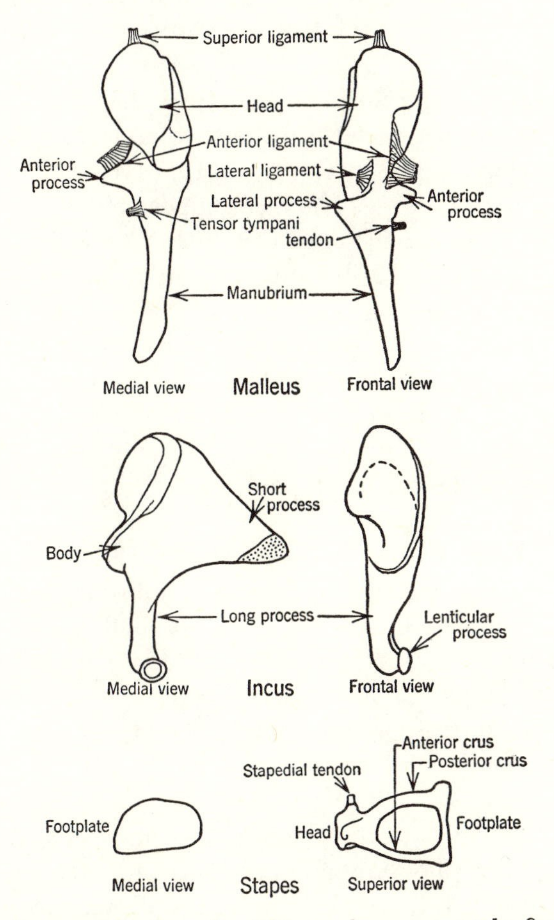

Fig. 1. The auditory ossicles from the right ear, shown separately, five times natural size.

posterior wall of the epitympanum (Plate 2). The actions of these ligaments in determining the axes of motion of the ossicles will be considered later on, in Chapter 7.

The inner ear lies in the bony labyrinth, a space of complex form in the deeper, petrous part of the temporal bone. There are three divisions of the labyrinth: the cochlea, vestibule, and semicircular canals, as shown in the frontispiece. The cavities of these parts contain a fluid, the perilymph, in which is suspended a complex system of membranous ducts and sacs, the membranous labyrinth. The cavities of the membranous labyrinth contain a separate fluid, the endolymph. Only the cochlea is concerned with hearing; the other parts of the labyrinth serve the function of bodily orientation.

The cochlea is a spiral-shaped cavity that is divided longitudinally into three tubes, as seen in the sectional drawing of Plate 3. The divisions are formed by two membranes, Reissner's membrane and the basilar membrane. The three tubes are the scala vestibuli, the scala tympani, and the cochlear duct. The scala vestibuli and scala tympani contain perilymph and communicate with one another through an opening, the helicotrema, located at the apical end of the cochlea. At the basal end of the cochlea the scala vestibuli communicates with the vestibule, in the outer wall of which the oval window is placed. As already mentioned, this window is closed by the footplate of the stapes. A corresponding opening in the bony wall of the scala tympani, the round window, is closed simply by a thin membrane.

The cochlear duct runs between the vestibular and tympanic scalae. Within it, lying on the basilar membrane, is the organ of Corti, a complex structure containing the final receptor cells, the hair cells. These cells are served by the dendrites of nerve fibers that originate in ganglion cells lying in clumps a little distance toward the axis of the cochlear spiral. The axon processes of these cells run farther inward and then follow the cochlear axis as the cochlear nerve.

That the hair cells are the essential receptor cells, and all other sensory structures play only an accessory role in conveying vibrations to these cells, is shown by three lines of evidence: (1) as just mentioned, these are the cells to which the dendrites of the cochlear nerve fibers are applied; (2) these cells are absent in certain congenitally deaf animals, like albinotic cats and a strain of Dalmatian dogs, while other aural structures are not very seriously affected; (3) these cells may be selectively damaged and caused to atrophy by overstimulation with sounds.

We can see reason, apart from the imperatives of evolution, in the deep and protected location of the cochlear hair cells. To develop a sensitivity to the most delicate vibrations, they have to be shielded from accidental injury and from disturbance by the ordinary mechanical processes of the body. To this end they ar floated in fluid and encased in a capsule formed by the petrous bone, the densest bone in the body. They have no immediate blood supply: no blood vessels are to be found around them or in any part of the organ of Corti proper. They derive their nourishment from the endolymph, the fluid with which they are bathed, and this fluid in turn is enriched by the capillary bed of the stria vascularis at the outer border of the cochlear duct. By this arrangement the hair cells are protected against circulatory pulsations that would form an intolerable background of disturbance. As will be brought out later, a further condition contributing to the isolation of the sensory cells is their particular mode of stimulation through mass displacements of the fluid in which they lie. It has therefore been possible for the evolutionary process to carry these cells to a remarkable level of vibratory sensitivity.

2

The Nature of Sound

As we have seen, the early students of the ear were handicapped because of a meager knowledge about sound and its principles of action. The science of acoustics reached a high level of development only in the latter part of the nineteenth century. That development has continued, with notable acceleration during the decades just past. We now consider some of the principles of this science as a basis for our dealing with the special acoustic device which is the ear.

Sound, as everyone knows, is a vibratory motion. Many physical objects, when energy is imparted to them, will give up this energy in the form of back-and-forth movements. Thus a tuning fork when struck a sharp blow, or a violin string when bowed, or a bottle when we blow a stream of air across its mouth, all have the special physical properties that cause them to convert the simple applied energy to a new form, the vibratory form, that constitutes a sound, and when of suitable frequency and intensity stimulates our ears.

The physical properties that all sound-producing bodies must have are two in number: inertia and elasticity. (Actually they always have also a third property, dissipation, as we shall see.)

Inertia is one of the characteristics of mass, and all physical bodies possess it. It is the property by which a body resists all change of motion: the body resists being set into motion when at rest and it resists being slowed down or changed in direction when once in motion.

Elasticity is the tendency of a body to keep its original size and shape. We often speak of this property of an object as "springiness" because a coiled spring has it to a marked extent. If we pull on the spring and expand it and then let it go it returns to its original shape, and if we push on it and compress it the same is true. Even liquids and gases have this property in some degree.

Now let us consider how these two properties operate so as to produce a vibratory motion, as for example in a tuning fork as shown in Fig. 2. If we strike a prong of the tuning fork, the force of the blow pushes the prong out of its original position, moving it to one side as pictured in part *b* of this figure. This displacement is limited in extent by the force of the blow and by the elastic resistance imposed by the prong. This elastic resistance or stiffness inheres in the tendency of the displaced and distorted molecules of the fork to regain their former positions. The energy imparted by the blow is now stored in these distorted molecules, mostly in

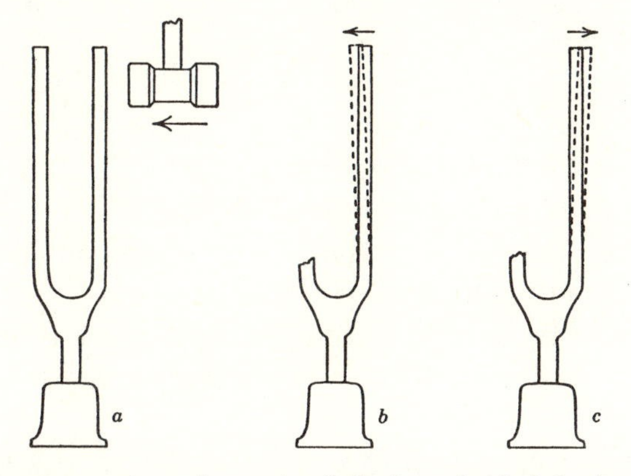

Fig. 2. The vibratory motions of a tuning fork. In *a* the fork is about to be struck. In *b* the right-hand prong has moved to the left, to a position indicated by the broken outline. Somewhat later this prong moves to the right of the equilibrium position, as shown by the broken outline in *c*. The left-hand prong, not shown in the last two views, executes symmetrical movements.

the molecules making up the basal part of the fork. This energy is responsible for all the action that follows.

The elastic force now acts upon the displaced prong so as to move it back toward its equilibrium position. This force is a maximum at the position of greatest displacement and it diminishes as the prong moves toward the equilibrium position, becoming zero when this position is reached.

However, the prong does not come to rest at the equilibrium position. In its movement from its lateral position it has been picking up velocity and it swiftly moves past the equilibrium position. It does so because of inertia. The energy that was stored in the form of molecular distortions (which we call potential energy) has been extracted and is now stored as molecules in motion (which we call kinetic energy).

The inertia causes the motion to continue to the right (as Fig. 2*c* shows), but the displacement of the prong in this direction encounters the same elastic resistance that we found for the original leftward displacement caused by the blow. Just as in that instance, the displacement continues until the motional energy is used up in overcoming the elastic resistance. The prong reaches an extreme right-hand position, at which point the kinetic energy has disappeared and we have potential energy once more, now in the same form as before except that the molecular distortions are the reverse of what they were.

If the tuning fork were in a vacuum and if there were no internal friction in the movements of the molecules, this cyclic transformation of potential energy into kinetic energy and back again would continue indefinitely. The kinetic energy at its maximum would equal the potential energy at its maximum, and at any point in the transformation cycle the sum of these two forms of energy would be the same. Actually, under practical conditions, a little energy is dissipated at every transformation and the fork ultimately comes to rest.

When the fork is vibrating in the air the energy is dissipated in two ways. There is first a frictional loss, caused by the rubbing of molecules of the fork on one another and also by contacts between the prong and air particles in its path. Through this friction a part of the energy in the fork is converted into heat and is of no further use.

The second form of energy dissipation is the useful one: this is a dissipation in the form of sound. As the prong of the fork moves from left to right it pushes the air particles before it and produces a condensation of these particles. Then as the prong moves back from right to left it has the effect of separating the particles in the same region, and produces a rarefaction. These alternate condensations and rarefactions constitute aerial sound.

MECHANICAL IMPEDANCE

A vibrating body, like the tuning fork in our example, gives only a limited movement in response to a force. The limitation of the movement reflects what we call the mechanical impedance of the body, and we can state as a general principle that the amplitude of displacement is directly proportional to the applied force and inversely proportional to the mechanical impedance. With a little amplification this last statement might serve as a formal definition of mechanical impedance. We can also define the impedance in terms of the velocity of motion, as the force required to produce one unit of velocity, or, in symbols, $Z = F/u$. Then, if the force is in dynes and the velocity is in centimeters per second, the impedance is in mechanical ohms.*

The mechanical impedance is contributed to by all three of the basic properties of vibrating bodies, namely, the inertia, elasticity, and dissipation already referred to. These contributions are indicated by the special terms of mass reactance, elastic reactance, and frictional resistance.

Mass reactance is the effect of inertia in opposing every change that the applied force tends to make in the velocity of motion. Its value in this opposition naturally increases with the mass of the vibrating body. Its value also increases as a function of the frequency of the vibration, because as the frequency rises the changes of velocity are more rapid. Or, more strictly, the mass reactance is proportional to what is called the angular frequency, which is the frequency multiplied by 2π. Therefore the mass reactance (in mechanical ohms) is equal to the mass (in grams) multiplied by the angular frequency (which is 2π times the frequency f), or, in symbols, $X_m = M \times 2\pi f$.

Elastic reactance is the effect of the elasticity or stiffness of the object in opposing the displacements resulting from the action of the applied force. Its value depends, of course, upon the magnitude of the elasticity, which we measure as the force required to produce a given amount of deflection and hence express in dynes per centimeter. Its value also depends upon the frequency, but in an inverse manner: the higher the frequency, the smaller the elastic reactance. The elastic reactance (in mechanical ohms) is equal to the elasticity (in dynes per centimeter) divided by the angular frequency or, in symbols,

$$X_s = \frac{S}{2\pi f}$$

* The mechanical ohm has been named in recognition of a direct analogy with the electrical ohm, which is defined as the electromotive force, in volts, acting on some element of an electrical circuit to produce a current of 1 ampere, or $R_e = e/i$.

Frictional resistance imposes a limitation on any vibratory action by the continuous withdrawal of energy and its conversion into heat. The magnitude of this resistance (in mechanical ohms) is simply proportional to the friction and does not vary with frequency as the two kinds of reactance do. These three constituents, mass reactance, stiffness reactance, and frictional resistance, when combined in a special way give us the total mechanical impedance.*

THE PROPAGATION OF SOUND

Aerial sound is propagated in all directions away from a vibrating body. The propagation comes about through the fact that the air particles communicate their motions to one another in progressive fashion throughout the extent of the medium. The propagation occurs at a constant speed, so that in free air a spherical wave results.

The propagated disturbance represents all the changing conditions of the sound source—every condition of condensation and rarefaction produced as the vibrating body, such as the tuning fork already mentioned, moves back and forth. As these conditions are repeated regularly the spherical waves are periodic. For the moment let us consider one particular condition, say, the maximum condensation produced as the fork moves farthest to the right. We now picture a spherical wave of condensation, which is something like a soap bubble originating at the source and blown larger and larger at a constant rate. A surface of this spherical wave, representing some one condition such as the maximum condensation, is known as a wave front. When the wave moves so far away from the source that the curvature becomes negligible we speak of such an advancing wave front as a plane wave.

Let us continue our example. When the fork moves to the right a second time it is as though a second soap bubble were blown within the first. Both bubbles will now be expanding, but because we have assumed a constant rate of expansion they will always

* The manner of combining is as follows: we find the difference between the two kinds of reactance, square this difference, add it to the square of the resistance, and take the square root of the result; or, symbolically,

$$Z = \sqrt{\left(2\pi f M - \frac{S}{2\pi f}\right)^2 + R_m^2}$$

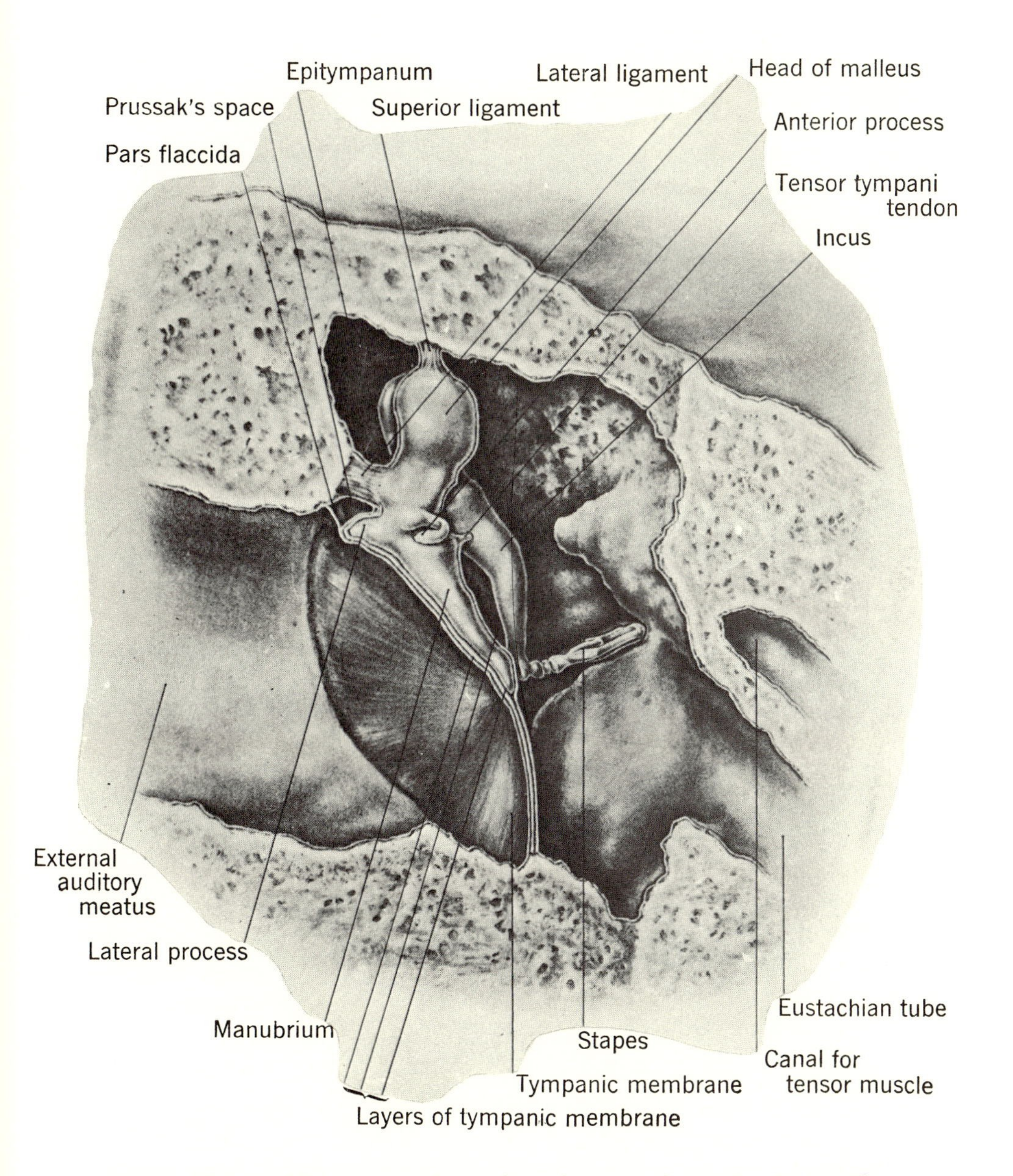

Plate 1. The middle ear on the right side, seen from the front. The tensor tympani muscle and most of its tendon have been removed. From John B. Deaver, *Surgical Anatomy of the Human Body*, 1926, The Blakiston Company.

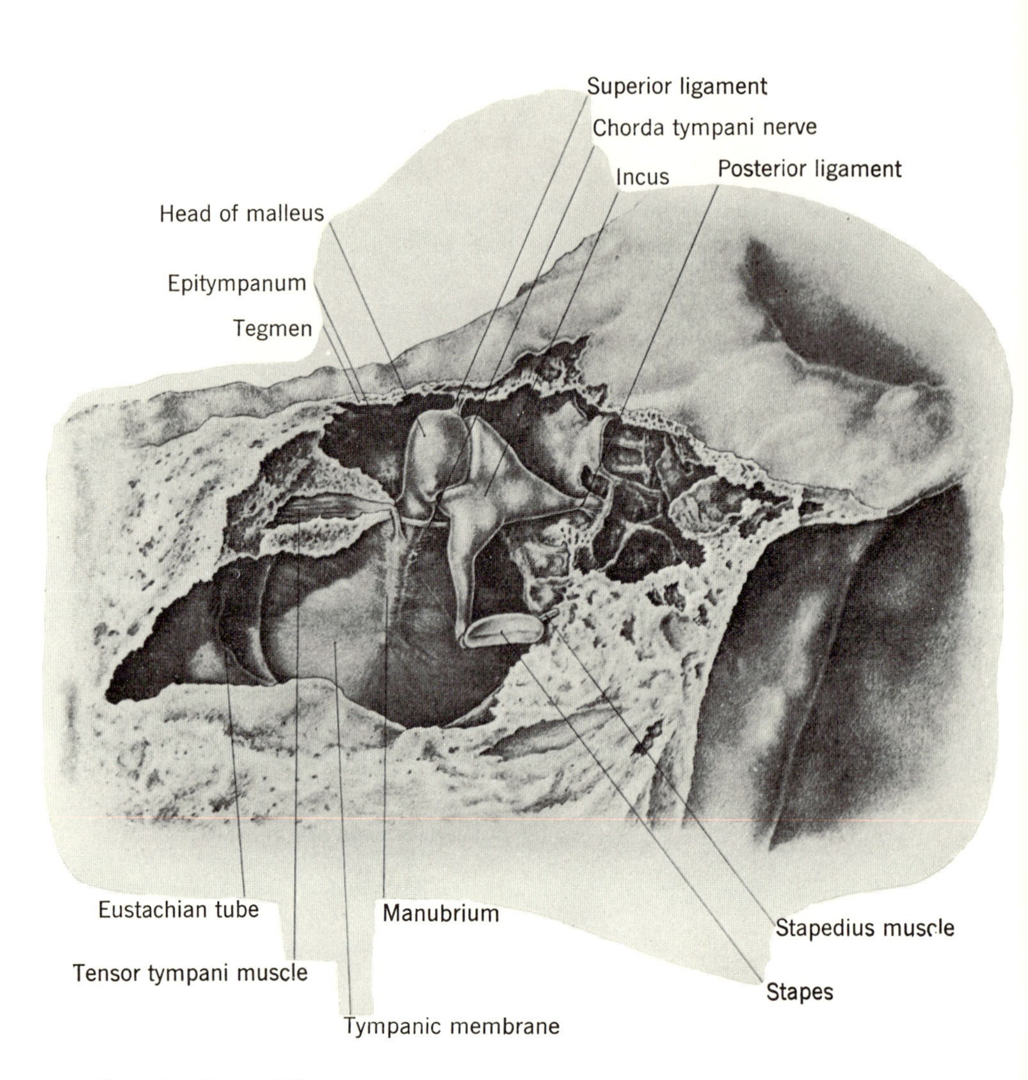

Plate 2. The middle ear on the right side, seen from within. From John B. Deaver, *Surgical Anatomy of the Human Body*, 1926, The Blakiston Company.

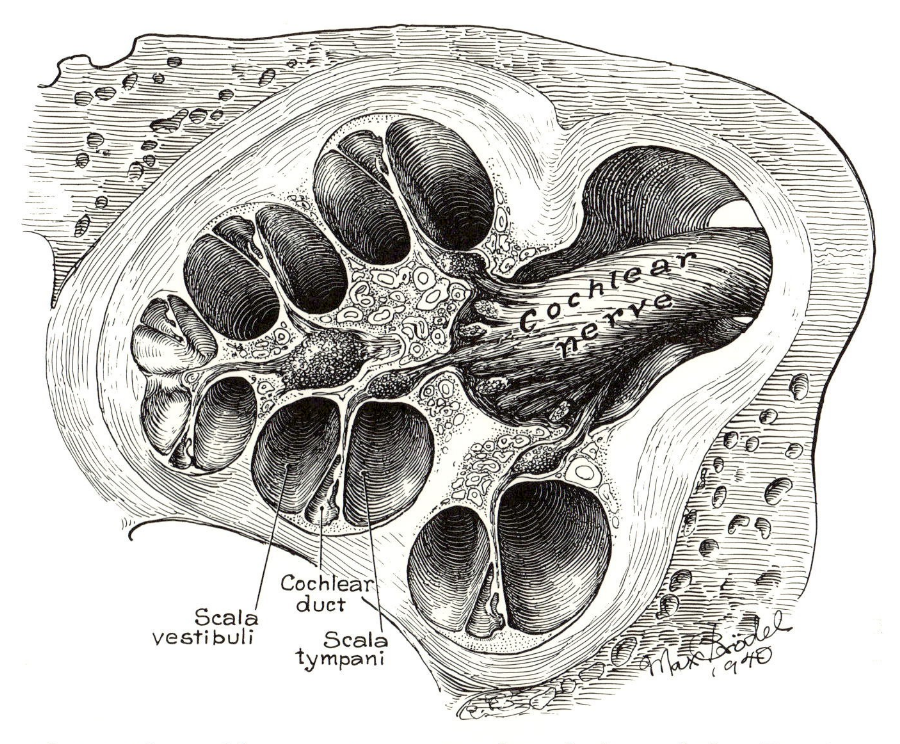

Plate 3. The cochlea, seen in a section through the modiolus. Drawing by Max Brödel, from the *1940 Year Book of the Eye, Ear, Nose and Throat.*

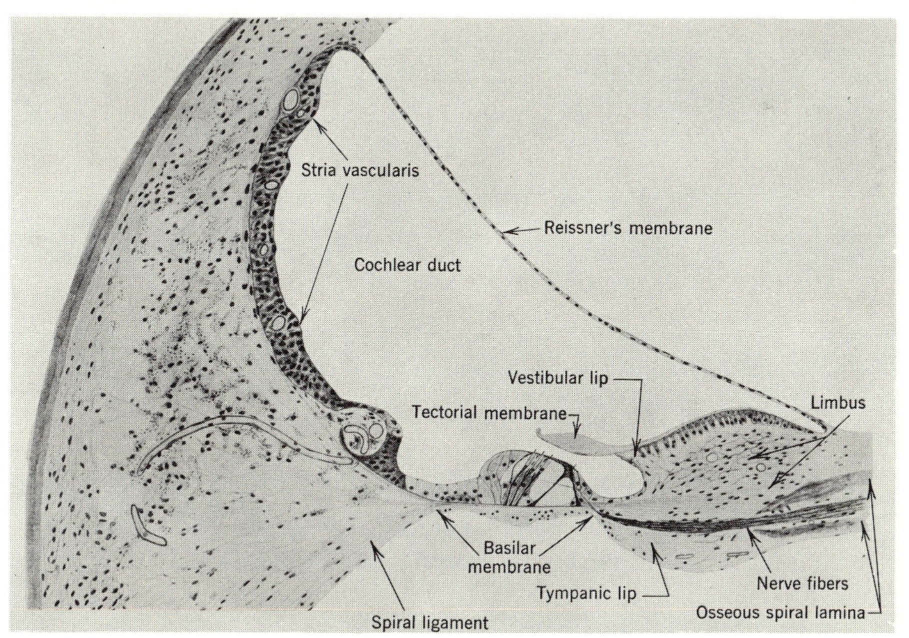

Plate 4. The cochlear duct, in cross section. From H. Held, in Bethe's *Handbuch der normalen und pathologischen Physiologie,* Bd. 11, *Receptionsorgane,* 1926, I, Julius Springer, Berlin.

remain a constant distance from one another. This distance between corresponding positions in periodic waves is the wave length.

If our expanding wave finds an obstacle in its path it does three things: it scatters about the obstacle, it penetrates the obstacle, and it is reflected back from the obstacle. These three processes are known as diffraction, transmission, and reflection. Their relative importance in the situation depends upon the particular conditions.

If the obstacle is small the energy in that part of the wave striking the obstacle is largely diffracted. This portion of the energy is radiated in all directions and becomes a minor disturbance of the general advancing wave front. The loss from the diffraction is hardly noticeable as long as the obstacle is small, for in effect the wave front closes up beyond the obstacle. A small obstacle thus does not cast a sound shadow. This is a fortunate circumstance, for by reason of it we are able to hear sounds around corners and despite many obstructions. "Small" in this connection refers to the size of the object relative to the wave length. As long as the linear dimensions of the object are below the wave length of the sound, or at least are not much greater than the wave length, the wave continues forward and maintains its form.

If the obstacle is large the portion of the advancing wave intercepted by it will be reflected back from the surface to form a wave in the opposite direction. Generally speaking, a surface needs to be at least five times as large as the wave length to produce an appreciable reflection, and its efficiency as a reflector continues to rise with its further enlargement. A familiar example of a reflected wave is the echo, and probably everyone has noted that a large surface is required if the echo is to be prominent.

Some of the sound energy striking the obstacle is neither diffracted nor reflected but is transmitted into the substance of the obstacle. For a large obstacle, the quantity of transmitted and reflected energy can be determined from the density and elasticity of the obstacle relative to the density and elasticity of the medium. The formula is given here because it will be of interest later on. Let ρ_1 represent the density of the medium and ρ_2 the density of the obstacle, and let S_1, S_2 correspondingly be their elasticities. Then $R_1 = \sqrt{\rho_1 S_1}$ is the specific acoustic resistance of the medium and $R_2 = \sqrt{\rho_2 S_2}$ is that of the obstacle. We find the ratio of these acoustic resistances as $r = R_2/R_1$. Then

$$T = \frac{4r}{(r+1)^2}$$

where T is the transmitted energy. The reflected energy is the remainder, or $1 - T$. We see here that when the properties of the two substances are similar, or R_2 is close to R_1 and r approaches 1, the value of T approaches 1, which means that most of the energy passes into the obstacle and little is reflected. On the other hand, as R_2 and R_1 become different and r becomes large, T becomes small; then the energy is mostly reflected.

The speed of propagation of a sound depends upon the density and elasticity of the medium, and for air under ordinary conditions it is 1129 ft per second (344 meters per second). For water the speed is more than four times as great, or 4714 ft per second (1437 meters per second). In rigid solids the speed is greater still. In ivory, which is a useful example because it is closely similar to the petrous bone about the cochlea, the speed is 9886 ft per second (3013 meters per second). These speeds vary somewhat with temperature.

SOUND WAVES AND THEIR DIMENSIONS

Vibratory movements and the resulting sound waves may vary in four ways: in amplitude, in frequency, in wave form, and in phase. The tuning fork discussed above varies in the amplitude of its movements according to the force of the blow that is delivered to it. The maximum amplitude is measured from the resting position to the farthest displacement to one side or the other. The amplitudes of aerial movements that are of practical importance cover a very wide range, from 10^{-9} cm, which is barely perceptible under ideal conditions, to 10^{-2} cm, which is approaching a magnitude dangerous to the ear.

The frequency of a sound is the rate of repetition of the vibrations. It is measured in cycles per second, and one cycle (for which we use the sign $\sim$) consists of a back-and-forth movement, that is, a movement from left to right and back again to the left. The time required by one cycle is called the period of the vibrations, and its value is the reciprocal of the frequency. As a tuning fork continues to vibrate it dies away in amplitude, but the frequency remains constant. Audible frequencies giving a simple pitch vary from 15 to 24,000~. Still lower frequencies are audible but are heard as noisy.

The simplest kind of vibration has a particular form called simple harmonic or sinusoidal. This form of vibration is produced by the tuning fork when lightly struck. If we were able to magnify the motions of the prong and to draw them out in time we should obtain a curve like that of Fig. 3*A*. Vibrations of this form seem

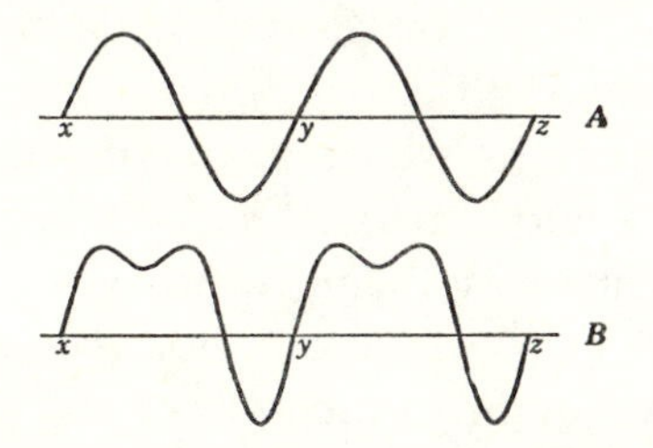

Fig. 3. Wave forms. At *A* is a sinusoidal wave as produced by a tuning fork when lightly struck, and at *B* is a complex wave as produced by an organ pipe when blown vigorously. For each wave, amplitude is represented on the vertical axis and time on the horizontal axis. The portion xy or yz is one cycle.

simple to the ear and are characterized as pure tones, whereas vibrations of any other form seem complicated and are capable of analysis into two or more component tones of the simple type. Most ordinary sounds, including the notes used in music, are of the complex form. They are complex because sounding bodies in general are capable of more than one frequency of vibration, and in effect they execute two or more of these frequencies simultaneously. For example, a stretched string may vibrate as a whole, by halves, by thirds, and so on. A musical observer can identify the component frequencies, or at least the more prominent of them. The note of an organ pipe has the form shown in Fig. 3*B*, and a trained observer will recognize the presence of a fundamental tone (the tone of lowest frequency) and two or three overtones. The overtones have frequencies that are simple multiples of the fundamental frequency. Thus if the fundamental frequency is 1000~, the first overtone is 2000~, the second overtone is 3000~, and so on. This relationship follows from the fact just mentioned that typically a vibrating body moves as a whole or in its motion breaks up into two, three, or more integral parts.

A further characteristic of tones is their phase, which is a particular kind of temporal relation. If we have two identical tuning forks and strike one just a little ahead of the other they will execute movements that are alike except that one series will always be slightly advanced in time. The sound waves that the forks produce will likewise differ in this respect. This difference we refer to as a phase difference. Ordinarily, because the movements of the forks are periodic, we are only concerned with phase differences that are less than the period of the vibration. If the time between our striking the two forks is just one period, or any whole number of periods, the movements of the forks will be strictly synchronous and thus without any phase difference. We refer to such movements as being in phase. If the time between striking the forks just exceeds one period we usually need to consider only the excess in our indication of the phase.

A phase difference between two sounds can be indicated in a number of ways. The simplest way is in terms of the time between corresponding positions reached by their waves. We can also express the phase as a fraction of a period. Further, for two progressive waves we can express a phase difference as the distance between corresponding points in space reached by the waves or as a fraction of a wave length. Most commonly, however, a phase difference is expressed in angular units, in degrees or radians. A wave period is then considered as made up of 360° or 2π radians. A phase difference of one fourth of a period, for example, then becomes 90° or $\pi/2$ radians. This method is carried over from circular measure and reflects the fact that a sine wave is a kind of transformation of uniform motion around a circle. It is a linear projection of uniform rotation, and anyone can demonstrate the fact for himself by moving his finger uniformly in a horizontal circle while observing the shadow cast by the finger on a wall.

It is usually necessary not only to state the amount of a phase difference but also to indicate its direction—to say which motion is leading or lagging with respect to the other. In this relation we need to bear in mind the peculiar conditions arising from the nature of wave motion. When a phase difference reaches 180° or π radians there is no leading or lagging but only a condition of opposition. The motions then are in contrary phase or are "out of phase." When one motion leads another by a little more than 180° it effectively is lagging by the excess over 180°.

We can best picture these phase relations in a simple analogy. Consider two runners *A* and *B* on a circular track. They start off together (zero phase relation) but *A* is the fleeter and he soon gains a few paces over *B*. We can say that *A* leads in phase by the angular distance between them. Now suppose that after a few circuits of the track *A* gains just half a lap over *B* (180° phase relation). Anyone who has not kept account of the whole performance will be at a loss to say which runner is ahead. When *A* has gained a little more (say to the 200° position) he will seem to our distracted observer as being the one behind. Then, when *A* has gained a full lap on *B* (360° relation), the observer who takes the instantaneous view will see the two runners as on even terms. Ordinarily in considering the phase relations of two waves we take simply this instantaneous view.

Phase relations are important when we combine simple waves to form complex waves. Two waves that are progressing along the same path give different sums according to their phase relations. If these waves are of the same frequency the phase relation is the same all along the path and we can express their sum in a simple manner. Consider the summation of two waves of equal amplitude; for the zero phase relation they add to give a wave of double amplitude, for the 180° relation they add to give zero amplitude, and for other phase relations they give sums of intermediate value. Note that always the amplitudes of the two waves are added point by point, taking account of sign as well as magnitude.

THE MEASUREMENT OF SOUND INTENSITY

In nearly all exact studies of the process of hearing we need to know the magnitudes of the stimulating sounds. Theoretically we might use any one of the three aspects of the sound that have already been mentioned: the amplitude of displacement of the air particles, the velocity of their back-and-forth movements, or the kinetic and potential energy contained in the motion. However, we do not have methods for dealing satisfactorily with these dimensions in the small magnitudes in which they exist in ordinary sounds. Fortunately there is another aspect of sound that more readily lends itself to measurement. It is the varying pressure that the vibrating air particles exert upon a surface that stands in the path of their movements. These variations of pressure are small in relation to those that we ordinarily deal with, and often amount to only a fraction of a dyne per square centimeter. In comparison, the normal, steady pressure of the atmosphere in which we live is about 10^6 dynes per sq cm.

For the measurement of sound pressure a condenser microphone is often used. It consists of a thin diaphragm stretched in front of a fixed backplate. These two parts are made of metal and are insulated from one another, and thus they constitute the two plates of an electrical condenser. Also, because the capacity of a condenser depends, among other things, upon the spacing of its plates, this capacity varies as the diaphragm moves back and forth under the influence of a sound. When a polarizing voltage is impressed across the condenser the variations of capacity produce a varying electric current through the condenser. Figure 4 is a

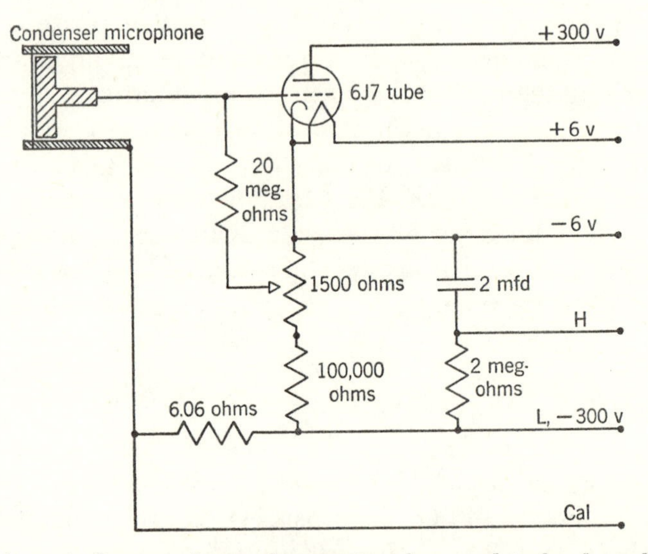

Fig. 4. Schematic diagram of a condenser microphone and its head amplifier.

schematic diagram of a condenser microphone together with the circuit that is used for applying the polarizing voltage and amplifying the resulting microphone currents.

The two ways of calibrating a microphone for use in sound pressure measurements are known as pressure and free-field calibrations. When given a pressure calibration the microphone simply shows the sound pressure that is active on its diaphragm at the moment that the readings are made. When given a free-field calibration the microphone shows the pressure that had existed at the position of the microphone before the microphone was introduced into the sound field and will again be present there when it is removed. These two calibrations are different insofar as the presence of the microphone alters the sound field.

As already mentioned, any object alters a sound field by absorbing, reflecting, and scattering the sound waves, but if its dimensions are small relative to the wave length of the sound these effects are negligible. Hence it is usually desirable for a microphone to be as small as possible within limits imposed by constructional difficulties and by the required sensitivity. A condenser microphone in common use for sound measurements is 1 inch long and ⅞ inches in diameter, a size that produces only slight disturbances of the sound field for all frequencies below about 1000~.

At times it is necessary to have measurements of sound pressure at points that are inaccessible with even a miniature microphone. The sound probe is an accessory that has been developed for this purpose. It consists of a cap that is fitted tightly over the diaphragm of the microphone and that leads to a slender probe tube, as shown in Fig. 5. This combination of microphone and probe tube is now calibrated to indicate the sound pressures present at

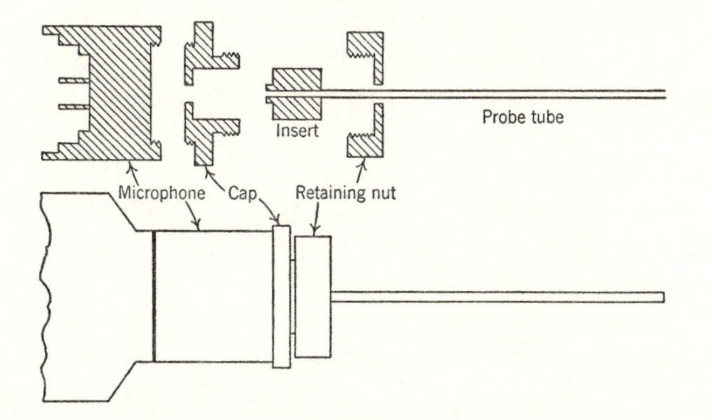

Fig. 5. Condenser microphone and sound probe. About two-thirds natural size.

the tip of the probe tube. For example, a probe tube may be inserted deep into the external auditory meatus close to the drum membrane, and will then reveal the sound pressures existing there.

Figure 6 gives typical calibration curves for a condenser microphone used alone and also as used with a probe tube. It will be noted that the microphone curve is fairly uniform, whereas the curve for microphone and probe shows several undulations. These undulations are due to the resonance characteristics of the probe

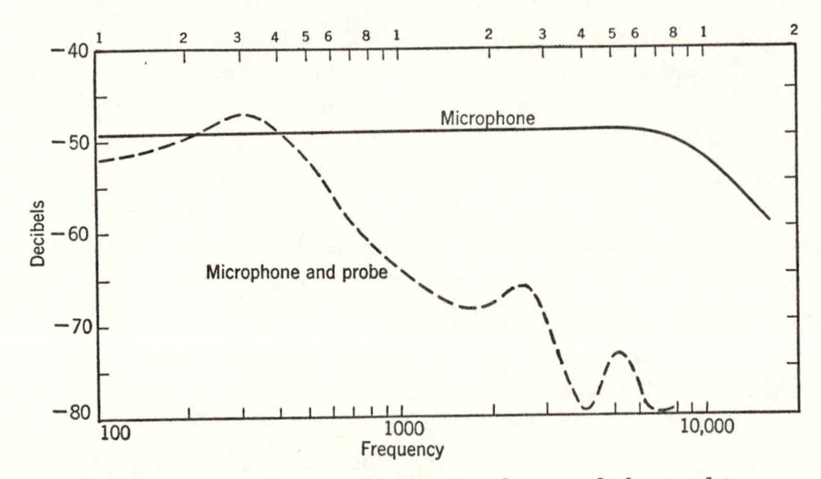

Fig. 6. Calibration curves for a condenser microphone used alone and in connection with a sound probe. The response produced by a sound of 1 dyne per sq cm is represented in decibels relative to 1 volt.

tube. The uniform character of the microphone curve makes this instrument suitable for measuring complex sounds as well as pure tones. The microphone and probe, however, will not give accurate results with complex sounds because certain component frequencies will be emphasized relative to others.

THE UNITS OF MEASUREMENT

In these operations we need to give attention to the units of measurement. For sine waves we can express the magnitude in terms of the maximum pressure reached in the course of the wave (peak pressure) or in terms of the root-mean-square pressure (rms pressure). The latter represents a particular sort of average, which we might obtain by taking all the pressures produced during a cycle, squaring them, adding these squares, dividing by the number, and finally taking the square root. Actually we do not need to go through this process, for in a sine wave the root-mean-square amplitude (for pressure or for any other magnitude of the wave) always equals the peak amplitude divided by the square root of 2. That is,

$$P_{\text{rms}} = \frac{P_{\text{max}}}{\sqrt{2}} = 0.707 P_{\text{max}}$$

Most measuring instruments are calibrated to read root-mean-square values. For complex waves there is no simple relation between peak values and root-mean-square values. For such waves the peak amplitude varies greatly, often from moment to moment, according to the phase relations among the components. The root-mean-square amplitude is more stable and therefore is the preferred value. In all numerical considerations throughout this book we shall use root-mean-square values except when stated otherwise.

Though measurements of sound intensity are nearly always made in terms of pressure it is sometimes convenient to convert the readings into other units. For such conversions we need to know the wave frequency and certain properties of the medium in which the wave motion occurs. The pertinent properties of the medium are the density and elasticity which determine the specific acoustic resistance, already defined as $R = \sqrt{\rho S}$. It is necessary to add that the conversions are limited to extended gases and liquids and to plane progressive waves in solids.

If we have the sound pressure P and the specific acoustic resistance R we can find the particle velocity u of the wave motion, which is $u = P/R$ and is measured in centimeters per second.

Here we should note that particle velocity refers to the back-and-forth motion of the molecules of the medium and is not to be confused with speed of propagation, which is the forward progression of the wave through the medium.

For plane waves we can find the displacement d of the molecules away from their equilibrium positions as $d = P/2\pi fR$, where f is the frequency and the other symbols are as defined above. The value of d is in centimeters.

The power present in the sound wave is $J = P^2/R$ in ergs per second per square centimeter or, because 10 ergs per second =

1 microwatt, $J = P^2/10R$ microwatts per square centimeter or $P^2/10^7R$ watts per square centimeter. We see in the relations expressed here that sounds represent exceedingly small amounts of power as measured by ordinary standards. Thus a sound in air with the dangerously high pressure of 1000 dynes per sq cm represents a power of only 0.0024 watts, and would have to be made ten thousand times as great to equal the power that we use in the form of electricity to operate a small electric light bulb.

To find the energy contained in a sound wave we need to consider the acoustic power as integrated over an interval of time.

The Decibel Notation. Because the ear operates over an enormous range of intensities and also because we find that perceptually this range is much condensed, it is convenient to employ a condensed scale for our physical measurements. This situation prevails in all the senses and has led to the general use of logarithmic scales. On such a scale we deal with exponents of some constant (which is the base of the system of logarithms) instead of the numbers themselves. This logarithmic scale has one notable drawback: it lacks a zero point. Strictly speaking, therefore, this system does not represent a true scale but is only a form of notation; we speak of it as a scale only loosely. Its values are ratios, and we must always know what the quantity is with which our stated quantity is being compared. When we are interested only in the relation between two intensities, then one of them becomes the reference for the other. When we wish to represent absolute values, however, it is necessary that we first choose a reference level and compare all other intensities with this reference. A number on our logarithmic scale tells us how many logarithmic units a given intensity is above or below the reference level.

There are any number of logarithmic scales that might be used, but the one in common use is the decibel scale. Originally this scale was worked out in the belief that the ear obeyed Fechner's law strictly and that psychological loudness varied as a logarithmic function of the physical intensity. The decibel was then considered as a just noticeable difference. Actually, loudness does not follow any simple function and it is better to treat the decibel notation simply as a mathematical transformation of stimulus magnitudes.

The decibel formula is

$$N = 10 \log_{10} \frac{J}{J_0}$$

where N is the number of decibels, J is the sound power in question, J_0 is the reference level, and the ordinary decimal logarithms are used.

A good many reference levels have been employed from time to time, and it is important that the particular one be stated in every case. Telephone engineers have used a "phonic level" of 1 microwatt per sq cm. Others in making pressure measurements have chosen a "zero level" of 1 dyne per sq cm. In auditory experiments it is often convenient to have a reference level that approximates the human threshold. Unfortunately, however, this threshold varies for different tones. The threshold for 1000~ has been used, as well as a rough average of threshold values in the middle range, which happens to come out about 0.001 dynes per sq cm. Again, many experimenters have taken as a reference point the threshold intensity of the particular tone being studied, which has the advantage of avoiding the necessity for absolute calibration but carries the disadvantage of being different for every tone and for every individual subject. Finally, in an effort to place the reference level sufficiently low that negative numbers could be avoided a reference level of 10^{-16} watts was proposed. This level was adopted by a standardization committee of the Acoustical Society of America and has been generally accepted.

An aerial sound whose power is 10^{-16} watts per sq cm has a pressure of 0.000204 dynes per sq cm under standard conditions (a temperature of 20°C and a barometric pressure of 760 mm of mercury). This figure is often rounded off to 0.0002 dynes per sq cm and this value is taken as the reference level for pressure measurements.

A table showing many of the reference levels and their relations to one another is given in Appendix A. Appendix B brings together the formulas relating acoustic power, pressure, particle velocity, and displacement.

The decibel formula as given deals with acoustic power, but in actual practice we are more often concerned with the electrical power that is sent into some sound-producing device like a loudspeaker. If the device is linear—if its output of sound is proportional to the electrical power sent into it—then the same ratios hold and electrical powers may be used for J and J_0. The value of J_0 is now the electrical power that gives some reference amount of sound (10^{-16} watts of sound if the ASA standard is being used).

If the situation is such that acoustic power is proportional to the square of the acoustic pressure, as is true when the acoustic impedance is constant, then we can use pressure in the decibel formula, which becomes

$$N = 10 \log_{10} \frac{P^2}{P_0^2}$$

or can be rewritten in the equivalent form

$$N = 20 \log_{10} \frac{P}{P_0}$$

When we are dealing with sounds in a given medium, such as air, the use of the pressure formula is proper. On the other hand, if we are comparing two sounds in different media, say one in air and another in water, we can use this formula only with a certain reservation, for now it will express only the pressure ratio and will not represent the power ratio. To obtain the power ratio we have to take account of the different impedances of the two media, and should use the power formula with

$$J = \frac{P^2}{R} \qquad \text{and} \qquad J_0 = \frac{P_0^2}{R_0}$$

When impedances are constant we can use the decibel formula for particle velocities as well as for pressures, as

$$N = 20 \log_{10} \frac{u}{u_0}$$

Also, when we are dealing with linear electroacoustical systems, and sound output varies as the square of the voltage or of the current, we can use the formulas

$$N = 20 \log_{10} \frac{e}{e_0} \qquad \text{and} \qquad N = 20 \log_{10} \frac{i}{i_0}$$

where e, e_0 are voltages and i, i_0 are currents. In practice our measurements are nearly always in terms of voltages.

The values of N in these formulas can be obtained by looking up the logarithm of the ratio in a table of logarithms and multiplying by the appropriate factor (10 for power ratios and 20 for voltage, current, pressure, or velocity ratios). Likewise, if we know N we can find the ratio by reversing this operation. It is more convenient to use a decibel table in which the values are read directly. Such a table is to be found, with directions for its use, in Appendix E.

Many acoustical and electroacoustical systems contain two or more stages at which power changes occur. As a simple example, suppose we have sound traveling down a tube with nonabsorbent walls and with two partial obstructions, one obstruction permitting ½ of the power to pass through and the other permitting ¼ to pass through. The transmission is then $\frac{1}{2} \times \frac{1}{4} = \frac{1}{8}$. This calculation is easy to make if we express the power changes in decibels, for the separate losses in decibels have only to be added to obtain the total loss. The first obstruction gives a 3 db loss ($J/J_0 = 0.5$) and the second gives a 6 db loss ($J/J_0 = 0.25$) for a total of 9 db. When gains and losses are both present the same procedure holds, if the individual decibel changes are added with proper attention to algebraic signs.

This example clearly illustrates a principle that must always be kept in mind: adding decibel values corresponds to multiplying the ratios that these values represent. A possibility of confusion arises when we are using decibels (above a reference level) to represent actual sounds and wish to know the effect of combining the sounds. Thus we may wish to know the total intensity of the sound of a typewriter and a background noise, when the two are measured separately as 45 and 32 db above 10^{-16} watts per sq cm. The sum is not 77 db. In fact, to obtain the sum we have to convert the readings to actual powers and add these. The powers are 3.162×10^{-12} and 0.158×10^{-12} watts per sq cm and their sum is 3.32×10^{-12} watts per sq cm, which is 45.2 db above the reference level. The background noise thus adds only 0.2 db to the total. In general, the combined intensity of two sounds that differ by more than 10 db only slightly exceeds the intensity of the stronger.

In electrical systems we often make use of a special kind of voltage divider called an attenuator, in which the dial readings are in decibels. Such an instrument delivers a voltage at the output end that is less than the input voltage by a certain ratio corresponding to the decibel reading to which the dial is set. For example, if we set for an attenuation of 20 db the output voltage will be one-tenth of the input voltage, regardless of the absolute values of the input. If the system works into a loudspeaker that converts the electric currents into sound, the sound intensity likewise will be reduced by 20 db, provided that impedance requirements of the circuit are satisfied and all parts of the system are linear.

In electroacoustical measurements it is always necessary to establish the linearity of the system for the conditions of interest. The best way to do this is to pick up the aerial sounds with a microphone and determine the change in sound pressure for some change in the attenuator setting. This is necessarily done at a high level of sound pressures because only at such a level is the microphone equipment sufficiently sensitive and free of background disturbances. Fortunately, if we establish linearity of operation at the high levels we can reasonably be assured of linearity at low levels.

3

Experimental Methods in the Study of the Ear

MODERN EXPERIMENTATION on the ear has reached a high level of advancement by the use of many new methods and techniques. Two developments are of special importance for the content of this book. One comes from the science of electronics, and is the elaboration of convenient and precise instruments and procedures for producing, measuring, and analyzing sounds. Another development comes from electrophysiology, and is the discovery of the electrical potentials of the ear and their use as a means of understanding auditory function.

In the following sections we shall describe a number of these new technical methods both for their immediate interest and for their experimental applications. We shall often refer to the particular procedures that we have employed in our own experiments, several of which are described later on. It should be mentioned, however, that in many instances there are other procedures that will serve the same purposes equally well.

THE PRODUCTION AND MEASUREMENT OF SOUNDS

The sounds used for experimental purposes are produced nowadays almost exclusively by electronic means. An arrangement of apparatus is shown on the left of Fig. 7. Here are five instruments: an oscillator, an electric filter, an attenuator, a power amplifier, and a loudspeaker. The currents produced by the oscillator are modified and controlled by the next three instruments and finally a portion of the electrical energy is converted into sound by the loudspeaker.

The most useful type of oscillator is one that produces sinusoidal currents. Many circuits are suitable for this purpose, all utilizing the same fundamental principle. A vacuum tube is arranged so that its output is partly returned to its input, and whatever activity is going on is amplified to a limit imposed by the available power. A tuned circuit is provided, consisting of a combination of an inductance, a condenser, and a resistor, or sometimes only two of these elements, and this circuit causes the system to oscillate at a particular frequency according to the magnitudes of the elements. By adjusting these magnitudes the frequency can be varied

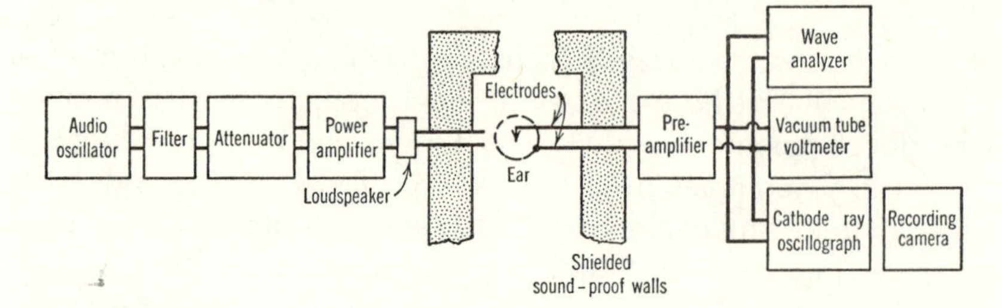

Fig. 7. Stimulating and recording apparatus for the study of the electrical potentials of the ear.

as desired. Audio oscillators are so called because they have adjustments for some part at least of the range of frequencies that are audible as sound.

The choice of an oscillator circuit depends upon such things as stability of operation, purity of wave form, and convenience of adjustment. The choice of an individual instrument depends also upon the range of frequencies provided. Many excellent instruments are commercially available. In our experience the most satisfactory for general experimental use is a beat-frequency oscillator with a continuous range from 20 to 20,000~ that is covered in logarithmic steps on a single dial. Such an instrument gives convenient and rapid settings of frequency and the accuracy and stability are sufficient for all ordinary purposes.

Another source of electric currents is the noise generator. The most common form is the thermal noise generator, which makes use of the fact that any vacuum tube produces a background of random current variations, mainly on account of the thermal agitations in the elements of its input circuit. When these variations are greatly amplified they produce a noise in a loudspeaker that is random except for modifications introduced by the circuit and peculiarities imposed by the loudspeaker. By the use of this generator the ear can be stimulated by a uniform noise or, by the addition of appropriate filter circuits, with certain bands of noise as desired.

Most electric generators incorporate one or more amplifier stages that follow the oscillator stage and increase the current to a suitably high level. They often include controls for reducing the output, but these controls usually are not intended for exact work. They are ordinarily used only for setting the output voltage to some desired level as indicated on an output meter, which is either incorporated in the generator or is added externally. This meter must have a wide frequency range, and satisfactory types are the rectifier meter and the vacuum tube voltmeter.

An electric filter is used to improve the wave form of the oscillating currents by removing any harmonic frequencies or any background noise. If only harmonics are present a low-pass filter will serve, and is selected to pass all frequencies up to the desired frequency and to cut off higher frequencies. In the presence of both harmonics and low-frequency noise, such as hum from the power lines, we must use a combination of low-pass and high-pass filters or the equivalent, a band-pass filter. Different filter sections are required for the various frequencies to be passed.

For precise control of intensity an attenuator is inserted between the oscillator and the loudspeaker. The attenuator consists of a network of resistances selected to reduce the output voltage to some desired fraction of the input voltage. Most instruments provide a number of steps, usually indicated in decibels. A convenient arrangement includes two decades of attenuation, one decade giving ten steps of 1 db and another giving ten steps of 10 db, for a total of 110 db. Additional decades of 10 db steps are added as needed to extend the range further. Adjustments below 1 db are not usually needed, but some of the phase studies described later required an additional decade of 0.1 db steps.

The power amplifier produces the currents necessary to operate the loudspeaker. It should provide the maximum power that can be handled by the loudspeaker and with a minimum of distortion. In practice it is well to select an instrument with a manufacturer's rating above the maximum of power needed so that the output tubes will always operate below their limits. A rating of 30 watts will meet usual requirements. Most power amplifiers also provide a moderate amount of voltage amplification, but more than about 20 db of such amplification should be avoided because it is likely to be accompanied by excessive tube noise and hum from the alternating current supply. If greater amplification is needed it should be obtained by a supplementary amplifier placed between the oscillator and the attenuator, for then any noise that this amplifier introduces will be reduced by the attenuation factor.

The terminal instrument in this system is the loudspeaker. It is nearly always of the electrodynamic type, and consists of a small coil which is suspended in a strong magnetic field and to which a conical diaphragm is attached. There are two general forms, one with a large diaphragm, designed to radiate sounds into the open air, and another with a small diaphragm, designed to operate into a horn.

The loudspeaker is the weakest element in the electroacoustic system. The other elements can be designed to give uniform output at all frequencies and at practically any desired level. The loudspeaker, however, is limited in its output and it varies enormously and irregularly in efficiency over the frequency range. It is difficult to cover the whole auditory range with one instrument and therefore it is customary to use two instruments, one for the low-frequency range, usually below about 400~, and another for the high-frequency range. The loudspeaker is subject to nonlinearity and wave-form distortion when driven hard, and it is especially difficult to obtain a substantial amount of sound at frequencies below 100~.

For experimental purposes it is often preferable to employ a closed system in which the sound is not radiated into the air but is conveyed directly from a horn type of loudspeaker to the ear. The amount of sound that must be generated is then less and the loading of the diaphragm is more suitable, so that distortion is much reduced. When only a moderate amount of sound is needed a telephone receiver may be used. Such a receiver is designed to work into a small air cavity and for best results should fit tightly over the ear. Soft rubber cushions are made for use with receivers, or sometimes are an integral part of the case, and serve both to give a good fit and to make the receivers more comfortable.

So far we have discussed only the production of aerial sounds, but there are occasions on which it is desired to stimulate the ear by bone conduction. For this purpose we require a bone-conduction receiver, a device that will communicate vibrations to a heavy mass such as the head. A number of these instruments are available, mostly developed for use in audiometry. Some of them work on the same electrodynamic principle as the loudspeaker but instead of the usual diaphragm they actuate a small plunger that is applied to the head. Others are based on the piezoelectric principle; a crystal of Rochelle salt or some similarly acting substance expands and contracts along one of its axes when an electric current is passed through it. We have had the most success with this crystal type of receiver in a special form. Our instrument consists of a series of laminations $\frac{5}{8} \times 3 \times \frac{1}{16}$ inches in size, cemented together with interleaving sheets of metal foil to give a final thickness of $\frac{5}{8}$ inch. The crystal plates are cut and interleaved in such a way that the finished crystal expands and contracts longitudinally when an alternating current is sent through it. The crystal is mounted on one end inside a heavy metal case, which serves both as a reaction mass and a shield, and to its other end is cemented a tapering plastic rod whose end protrudes through the case and forms the plunger to be applied to the head. This receiver gives a useful output over the whole audiofrequency range.

A complication in the use of any bone-conduction receiver is the varying pressure involved in usual methods of applying it to the head. With most receivers the vibratory output varies with this pressure. In the model just described this variation is a minimum because the natural stiffness of the crystal is very great. A valuable accessory to this instrument is a vibration pickup that serves as a monitor to the receiver. A crystal type of phonograph pickup is mounted so that its needle rides in a conical hole drilled in the side of the plastic rod. The output of this pickup is led to a sensitive vacuum tube voltmeter and reports the relative amplitude of movement of the plunger.

The electroacoustical system as described above for pure aerial tones under closed-tube conditions will operate with negligible distortion at low levels and then as the intensity is raised the distortion increases gradually for a time and finally, at a rather high level, it increases with great rapidity as the loudspeaker overloads. The intensities at which these changes occur vary with frequency, and in general are less for the low tones. The points of rapid overloading are readily detected on listening to the sound, on looking at oscillographic traces of the wave form, or on noting the change of output for some given increase in input voltage.

The electric filter in Fig. 7 is shown as preceding the power amplifier. This is so because most filters will not withstand the heavy currents produced by this amplifier. If, however, a filter is available with the necessary current capacity it is possible to place it between the power amplifier and the loudspeaker. The filter then has the advantage of reducing the distortion and extraneous noise arising in the amplifier. However, this arrangement makes more difficult the proper matching of impedances and the damping of the loudspeaker. We have used it successfully only for producing steady pure tones of low and moderate intensities.

For especially high degrees of purity it is necessary to interpose an acoustic filter between the loudspeaker and the ear. The simplest of such filters is the interference tube, made in the form of a side branch in the conduit from loudspeaker to ear. This side branch contains a tight-fitting plunger so that its effective length can be varied. This length is adjusted to approximately one-fourth of the wave length of the tone that is to be eliminated. Thus there must be as many side branches as components to be filtered out. Usually two will suffice, because only the second and third harmonics are of appreciable magnitude. The final adjustments of the tubes must be made under the actual conditions of the experiment, with the ear connected to the system. This method when added to careful design of the electrical circuit gives the high degree of purity necessary for studies on the distortion of the ear, as described later.

THE STUDY OF COCHLEAR POTENTIALS

The electrical potentials of the cochlea have proved of great value in the study of the conductive mechanism of the ear. Their use, of course, largely limits the investigation to experimental animals, of which the ones most used are the guinea pig and cat. Though the potentials are present in the human ear it is only rarely that circumstances permit their recording, and their systematic investigation is impracticable. However, the close resemblances among the ears of higher animals readily permit an extension to man of the general principles established in the laboratory animals. Naturally, consideration must always be given to possible differences of detail.

The electrical potentials are produced in the interior of the cochlea during stimulation of the ear with sounds and are recorded by means of electrodes in the vicinity of the cochlea. It is thus necessary to expose the cochlea by opening the middle ear cavity and to place an active electrode at some point on or in the cochlear capsule and an inactive electrode in other tissues near by. After suitable amplification the potentials picked up by these electrodes are observed and measured. We shall describe the steps of this procedure and the special precautions that are necessary.

Anesthesia. The choice of anesthetic depends upon the expected duration of the experiment and whether the animal is to be kept or disposed of at the end of it. If the animal is to be kept the operative procedure must be sterile and the experiment must not be too prolonged, preferably not beyond an hour or two. When the experiment does not require a subsequent examination of the animal it is preferable to avoid the complications of sterile procedure and the necessary after-care of the animal. The animal is then dispatched at the end of the tests. By far the greater part of our work is done in such acute experiments.

For acute experiments on guinea pigs we use ethyl carbamate (Urethane, U.S.P.) in 20 per cent aqueous solution, in a dosage of 7.5 cubic cm per kilogram of body weight, injected intraperitoneally. This drug in such doses produces a very deep anesthesia lasting many hours. Indeed, recovery from the anesthesia is unsatisfactory. The respiration is much depressed, yet in most animals is sufficient to maintain good physiological condition for half a day or so.

For chronic experiments on guinea pigs we use sodium pentobarbital (Nembutal—Abbott) intraperitoneally in a dosage ratio that varies with body weight. Careful records kept over several years by Alexander and Githler have shown that small animals up to 500 grams in weight require 0.6 cubic cm per kilogram, and animals over 500 grams require 0.5 cubic cm per kilogram. These doses produce deep surgical anesthesia over a period of about two hours. The animal must be watched and manual aid to respiration must be given if breathing becomes labored or light.

For acute experiments on cats we use a solution of diallyl-barbituric acid with Urethane (Dial–Ciba) intraperitoneally in a dosage of 1 cubic cm per kilogram of body weight. This dosage insures a deep anesthesia, but usually, to prevent activity of the tympanic muscles, it is supplemented with curare (d-Tubocurarine chloride–Parke, Davis, and Co.) in a dosage of 0.5 cubic cm per kilogram injected into the femoral vein, after which artificial respiration is required throughout the experiment. This procedure gives a very uniform physiological level, such that we have been able to obtain a cochlear response that is constant within 1 db over a period of two or three hours while stimulating with a constant sound.

For chronic experiments on cats and monkeys we use Nembutal intraperitoneally or intravenously in a dosage of 0.7 cubic cm per kilogram of body weight. In intraperitoneal injection the complete dose is given at one time unless the animal is in a highly excited state, as untamed animals often become in the process of being caught. With such excited animals it is wise to give about one-third of the dose first and wait a few minutes until the pulse rate is down to normal before giving the remainder of the dose. A full dose in excited animals may be fatal. In intravenous injection the rate of injection should be slow, so that about a minute is taken for the full dose. The animal is carefully observed during the injection and the injection is interrupted if the pulse and respiration become seriously impaired.

Introduction of the Stimulus. Aerial sounds are introduced into the ear by two methods, the open-tube and the closed-tube methods. In the open-tube method the animal's ear is placed at a standard distance (say, 2.5 cm) from the end of a tube leading from a loudspeaker. The intensity of this sound can be measured by removing the animal and placing a microphone at the same position that was occupied by the ear or else by keeping the animal in position and introducing an acoustic probe tube into the orifice of the ear. The latter method is the more exact. This open-tube method has the advantage that the animal's ear is left intact and is unmodified by the stimulating apparatus.

In the closed-tube method the tube from the loudspeaker is attached to a rubber tube that ends in a cannula inserted into the external auditory meatus and firmly secured there. For best results it is advisable to remove the pinna, leaving a short length of the cartilagenous portion of the meatus into which the cannula is inserted and to which it is tied with strong thread.

This closed-tube method has two advantages over the other. It gives greater intensities of sound and less distortion because it affords a better acoustic matching of loudspeaker and ear. Also the removal of the pinna permits a clear view of the deeper part of the meatus, which in many animals, especially in old cats, is more or less filled with cerumen and dirt, especially near the drum membrane. When this material is present it is necessary to remove it to permit a normal entrance of sounds and a free functioning of the drum membrane.

With the closed-tube method the acoustic probe must be used to measure the sound intensity. We employ a special cannula into which this probe is incorporated as shown in Fig. 8. The probe

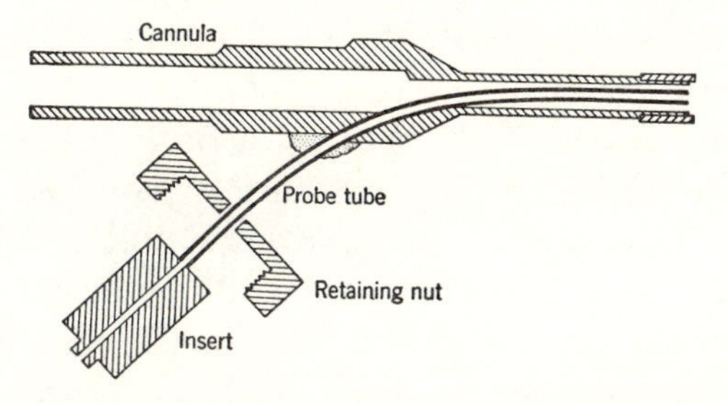

Fig. 8. Sound cannula and probe tube. The insert fits into a cap over the microphone as shown in Fig. 5.

tube enters the cannula through the side and then runs concentrically with it to end at the same level. Therefore the sound intensity as recorded is at the deep part of the meatus, within 2 or 3 mm of the drum membrane.

For stimulation by bone conduction we have used the crystal receiver described above, fitted with a special nosepiece that bears a heavy steel needle. This needle is applied so as to make contact with some portion of the cranium or sometimes with the cochlear capsule itself.

Operative Procedures. The particular surgical procedures vary according to the purposes of the experiment, but an exposure is always made of the round window region or of some other portion of the cochlear capsule. The approach varies a little in different animals.

The exposure of the round window in the guinea pig is particularly simple. An incision is made through the skin immediately behind the pinna and a part of the mastoid portion of the temporal bone is thus exposed just forward of the lambdoidal ridge. A hole is then drilled with a small dental burr until the middle ear cavity is reached. The proper place of drilling is behind the point of exit of the facial nerve through the stylomastoid foramen, or, more exactly, midway between this point and the lambdoidal ridge. The bone is usually less than half a millimeter thick in this region. The opening is enlarged a little downward until on looking in an anterolateral direction the round window and the basal end of the cochlea can be seen. This window lies below and beyond a ridge of bone that marks the location of the lateral semicircular canal. An electrode can easily be placed upon the round window membrane through the opening. A binocular operating microscope or loupe and a light that can be directed into the opening are essential for this work.

The round window in the cat has a position corresponding to that in the guinea pig but its exposure is more difficult. In acute experiments on the cat the procedure is as follows. A skin incision is made in a line between the ear and the angle of the mouth directly over the parotid gland. The facial vein lying over the parotid gland is then cut between double ligatures. The gland is elevated and dissected out to expose its blood supply. To expose the arterial supply it is often necessary first to free and elevate the digastric muscle, which runs from below the angle of the lower jaw to the mastoid region. Then the parotid, external maxillary, and external carotid arteries are identified. Usually the external carotid artery gives off the parotid artery and then the parotid artery immediately gives off the external maxillary artery. Sometimes the parotid and external maxillary arteries arise from the external carotid artery independently and at the same place. The external maxillary artery is ligated as close to the jaw as possible. The external carotid artery is ligated above and below the region of origin of parotid and external maxillary arteries. The parotid gland is now removed. The digastric muscle is removed by cutting it as close to the mandible as possible and then freeing its upper end from its origin at the jugular process of the occipital bone. The external carotid artery is cut between its ligatures. The styloglossus muscle will now be seen lying over the stylohyal bone of the hyoid chain. Also in view is the small, flat jugulohyoideus muscle lying on the lateral surface of the auditory bulla. The styloglossus muscle is cut and the hyoid chain disarticulated between the stylohyal and epihyal bones. The styloglossus and jugulohyoideus muscles are removed together with the upper end of the hyoid chain, which includes the tympanohyal bone. The auditory bulla is now exposed, and is opened with a dental burr. This procedure has the advantage of giving a very flat field.

In experiments on cats in which the animal is to be saved and on guinea pigs in which access to the apical portion of the cochlea is desired, it is best to use a ventral approach through the region of the throat. After orientation by palpating the auditory bullas, a 6 cm midline incision is made through the skin and superficial muscle layers. Then by blunt dissection and wide retraction the bulla is exposed on either side. This method has the advantage that no blood vessels need be tied or other parts dissected away. Also it gives an approach to both bullas through one incision. However, the working field is rather deeply situated.

Recording Methods. The electrode that is applied to the round window membrane consists of a piece of platinum foil 2.5 micra in thickness, cut in the form of an isosceles triangle about 1.5 to 2 mm wide at the base and 3 mm high, and soldered to the end of a piece of No. 36 gauge enameled copper wire (diameter 0.013 cm). The wire is held in a manipulator and placed so that the tip of the foil is in contact with the midportion of the round window membrane. Contact between the foil or the uninsulated end of the wire and the rim of the round window or other tissues in the region should be avoided whenever possible, as such contact causes a partial short-circuiting of the potentials and therefore reduces the readings, usually by about 3 db. In guinea pigs, especially young ones in which the opening must be small, it is often more convenient to place the electrode manually.

For recording from the surface of the cochlear capsule we usually use a fine steel needle mounted in a thin brass handle, to the upper end of which an enameled wire is soldered.

For recording from the interior of the cochlea a minute hole is drilled barely through the bone and a fine wire is inserted. It is practically impossible to carry out this procedure without doing some damage to the sensory structures.

The indifferent electrode is usually a steel hypodermic needle placed in the masseter muscle.

The potentials that may be recorded from the cochlea vary in magnitude from a little over 1 millivolt to something less than 0.1 microvolt. The small magnitudes make it necessary to use considerable amplification before these voltages can be recorded. It is also essential that the amplification occur with a minimum of spurious noise, for the lower limit of the measurements is determined in part by the level of this noise. Moreover, the amplifier should operate over the whole audiofrequency range. The amount of amplification depends upon the recording method.

Our arrangement of amplifying and recording apparatus is indicated on the right of Fig. 7. The preamplifier is designed for wide range and freedom from internal noise and is adjusted to give a voltage gain of 1000-fold (60 db). Further amplification is incorporated in the recording instruments, of which three types are shown.

For measurements on sinusoidal responses we use a wave analyzer, which is a highly selective voltmeter. This instrument has the special advantage of minimizing the background noise and is particularly useful in measuring the weakest potentials. Its dial is set to the frequency of the response being measured. For complex responses it gives the magnitudes of the various components present when its dial is set to these frequencies in turn. When many components are present this is a rather laborious procedure, and if only the root-mean-square magnitude of the complex response is needed a more convenient instrument is a vacuum tube voltmeter designed for this sort of reading. For visual observations of wave forms a cathode ray oscillograph is used. It has the particular advantage of allowing a photograph to be made of its trace, giving a permanent record.

As the figure shows, the animal is placed in a soundproofed, electrically shielded room. It is essential that the shielding be sufficient to exclude any extraneous currents, and no electrical conductor must be allowed to pass through the shield. The adequacy of the shielding and the absence of artifacts from any source should be checked by carrying out the regular procedure on a dead animal. Special care is necessary in checking the responses to intense high tones.

4

The Sensitivity of the Ear

In our consideration of the sense of hearing the first and most fundamental problem is that of sensitivity: the delicacy with which our ears inform us of the presence of vibratory energy in our environment. Because this sensitivity is great—because under good conditions the ear is able to detect incredibly small amounts of energy—this sense has attained the highest biological status as the protective sense, as the one that most promptly and diligently acquaints us of those happenings in the world outside that are of most urgent concern to us. At the same time this sense stands in the first rank alongside of vision in providing us with information about the finer details of the world and its changes. We are, of course, most acutely conscious of this service of the ear only after it is denied to us, when accident or disease has seriously impaired the sensitivity.

THE EXPERIMENTAL METHODS

In dealing with the problem of sensitivity we have to recognize two different though always related approaches, one general, in which we take a broadly functional view and treat the ear as an integral part of the responding system, and one specific, in which we take a narrower view and treat the acoustic mechanism by itself. In the first approach we determine the sensitivity of the whole organism by measuring the strength of the sound field that has to be present in an environment in order to be detected by the person when he is introduced into the environment. In the second approach we determine the sensitivity of the ear by introducing the sound at some designated place within the ear and measuring it there. Our choice of these two approaches is determined by practical considerations, for the first gives us information about suitable listening conditions and the second gives more specific aid to our understanding of the operation of the ear.

The results obtained under these two procedures are different because of a peculiarity of sound already referred to: it is a labile sort of energy and its magnitude in an environment is altered, often considerably, by the introduction of new objects like the human body. When a person enters a sound field his presence may raise or lower the energy as a whole and in any event will cause a redistribution of the energy in his vicinity. Therefore the energy playing upon his ears will not ordinarily be the same as the energy that formerly was passing through the same points in space. A fairly rigid object like the head, for those sounds whose wave lengths are small with respect to its dimensions, has the general effect of raising the sound pressure close to its surface on that side from which the waves come. We have already met these conditions on a smaller scale in connection with the calibration of microphones.

The Free-Field Method. The first general method of measuring sensitivity we identify as the free-field method. The procedure is to introduce the person into the sound field and to vary the field intensity until the person is barely able to perceive the sound, and then to remove the person from the field and to measure the intensity existing there. The person's effects upon the field—the absorptions of his clothing and skin, the reflections and diffractions by his body and especially the head—are all taken as aspects of his hearing capacity. It is further necessary to specify the kind of sound waves present and the person's position with respect to them.

The simplest kind of sound waves are plane progressive waves. These are waves that are coming from some one direction in a regular manner, undisturbed by reflections from intervening objects and therefore with wave fronts that are perpendicular to their paths of flow. Such waves are easy to work with because their energy distribution is uniform along the wave front and we do not have to be concerned about slight changes in position of our measuring instrument. It is difficult in practice to obtain a wave that is truly plane, but a close approximation is obtained if we work at a moderate distance (say, 2 meters or so) from a small source like a tube with a 2 or 3 cm opening from which the sound comes. The waves from such a source spread out with spherical wave fronts, but after they have traveled a little distance their curvature is small enough to be neglected. It is necessary to guard against any disturbance of the sound field by extraneous objects. The measuring apparatus must be kept small and reflections from the walls of the experimental room must be minimized by lining them with sound-absorbing material. It is usually specified that reflected waves should be at least 20 db weaker than the waves received directly from the sound source. To obtain this much reduction of reflection is difficult at low frequencies, and elaborate treatment is necessary to secure free-space conditions down to 100~. Beranek and Sleeper described an "anechoic" chamber that fulfilled this requirement. Its walls were lined with wedges of compressed mineral wool, which provided an energy absorption of 99.99 per cent above 100~.

The most natural orientation of the listener is facing the source with head and trunk erect and both ears open. Results of Sivian and White obtained by the free-field method, and with this natural orientation, are shown as curve *a* of Fig. 9.

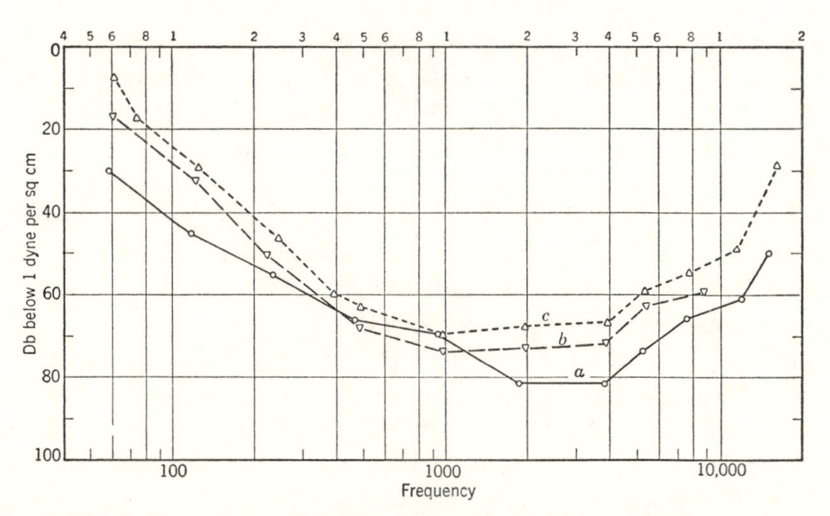

Fig. 9. Threshold sensitivity of the human ear, as measured under three conditions. For curve *a*: binaural stimulation, free-field method; average of results for 10 subjects. For curve *b*: binaural stimulation, closed-ear procedure; results for 22 subjects. For curve *c*: uniaural stimulation, closed-ear procedure; results for the 22 ears of 11 subjects. The data were obtained by W. A. Munson, and most of them have been reported by Sivian and White.

The Intra-aural Method. The second way of measuring sensitivity may be called the intra-aural method because the measurements are made at some designated place in the ear, as for example at the entrance to the meatus or immediately adjacent to the drum membrane. The information wanted is the sound pressure actually present at this place when the sound is just audible. To obtain it two procedures have been used.

Closed-Ear Procedure. In one procedure the ear is more or less tightly closed by the sound-producing instrument, which is usually a telephone receiver, so that a volume of air is trapped inside. The sound pressure established in this enclosed space when a given current is sent into the receiver is then determined from the receiver characteristics or is measured by the sound-probe method. The calculation procedure, though much used in the past, has drawbacks arising from the difficulty of determining the receiver characteristics under the particular conditions and from uncertainty about the volume of the enclosed air. Also it is necessary to assume uniformity of pressure in the enclosed space, which is valid enough for low-frequency sounds but ceases to be so for high-frequency sounds whose wave lengths come close to the dimensions of the cavity. The sound-probe method makes use of a microphone fitted with a front cap and an extending probe tube, as described earlier (page 27). The tube is passed through an opening made at the rim of the receiver so that its tip lies within the enclosed space. The microphone and probe tube are calibrated to indicate the sound pressures existing at the tube's tip. Results obtained by this method for binaural stimulation are presented in curve *b* and for uniaural stimulation by curve *c* of Fig. 9.

Open-Ear Procedure. In a second procedure the ear remains open and the sound probe is used to measure the pressures set up in the ear by an external sound field. This method has a number of advantages over the other. It is the more natural way of listening and usually is the more comfortable. Most persons find that the wearing of a receiver on the ear becomes unpleasant and even painful when continued for the extended periods often necessary for threshold determinations. Only limited data have so far been reported by this method.

ANALYSIS OF CONDITIONS AFFECTING SENSITIVITY

We now attempt an analysis of the various experimental conditions affecting determinations of the ear's sensitivity.

Obstacle Effect of Head and Body. As the foregoing discussion has suggested, an outstanding difference between the free-field and the intra-aural methods of listening is the presence of the listener as an obstacle in the free field. In the open-ear form of the intra-aural method the presence of the listener also modifies the sound field, but because the measurements are made at the ear these modifications are of no account.

The effects of the listener as an obstacle are shown in Fig. 10.

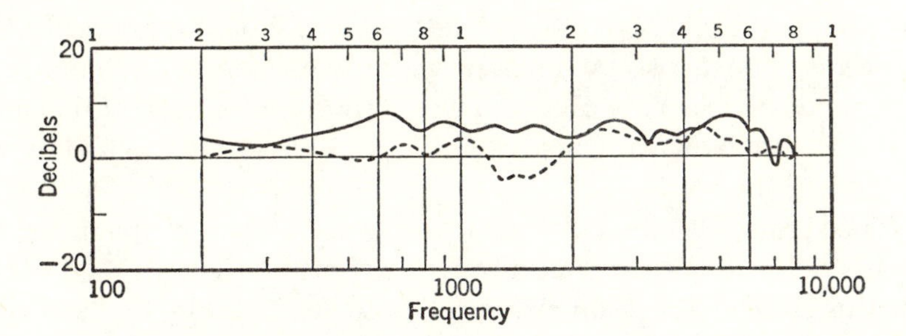

Fig. 10. Obstacle effects of head and body. The sound pressure at the entrance to the blocked external auditory meatus is compared with the pressure at the same point in the sound field in the subject's absence. A positive difference represents an increase in pressure due to the presence of the subject. For the broken curve the subject faced the source of sound, and for the solid curve he directed one ear to the source and the measurements were made in this ear. After Wiener and Ross.

These results represent differences between pressures existing in an empty sound field and pressures existing at the entrance of the external auditory meatus on one side after a person was introduced into the field. In order to isolate the obstacle effect from the resonance of the meatus, which as we shall see has an effect of its own, the meatus was filled with a rubber and wax plug. The broken curve of this figure was obtained when the person faced the source of sound. Even larger effects were found, as the solid curve shows, when the head was turned so that one ear was directed toward the source and the measurements were made at the entrance to this ear.

As will be seen, the obstacle effects are small and when the person faces the source they never exceed 5 db, but they are generally positive, representing increases of sound pressure, except for a narrow region around 1500~. The more marked effects are found in the regions about 2500 and 4500-5000~. When the head is turned so that the ear is toward the source the effects are larger at nearly all frequencies, and herein lies the reason why in listening carefully to a faint sound, as everyone knows from experience, it is advantageous to turn one ear directly toward the sound. Presently we shall consider this effect of angular position of the head in more detail.

Resonance of the Meatus. In both the free-field method and the open-ear form of the intra-aural method there is a resonance effect of the meatus. Because the measurements are made in the air outside, this effect has to be considered as a separate factor. In the closed-ear form of the intra-aural method there is similarly a resonance effect of the meatal cavity (for high tones), but because the measurements are made within the cavity this effect is included in them.

The resonance effect of the meatus is seen when probe measurements are made of the sound pressures both at the entrance to the meatus and at its termination as close as possible to the drum membrane. The differences in these measurements represent the effects of the meatus. Results obtained by Wiener and Ross are given in Fig. 11, and show a prominent peak near 4000~, where

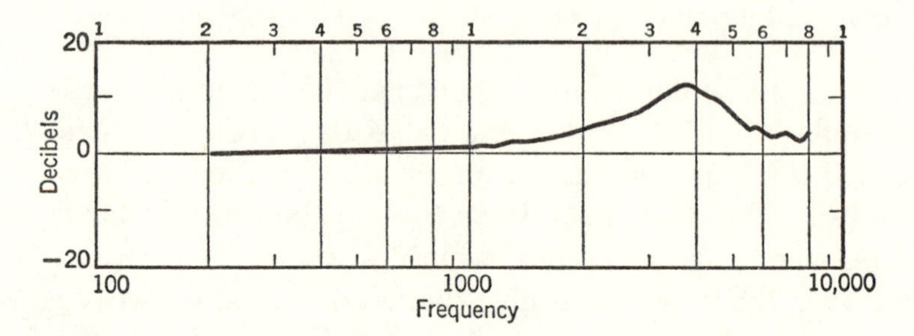

Fig. 11. Resonance of the meatus. After Wiener and Ross.

the gain in pressure at the drum membrane amounts to about 12 db. This is clearly a resonance effect. The meatus is acting like a closed tube, which resonates best when its length is one-fourth the wave length of the applied sound. Wiener and Ross found the average length of the meatus in their subjects to be 2.3 cm, which would give 9.2 cm as the wave length of the resonated sounds and hence a frequency of 3800~, which is close to the one observed.*

* At the meatus temperature, estimated as 30° C, the speed of sound is 350 meters per second.

The resonance is not very sharp because the tube is not closed by a firm wall but by the drum membrane which transmits a large portion of the sound inward and so has a damping effect.

Effects of Azimuth. The angular position of the listener and more particularly of his head with respect to the source of sound has often been indicated in terms of the "azimuth," a term carried over from astronomy where it refers to the location of a star on the arc of the horizon. In auditory usage, unfortunately, the azimuth has been defined variously by different authors. It seems wise to adhere to a definition that conforms as closely to the usage in astronomy as the altered conditions will permit. Therefore we shall refer to the azimuth as the angle, measured in a clockwise direction, between the listener's line of sight and a line drawn toward the source of sound, as illustrated in Fig. 12. For increas-

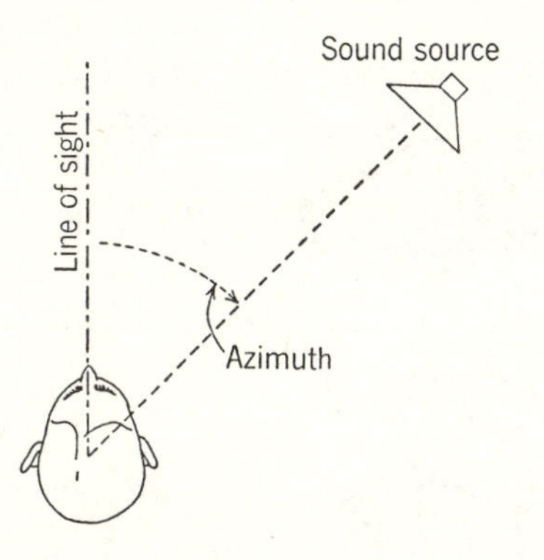

Fig. 12. The azimuth of a source of sound.

ing azimuths therefore we may think of the sound as moving clockwise around the observer, whose position remains fixed. Actually, of course, it may be found more convenient in an experimental situation to keep the sound source fixed and rotate the observer, but our designations of azimuth are still made as described. This definition gives zero azimuth for the normal listening position when the source is in front, a 90° azimuth when the source is opposite the right ear, a 180° azimuth when it is directly behind, and a 270° azimuth when it is opposite the left ear.

If listening in a free field is to be done with only one ear the other is carefully plugged, for example with oil-soaked cotton or a specially designed ear stopple. The azimuth conditions then become more important than when both ears are open, especially when the single functioning ear is turned away from the sound into the "shadow" cast by the head.

Figure 13 shows some of the effects of azimuth conditions on

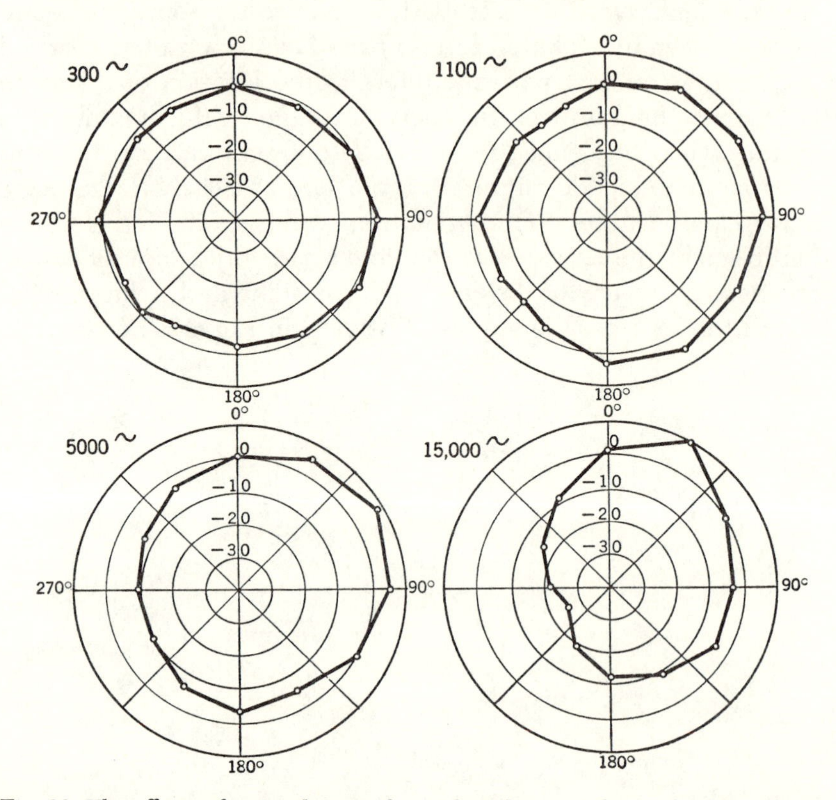

Fig. 13. The effects of azimuth upon four selected tones when only the right ear is used, the left being plugged. These are polar graphs, with azimuth plotted as angular displacement and the change in effective intensity plotted in decibels along the radius relative to the intensity at zero azimuth. Thus as the curve approaches the center of the graph it represents a diminution in the sound reaching the right ear. Data from Sivian and White.

tones of 300, 1100, 5000, and 15,000~. The results represent threshold observations by Sivian and White, made on three subjects with one ear open and the other plugged. The data as plotted refer to the right ear as the open one. Similar results were obtained by Wiener (*1*) by direct measurement, by means of a sound probe, of the sound intensities in the vicinity of the drum membrane for various azimuth positions.

Both the threshold observations and the physical measurements included the resonance effects of the meatus, but because this resonance does not vary with the azimuth the results as shown are due only to obstacle effects. It is clear that at 300~ the angle of the sound has little effect, but for the higher tones, and especially at 15,000~, the head casts an appreciable sound shadow.

In Fig. 14 we use an alternative method of presenting the

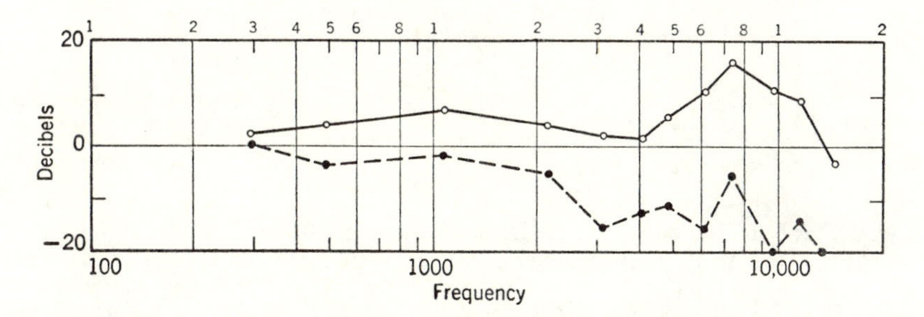

Fig. 14. The effects of 90° azimuth (solid line) and 270° azimuth (dashed line) over the frequency range from 300 to 15,000~. Data from Sivian and White.

Sivian and White results, a method that brings out in greater detail the variations with frequency for any given azimuth. This is a rectangular plot for two azimuth conditions, again represented relative to zero azimuth. The solid line is for 90° azimuth (listener with his right ear open and directed toward the sound) and the dashed line is for the 270° position (this ear directed away from the sound). The 90° position shows an appreciable gain in the region of 1000~ and a very marked advantage around 7000~. The 270° position, on the other hand, shows a loss that grows more serious with the higher frequencies, amounting to 20 db at the upper end of the scale. From these results we see clearly why it is that a person who is deaf in one ear has trouble in hearing us when we are on the wrong side of him.

Because the diffraction effect of an obstacle is rather well localized, reaching its largest values close to the surface of the obstacle, the changes shown in Figs. 13 and 14 are predominantly due to the head, and the trunk makes only a slight contribution to them. That this is true is shown by a comparison of the effects just described with those produced by a rigid sphere of about the size of the head, as represented in Fig. 15. Of particular importance here are conditions of asymmetry about the head: the location of the ears behind the center of the head and especially the presence of the auricles. Differences found between 45° and 225° azimuths are almost wholly due to the auricle. This structure has practically no effect upon hearing at low frequencies, but in the high frequencies, for which its dimensions are of the order of magnitude of the wave lengths, it has a barrier effect. An exag-

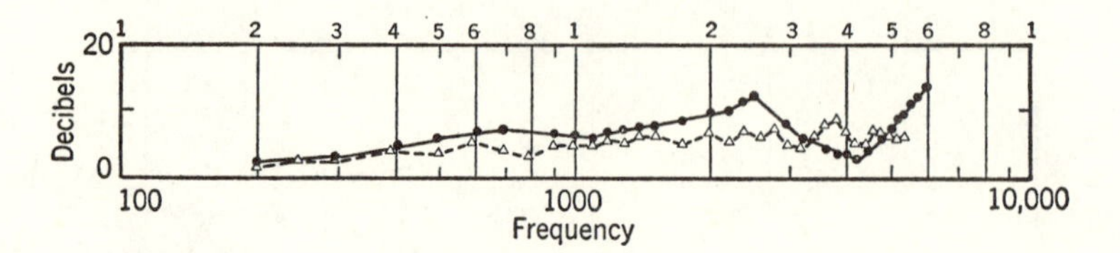

Fig. 15. The diffraction effects of the head (solid line) compared with those of a wooden sphere 19.4 cm in diameter (dashed line), measured for the head at the 90° azimuth position and for the sphere at the 0° position, and expressed in decibels relative to the free-field intensity. Data from Wiener (2) and Wiener and Ross.

geration of this effect is produced when in intent listening we place the cupped hand behind the ear (Wiener and Ross).

Binaural and Uniaural Listening. The free-field method as ordinarily employed has the advantage of including two ears, whereas the intra-aural method uses only one ear. This advantage is small, however, and under some conditions is obscured by experimental variations. Yet it is possible to show that two ears, if functionally about equal, are uniformly more acute than one. Shaw, Newman, and Hirsh, by a method in which they compensated for any differences between the two ears, got a binaural summation amounting on the average to 3.6 db. If one ear is considerably better than the other this summation is insignificant and the performance of two ears is not noticeably different from that of the better ear.

The Closed-Ear Effect. A significant difference has been observed in auditory sensitivity when tests are made with a loudspeaker and with a telephone receiver worn in the usual way. The loudspeaker sound seems louder than that of the receiver even though the two stimuli are equal as measured in terms of sound pressure in the auditory meatus. The difference is most prominent for the low tones, for which it may amount to 6 to

10 db. The effect has mainly been studied at levels well above threshold, but no doubt it holds for threshold sensitivity as well.

Munson and Wiener carried out a number of well-controlled experiments showing that the difference is due to the closing of the ear and is not an instrumental artifact, but they were unable to reach a final conclusion as to its cause. Our suggestion is that the difference in sensitivity reflects an alteration in the effective impedance of the ear. Evidently the drum membrane and its associated structures are better matched to the impedance of the open air than to the impedance of a limited volume of air entrapped in the meatus. Our experiments on cats support this hypothesis, for we have found that such things as extending the length of the meatus by means of a tube will alter the sensitivity as seen in the cochlear potentials, even though the pressure of the sound at the drum membrane is kept constant.

THE MECHANICAL IMPEDANCE OF THE EAR

So far in this chapter we have been concerned with sensitivity in a general sense, as the conditions that have to be established in the vicinity of the ear in order for a sound to be heard. We now take up the problem more specifically in a consideration of the energetics of the ear: we wish to determine the quantity of sound that flows into the ear and is utilized there. Our measurements of sound pressures or energy in the environment or even as far inward as the drum membrane are insufficient for this purpose. These measurements show the sound available, but they do not tell what part of this sound is made use of. What must be revealed further is a basic characteristic of the ear itself—its mechanical impedance—for it is this characteristic and its relation to the impedance of the air outside that determine the mechanical efficiency of the receptive process.

The study of the mechanical impedance of the human ear began with some simple observations by Wegel and Lane in 1923 and by Thuras in 1925 (reported by Wegel). The first systematic measurements were made by Tröger in 1930.

These investigators and others who have continued this study of the ear made use of the reflection method of measuring impedance. This method depends upon the principle that the acoustic properties of a mechanism determine its manner of reflection of sounds. When a sound wave is introduced, the mechanism accepts a part of the energy in the wave and reflects the remainder, and the relation between the incident and reflected waves depends upon the impedance of the mechanism. This principle has particular value in our dealing with mechanisms to which we have only limited access, for the observations are made outside the mechanism, in the pathways of the stimulating sounds.

Tröger introduced the sounds through a tube that was closely joined to the external auditory meatus. He observed the wave patterns near this junction and compared them with the ones produced when the end of the tube was closed with a rigid plug.

This comparison determines the nature of the ear's impedance. If the ear behaved just like the rigid plug it would reflect the sound waves completely. Then the impedance would be infinitely large and the absorption by the ear would be zero. On the other hand, if the ear had just the same impedance as the tube it would accept all of the sound without reflection. Actually the ear's characteristics vary with the frequency of the sound that is being used, but they always lie between the two extremes of total reflection and total absorption: the ear accepts a part of the applied energy and reflects the remainder. The experimental procedure reveals this proportionate absorption.

At the same time the procedure reveals further information of interest: it reveals the phase relation between the applied wave and the reflected wave. If in addition to the absorption we know this phase relation we can determine both the effectiveness of the transmission into the ear and the kind of reactance that the ear is presenting. If for any applied frequency the reflected wave has a phase lead over the applied wave we know that the ear's action is predominantly determined by stiffness, whereas if the reflected wave lags in phase with respect to the applied wave we know that the action is predominantly determined by mass. If the reflected wave has the same phase as the applied wave we know that stiffness and mass reactances have canceled one another and that the only limitation on the ear's action is that provided by friction.

If with any simple acoustic device we vary the frequency over a wide range we nearly always encounter all three of these conditions: at the low frequencies the stiffness dominates, at the high frequencies the mass dominates, and at some frequency in between the friction is left alone as the only limitation. What we are usually interested in knowing about the device is just where along the frequency scale the critical changes occur.

Tröger's study was carried out on a single ear stimulated with 28 tones over the range from 250 to 3000~. He found the measurements difficult and subject to large experimental errors. This was especially true for frequencies above the resonance point of the ear. These difficulties have continued to plague subsequent investigators despite numerous improvements of method.

Because Tröger's results and the others have been expressed in different ways we have carried out the calculations necessary to obtain a uniform measure of transmission, which is called the transmission factor. This factor expresses the amount of energy that the ear absorbs in comparison with what it would have absorbed if its impedance were the same as that of free air, a comparison that takes into account both the magnitude of the reflected wave and its phase with respect to the incident wave. Finally, for purposes of presentation here, we have expressed this factor as a transmission loss in decibels. Hence the zero line of our graph represents perfect transmission, and in general the smaller the values the greater is the ear's effectiveness.

Tröger's results are given by the solid circles of Fig. 16. They

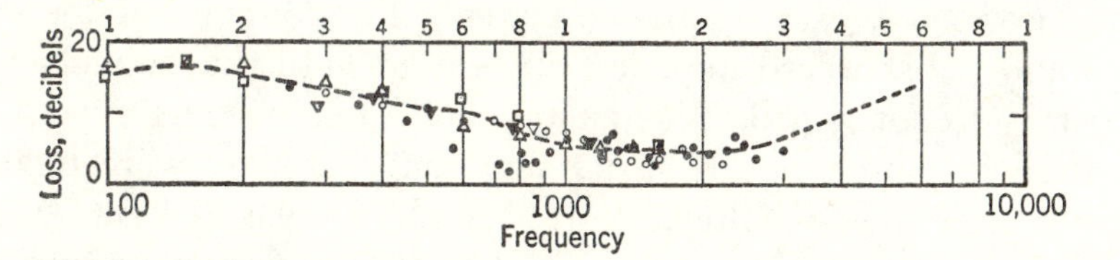

Fig. 16. The impedance of the ear, expressed as a transmission loss in decibels. From six sources: Tröger, solid circles; Geffcken, open circles; Waetzmann and Keibs, upright open triangles; Keibs, inverted open triangles; Kurtz, squares; and Metz, inverted filled triangles. The curve represents a weighted average of all the data.

show a progressively smaller loss, i.e., an improvement of transmission, as the frequency rises from 250~ to the region of 700 to 900~, and then a series of rapid fluctuations as the frequency rises further.

After Tröger's study came others in rapid succession, mostly carried out under the guidance of Erich Waetzmann in Breslau.

Geffcken in 1934 reported observations by a method similar to Tröger's, and also on a single ear. As the open circles of Fig. 16 show, they agree fairly well with Tröger's except that the smallest loss (most favorable transmission) occurs at higher frequencies.

Waetzmann and Keibs in 1935 gave results for two ears, as shown by the upright open triangles of Fig. 16. They corrected a fault in the previous tube method, which employs a tube so short that the wave reflected from the ear is again reflected by the sound producer at the other end of the tube and returns down the tube to disturb the wave patterns. They extended the length of the tube to several meters, which makes this secondary wave negligible. Unfortunately, however, this change also reduces the sound intensity available at the ear, which is a serious limitation. Keibs developed this method further and worked out some welcome simplifications of the mathematical treatment of the data. His results for two ears are given by the inverted open triangles of Fig. 16.

Kurtz in 1938 used the Waetzmann and Keibs method on 12 subjects, with results as shown by the squares of Fig. 16.

Waetzmann in 1938 introduced a new method which employs the acoustic bridge that Schuster had devised. In this method the ear is placed at one end of a tube and a standard impedance is placed at the other end, while sound is introduced at a place midway between. Observations are made at the midpoint of the system and the standard impedance is varied until the ear's characteristics are balanced. Waetzmann and later Menzel used this method for absorption measurements only, as they were unable to overcome the difficulties involved in phase measurements. Metz finally developed the method for complete determinations of impedance and carried out extensive measurements. His results, shown by the inverted filled triangles of Fig. 16, were obtained with 4 tones on a large group of ears, the number of ears varying for the different tones from 41 to 57.

We have averaged the results of the six studies represented here, interpolating when necessary and weighting the results according to the number of ears measured in each study, and the composite function is shown by the curve of Fig. 16. This curve shows a gradual decline in the transmission loss to a minimum that seems to be around 1350~. It is unfortunate that the curve's further course is not clearly indicated, though there seems little doubt that for the higher frequencies it rises as its short-dashed extension suggests.

The Measurement of Amplitude. A different approach to the problem of the ear's sensitivity was made by Wilska. He attached a slender rod to the moving coil of a dynamic loudspeaker and carefully brought the end of this rod in contact with the drum membrane. On sending a current through the loudspeaker coil he was able to drive the drum membrane mechanically. He determined the minimum current required for threshold perception over a range from 45 to 9000~. The measurements evidently were carried out only in a single ear, which is understandable in view of their hazardous nature.

It was possible to calibrate the apparatus directly by measuring the amplitude of movement with a microscope only over the lower portion of the tonal range, up to 270~. The amplitudes at higher frequencies were determined by an extrapolation procedure. This is an unfortunate limitation of the method. The results are given in Fig. 17, but are not to be taken very seriously above the calibration point of 270~.

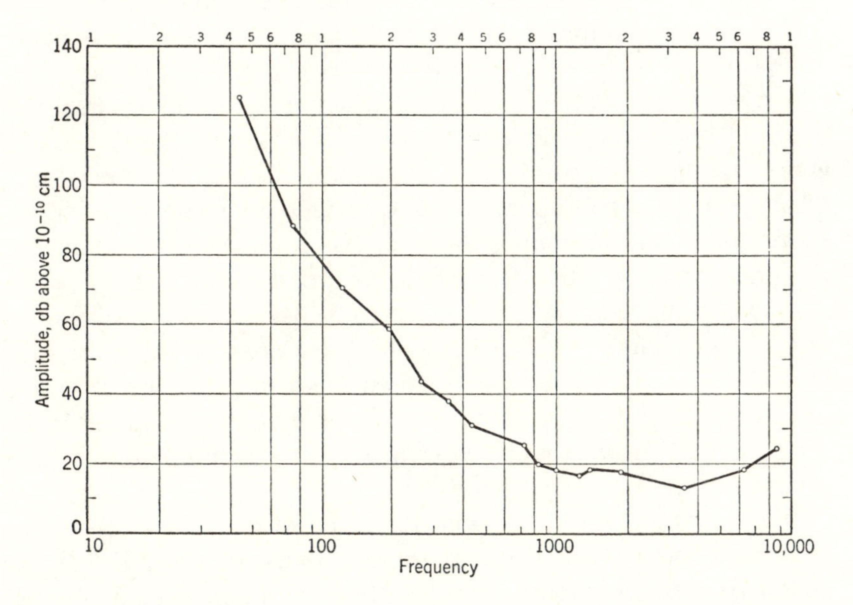

Fig. 17. Sensitivity in terms of the amplitude of displacement of the drum membrane. Data from Wilska.

This method of Wilska's is a promising one, especially if it can be carried out with direct calibration over a more extensive range. If it were used in conjunction with measurements of pressure sensitivity it would provide further information on the ear's impedance.

Energy Requirements. If we know the impedance of the ear it is possible to determine the actual rate of flow of energy into the ear under any condition of stimulation. Of particular interest is this rate of flow at the threshold of stimulation. The calculations involve further only the sound pressures present at the drum membrane at threshold and the effective area of the drum membrane.

The effective area of the drum membrane presents a little difficulty. This area is the one that a simple piston would need to have in order to produce the same volume displacement as the drum membrane when it moves back and forth the same distance as the center of the membrane. It is not the anatomical area of the drum membrane, but something less, because the surface of a membrane cannot move uniformly like a piston. Its edges are fixed and movements of the marginal portions are restrained. There is, therefore, a gradation of motion from the central portion, which has the greatest freedom, down to zero at the edges.

If the drum membrane were flat and uniformly flexible we could easily solve the problem. For a simple plane diaphragm the effective area is 0.306 of the geometrical area. Geffcken used this figure in his computations of threshold energy for the ear, but it is not the correct one. The drum membrane has a conical form, which stiffens its midportion, as discussed in more detail later. By reason of this stiffening the midportion moves as a whole and the gradation of motion is pushed toward the periphery.

It is certain that the effective area of the drum membrane is much more than a third of its geometrical area. Results obtained by Békésy on human ears and by ourselves on cat ears, reported in full in Chapter 6, indicate that the effective area varies in individual ears from about two-thirds to a maximum of perhaps three-fourths of the anatomical area.

We do not know of any extensive series of measurements of the anatomical area of the drum membrane in man. Schwalbe's figure of 0.695 sq cm, evidently representing the average of a number of ears, has already been cited. Keith gave a figure of 0.650 sq cm, which probably is for a single ear. One of us has measured two specimens as 0.598 and 0.630 sq cm. These four values give an average of 0.643 sq cm. There are measurements also of the two principal dimensions of the membrane, the greatest longitudinal diameter and the greatest transverse diameter, from which we can compute the area if we treat the form of the membrane as a true ellipse. On this basis, measurements reported by Gray give an area of 0.694, by Helmholtz (*3,4*) an area of 0.590, and by Bezold (see Siebenmann, *2*) an area of 0.615 sq cm. These figures give an average of 0.633 sq cm, in close agreement with the direct areal measurements just mentioned. If we take two-thirds of 0.643 we have 0.429 sq cm as the effective area of the drum membrane.

The rate of flow of sound energy into the ear for threshold stimulation is computed from the formula

$$\frac{E}{t} = \frac{ATP^2}{R} \qquad \text{ergs per second,}$$

where A is the effective area of the drum membrane, T is the transmission factor obtained from impedance measurements as described, P is the root-mean-square pressure in dynes per square centimeter at the drum membrane, and R is the specific acoustic resistance of air, which is 41.5.

In this calculation it is of course desirable that the values of P, representative of threshold sensitivity, be obtained on the same ears used for the impedance measurements. Such sensitivity determinations were made in the studies of Geffcken and of Waetzmann and Keibs. Kurtz in his study gave sensitivity data on 8 ears, only 4 of which were common to his group of 10 ears used for impedance measurements. Calculations from these data yield the threshold values of energy flow shown in Fig. 18. The values are plotted in decibels relative to an energy flow of 10^{-11} ergs per second. There is fairly good agreement among these values and a single curve is drawn as representative of them. It is clear that the energy flow required at threshold is greatest at the lowest frequencies and becomes progressively less as the frequency rises until the region of 800~ is reached. In this region the function levels off and then rises a little as the frequency increases to 2000~ The further course of the function remains to be determined but there is little doubt that it will continue to rise at the higher frequencies.

The values of energy flow represented in Fig. 18 are those required to produce a continuous threshold stimulation. It is of

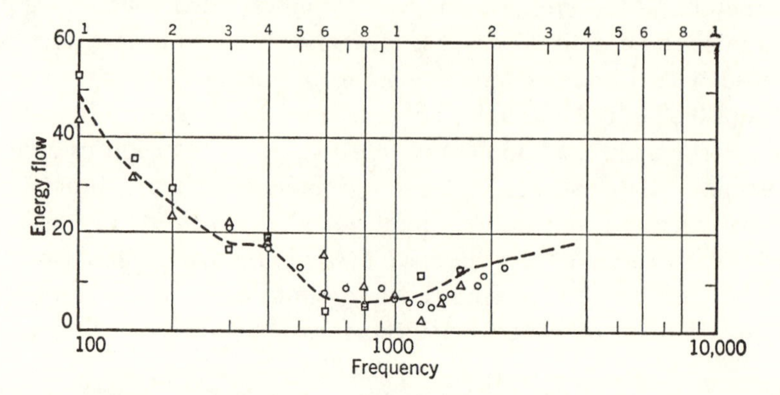

Fig. 18. Energy flow at threshold, in decibels relative to 10^{-11} ergs per second, based on impedance and pressure measurements by Geffcken (circles), Waetzmann and Keibs (triangles), and Kurtz (squares). The curve represents the weighted average.

interest to determine the minimum quantity of energy needed to give a single momentary sensation of tone. This determination involves the functional relationship between the loudness of a tone and its duration. This functional relationship is not fully known for tones near threshold, though it is clear that loudness increases progressively with duration up to some limit (Kucharski, Garner). Studies on tones well above threshold show further that the growth of loudness with duration practically attains its maximum at 0.2 second (Békésy, 2, Munson). This duration also permits a perception of the pitch of the tone (Turnbull). If we multiply the mean value of energy flow at threshold for a tone of 1000~, shown in Fig. 18 as 7 db above 10^{-11} ergs per second or 5×10^{-11} ergs per second, by this duration of 0.2 second we obtain 10^{-11} ergs as the threshold energy.*

THE LIMITS OF SENSITIVITY

Sivian and White raised the question whether threshold sensi-

* A figure for this threshold energy obtained by DeVries (*1*) is in good agreement with this one, but only fortuitously so. He used a duration of 0.4 second, which is about twice too long, and the energy flow data as reported by Geffcken, which involved a drum membrane area that is about half large enough.

tivity for tones is limited primarily by the ear's structural and physiological characteristics or by the background of external activity upon which hearing must occur. The molecules of the air are continually undergoing random thermal agitations, whose effects can be seen under high magnification as the Brownian movements, and these agitations impose fluctuating pressures upon the drum membrane. Theoretically these fluctuations extend over an infinite range of frequencies, but it is obvious that we can neglect the energy lying outside the audible range. Sivian and White considered only the range from 1000 to 6000~, which is the region of the ear's greatest sensitivity. They calculated the pressure level of the thermal fluctuations over this range and found it to be about a third of what any tone in this region must attain in order to be heard by a good average ear. The thermal noise level is thus 10 db below the 1000~ threshold. They pointed out that an exceptionally keen ear might hear this noise.

DeVries in 1948 in discussing this problem made reference to the observations of Wegel and Lane which showed that a tone is noticeably masked only by other tones in its immediate region of frequency. Therefore the effective level of thermal noise is less than that calculated by Sivian and White. DeVries estimated that for a 1000~ tone we need only consider the thermal noise energy from 900 to 1100~, or a band 200 cycles wide. On this basis he obtained a noise level about 25 db below the threshold. He concluded that this background is probably inaudible even to sensitive persons.

DeVries might have made use of the masking data obtained in 1938 by Fletcher and Munson (see Fletcher, 2), which showed more exactly the nature of the masking of tones by noise. These investigators found that in this form of masking it is only necessary to consider the noise energy lying within a critical band of frequencies about the particular tone. These critical bands are especially narrow in the middle frequencies, covering only about 40 cycles for a 1000~ tone (Munson, 2).

A calculation of the effective level of thermal noise in the band from 980 to 1020~ gives a pressure of 1.2×10^{-6} dynes per sq cm. This pressure is about 42 db below that necessary to give threshold stimulation at 1000~ according to the data used for Fig. 18 and even farther below the threshold indicated in Fig. 9. It is clear

PART II

The Middle Ear as a Mechanical Transformer

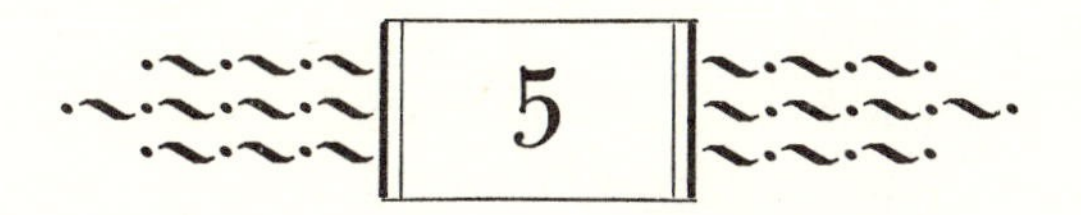

5

The Function of the Middle Ear

Over the long course of evolutionary history many things have happened to produce the highly efficient ear that is now possessed by man and his close relatives among the vertebrates. The original ear, as evolutionary study suggests, was present in some forerunner of the fishes as a simple fluid-filled sac bearing on its interior walls a layer of sensory cells and including also, in contact with these cells and loading them, a dense and somewhat mobile mass. Doubtless the primary function of this organ was not hear-

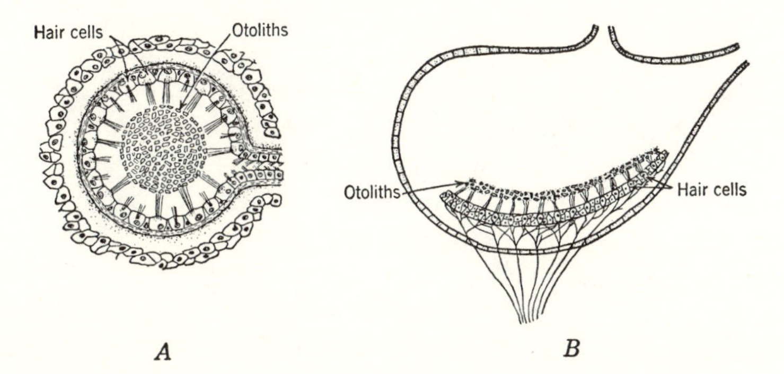

Fig. 19. *A*, the otolithic organ of a mollusc (*Pecten inflexus*), after Bütschli. *B*, the saccule of man.

ing at all but the determination of bodily position, for many animals now living, including ourselves, have static organs of this nature, as for example the macular endings of the vestibule. It happened, however, for simple physical reasons, that this structure was sensitive not only to bodily position but also to movements. It was sensitive to translatory movements of a sudden kind and to vibratory movements of slow rates. The same is true today of our own vestibular organs, which respond to jerky movements and also to sounds if these are of sufficient strength.

Though as an auditory receptor this organ was dull as to sensitivity and limited to the very low frequencies, we can well imagine that it was of great biological value to its possessor. Therefore it was elaborated and improved through the evolutionary process until some of the fishes came to be equipped with an ear of creditable sensitivity and range.

This was an aquatic ear. It responded to vibrations passing through the water, and so it signaled events of the fish's immediate environment. It must have been helpful in the escape from enemies and the avoidance of obstacles such as sharp rocks in swift water, and it may also have acquired a significance in mating.

DEVELOPMENT OF THE MIDDLE EAR

Now we come to an event that may well be regarded as the most remarkable of all in the ear's evolutionary development. This is its transformation into a receiver of aerial vibrations. It is believed to have happened in Devonian times, about midway in the Paleozoic era, some 260 million years ago. Great geologic changes were gradually taking place upon the earth. The earth's solid crust bent and buckled, the seas crept over the continental masses, lakes and marshes appeared and then were filled with volcanic ash and the debris of overtowering mountains. These changes of environment forced the living creatures to adapt or perish, and one of the fortunate adaptations was a foresaking of the inhospitable waters and the taking up of a new life on land. This transition had already been made a little earlier by invertebrates that became the insects and other arthropods of today, and now it was accomplished by some of the fishes.

From the little evidence that we have we can picture the evolutionary process as something like the following. When the seas flowed over the land they carried many marine animals along, and after the seas receded they left these animals behind in brackish pools and marshlands. Rainfall gradually diluted the waters and bore their salt out to sea, and some of the fishes were able to adjust to this change. So the freshwater fishes developed. As the pools became choked with ash and silt and as they grew overpopulated their capacity to supply their inhabitants with oxygen became diminished, and many of the fishes may have resorted to the expedient of gulping air at the surface. Some of these with the necessary powers of mutation developed an air pouch in the throat, the forerunner of the air bladder used as an organ of buoyancy in many fishes. When the walls of this pouch were supplied with a network of blood capillaries a new and efficient way of obtaining oxygen came into operation and so the lungfishes were born. A further supposition is that as the pools grew still less suited to life and finally dried up, these lungfishes were able to hobble out on land and to sustain themselves in a new way of life. They had many adjustments to make, but by slow stages they achieved the transformation into amphibians. In this process their fins changed into legs, their eyes altered in refractive powers to be suitable for aerial vision, and their deep-lying aquatic ears were fitted with a middle ear mechanism, an accessory structure reaching to the surface of the head for the reception and transmission of aerial sounds.

This aerial ear was further elaborated and improved in the reptiles and birds and it reached its highest level of development in the mammals. It proved so far superior to the simple aquatic ear that it has been retained even by those animals like the dolphin and the whale that have forsaken the land to spend most of their lives in the sea.

THE PROBLEM OF SOUND TRANSMISSION

The reason why a special mechanism, the middle ear, is necessary for the adequate reception of aerial sounds is to be found in an elementary principle relating to the transmission of sound between media. Sound waves in one medium do not readily pass to another medium of different acoustic properties. Instead, they are largely reflected back from the boundary. The physical property in consideration here is the acoustic resistance, which as already mentioned depends upon the density and elasticity of the medium. Thus sound waves in air do not enter the water of a pool to any great extent because air and water have vastly different acoustic resistances. Hence a fisherman—contrary to the usual angler's lore—may whistle and shout as much as he pleases without endangering his catch of fish. He may not unwarily splash about in the water, however, if the fish that he is after are of the sort with good hearing, for water and flesh are not greatly different in acoustic resistance and sound waves produced in the water will enter the body of the fish and penetrate to its ear with comparative ease. How effective this sound is in stimulating the sensory cells there depends upon local conditions of the structure, and these vary greatly among the fishes.

We can calculate the amount of sound energy that passes from the air to a body of water because for extended media like these the densities and elasticities and hence the acoustic resistances are known. In this calculation we shall use values of density and elasticity for air at average outside temperature and, because it aids the comparison with the ear later on, we shall use values for sea water at body temperature.

For air at 20°C the density $\rho_1 = 0.0012$ grams per cubic cm and the bulk modulus of elasticity $S_1 = 1.42 \times 10^6$ dynes per sq cm. For sea water at 37°C the density $\rho_2 = 1.024$ grams per cubic cm and the bulk modulus of elasticity $S_2 = 2.53 \times 10^{10}$ dynes per sq cm. The specific acoustic resistance of a medium is the square root of the product of its density and elasticity, or $R = \sqrt{\rho S}$, hence for air the specific acoustic resistance $R_1 = \sqrt{\rho_1 S_1} = 41.5$ grams per second per sq cm (or mechanical ohms per sq cm) and for sea water it is $R_2 = \sqrt{\rho_2 S_2} = 161{,}000$ mechanical ohms per sq cm. The ratio of these acoustic resistances therefore is $r = R_2/R_1 = 3880$.

To find the energy transmission from air to sea water we apply the formula $T = 4r/(r+1)^2$, and for $r = 3880$ the transmission $T = 0.001$. This figure means that only one-tenth of 1 per cent of the energy contained in the aerial waves will enter the water; the remaining 99.9 per cent of the energy is reflected back from the liquid surface. This is poor transmission indeed. The loss expressed in decibels is 30 db.

Now if sound waves in the air acted directly upon the cochlear fluid and this fluid had the same acoustic resistance as sea water we should expect the same transmission loss as just calculated. However, the value used for the elasticity of sea water is for a body of fluid so large that the boundaries of the enclosure have only negligible effects. Under these conditions the elasticity is determined only by relations among the molecules.

The molecular relations in a liquid are peculiar. The molecules are closely packed, with relatively little space between them, and are in a state of equilibrium such that any force tending to change their separation meets with opposition. An attempt to move any two molecules closer together meets a force of repulsion and an attempt to separate them meets a force of attraction. These forces increase in proportion to the molecular displacements. Therefore any deformations, such as the condensations and rarefactions of a sound wave, can arise only in the face of strong molecular forces. What happens in the condensation phase of a sound wave is as follows. The force exerted on a given fluid molecule tends to drive it forward, but such a movement, by bringing it closer to a neighboring molecule, encounters the opposing force just mentioned. The second molecule tends to give way, but in its movement it encounters a similar opposition from a third molecule farther along. If all these molecules could move with perfect freedom the molecular opposition would remain negligible, but this does not happen because every molecule has inertia. Because of this inertia the movement is confined to a limited number of molecules, and their combined inertias permit the molecular forces to exert themselves in opposition to the impressed force. The effective elasticity is relatively great, much exceeding that obtaining in a gas.

Now let us consider the situation in the ear. The cochlear fluid is limited in amount and is closely confined within the cochlear capsule. When the middle ear mechanism is absent, sound has free access to this fluid at two narrow openings, the oval and round windows. If the sound has equal access to the two windows, then at a condensation phase it presses inward on the fluid at both places. The fluid tends to move inward, but this movement is opposed not only by the inertia of the fluid molecules but also by the resistance of the confining bony walls. The fluid movement is largely prevented and the molecular forces rise to a large value. The effective elasticity is very great.

On the other hand, if we should arrange matters so that the sound is applied only to one window—say, to the oval window—and the other window is left free, the constraint imposed by the bony walls is largely removed. A little of this constraint may remain because of friction between the fluid and the walls, but apart from this friction the fluid particles can move freely in response to the pressure of the sound. Accordingly, the molecular forces are now minimized. The inertia of only a small quantity of fluid comes into play, and essentially there occurs a displacement of the whole fluid column extending from one window to the other, and with only a minimum change in the separations of the molecules. The elasticity from molecular forces will therefore be less than in an extended fluid.

If there is this sort of mass displacement of the fluid we must consider the restraints imposed by the two cochlear windows and by the membranes lying between them. The footplate of the stapes is held by its annular ligament and the membrane of the round window has an elasticity of its own. Reissner's membrane is weak and yielding and probably contributes little, but the basilar membrane must present an appreciable opposition to the displacement. The actions of all these membranes tend to restrain the fluid column and to restore it to an equilibrium position. The effective elasticity of this system thus must be considerable, and may easily approach or even exceed that of a boundless fluid. We can find out what this elasticity is only by measurement, and unfortunately such a measurement presents considerable technical difficulties. A further discussion of this problem will be entered into later, in Chapter 17, but for the present we need merely to bear in mind that the elasticity and hence the acoustic resistance of the cochlear fluid is great, and the problem of an efficient transfer of acoustic energy to this fluid from the air is a serious one.

In general, this problem is one of securing a mechanical advantage to the light and relatively inelastic air particles so that they may more effectively communicate their motions to the dense and highly elastic particles of liquid. The solution of this problem is an acoustical transformer.

This sort of problem is not peculiar to sound transmission. It finds a good analogy in our attempt to move a weight that far exceeds our muscular strength. When we push against a huge boulder it altogether resists our efforts and no appreciable energy is communicated to it. Yet when we use a lever we are able to move the weight; the mechanical advantage provided by the lever enables us to make an effective energy transfer. We accomplish this transfer by working on the long arm of the lever and applying our force over a distance much exceeding that through which the burden moves. We pay a price, of course, in the distance over which we must apply ourselves, but this is no real disadvantage. The energy at either end of the lever is the force times the distance moved and, neglecting friction, all the work that we do goes into moving the weight. A better analogy, though not as familiar in everyday experience, is found in the transfer of electrical energy from one circuit to another. If the circuits are not matched in electrical properties—if they differ in their electrical impedance, which is a complex combination of resistance and reactance—the electrical energy will largely be reflected at the junction point. To obtain a suitable transmission it is necessary to use an electrical transformer, which makes the impedance of each circuit effectively the same.

The middle ear apparatus is an acoustical transformer, and it solves the acoustical problem just as the lever and electrical transformer solve the analogous ones. It augments the alternating pressure of the sound waves and presents this higher pressure to the cochlear fluid, thereby effectively matching the external and internal impedances. The transformer ratio that is needed to make this impedance match is the square root of the ratio of the acoustic resistances of the two media. For air and sea water, whose ratio of acoustic resistances is 3880, the required transformer ratio is 63. It is, perhaps, of the same order of magnitude for air and cochlear fluid. In other words, for optimum sensitivity we must have a mechanism that increases the alternating pressure at the oval window over what it is in the aerial waves by some such factor as this.

The conception that the middle ear serves this physical purpose developed slowly during the nineteenth century. It was dimly perceived by Charles Bell in 1803 and by Eduard Weber in 1851, and finally was given a clear formulation by Helmholtz in 1863. Helmholtz's formulation met with immediate favor and has continued in acceptance as a general principle of middle ear function. It is only in recent years, however, that we have obtained experimental evidence of the validity of this principle and quantitative measurements of its effectiveness of operation.

ADVANTAGES OF THE MIDDLE EAR TRANSFORMER

The benefits to sensitivity that the middle ear mechanism provides are easily demonstrated clinically, in the impairments that follow an injury to this mechanism. An accident or disease that interrupts or damages the ossicular chain causes a serious loss of hearing. The person no longer hears such faint sounds as whispering and soft speech, though he will still be able to perceive the louder sounds if the inner ear has not been involved in the injury.

The amount of loss resulting from a middle ear injury varies greatly according to the particular conditions. An injury to the middle ear may not only prevent the transformer action but also may affect the transmission of sounds in other ways. For example, an infection that erodes away the ossicular chain may sometimes remove the drum membrane and sometimes may leave it intact though scarred and stiffened and thereby converted into an obstruction to sound. Such an infection may cause the growth of scar tissue in the middle ear cavity. The conditions may give the sounds various degrees of access to the round window as well as to the oval window, which as we shall see is a serious complication. These variations and their particular causes we shall examine in detail later on. We must make note of them now because they prevent our use of the clinical evidence for anything but a general indication of the transformer action.

The loss of hearing from middle ear damage usually amounts to 30 db and may be considerably greater. Békésy (*8*) reported the average results of tests on 5 ears in which a visual examination showed the drum membrane, malleus, and incus to be absent and in which no other abnormalities were obvious. The average impairments varied with frequency over a range of 15 to 65 db. Of particular interest are measurements on persons who have suffered the disease of otosclerosis and have lost the functioning of the middle ear mechanism and of the oval window, and who then have been treated by means of the fenestration operation to provide a substitute for the oval window (see below, Chapter 16). These persons are improved from a condition of extreme deafness to one of serviceable hearing, but because they no longer have the benefit of a middle ear mechanism they remain on the average at a level of about 30 db below normal.

Exact determinations of middle ear functioning can be made only in experimental animals in which the conditions are under immediate control and particular changes can be produced at will. The following experiments were carried out on cats by use of the electrical potentials of the cochlea to indicate the effectiveness of the ear's action (Wever, Lawrence, and Smith, *1*). These animals were deeply anesthetized and immobilized with curare and the auditory bulla was opened to give access to the round window membrane on which the recording electrode, a platinum foil, was placed. Pure tone stimuli were conducted through a tube to the animal's ear, and near the terminus of this tube and running concentrically with it to end at the same level at the ear was a probe tube fitted to a condenser microphone. This probe and microphone recorded the sound pressures prevailing at the terminus of the sound tube. The sound was applied at the external auditory meatus within about 3 mm of the drum membrane.

The ear was stimulated with various tones and readings were made of the sound pressure required to produce a standard response of 10 microvolts. A set of readings was first obtained for the "normal" condition, with the middle ear mechanism intact. Then the ossicular chain was broken at the incudostapedial joint and the peripheral structures, including the drum membrane, malleus, and incus, were removed. The lateral wall of the tympanic cavity was dissected away to give free access to the cochlea and then the crura of the stapes were carefully removed, leaving only the footplate in position in the oval window. The sound tube was then brought over the oval window, with care to prevent any contact with the stapedial footplate, and was sealed in this position with wax. This procedure was used to prevent access of the sounds to the round window. The ear was again stimulated with the same tones as before and the sound pressures were found that gave the standard response. Some representative results are presented in Fig. 20.

The differences between the two curves of Fig. 20 represent the losses of sensitivity from removal of the middle ear mechanism. These differences, expressed in decibels, are shown in Fig. 21. Alternatively these differences may be regarded as representing the advantage to sensitivity due to the action of the middle ear. This advantage varies with frequency but is considerable along the whole range. It is greatest for tones in two regions, around 500 to 2000~ and around 5000 to 7000~, for which it attains a magnitude of as much as 35 db, and is least for 3000~ and for the lowest tones, for which it is around 20 db. An average figure is 28 db.

The results given here are for a single ear selected as representative of our measurements on a group of 17 ears. Individual ears vary only slightly, usually no more than 2 or 3 db, except for

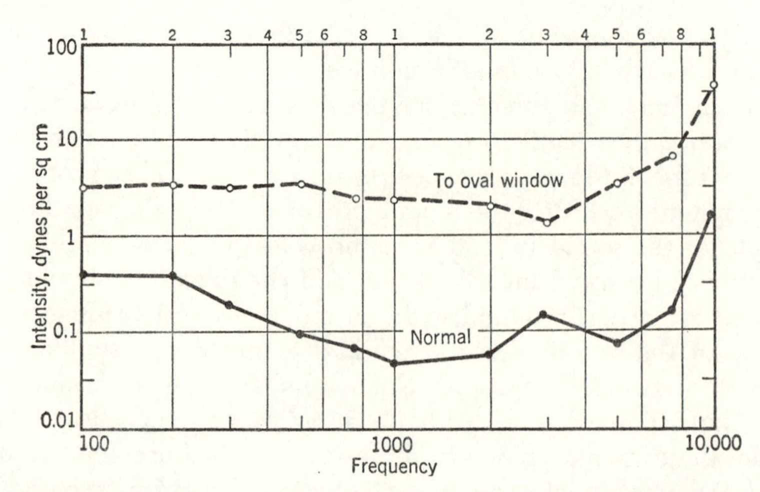

Fig. 20. Sensitivity in the cat, expressed as the sound intensity required to produce a cochlear response of 10 microvolts. The solid line is for an ear with the middle ear intact and the dashed line is for an ear with this structure removed and the sounds delivered directly to the oval window.

ears with clear signs of having suffered from middle ear disease.

These results are somewhat complicated by incidental resonances of the middle ear structures and especially, as we shall see in Chapter 17, by variations of the inner ear impedance as a func-

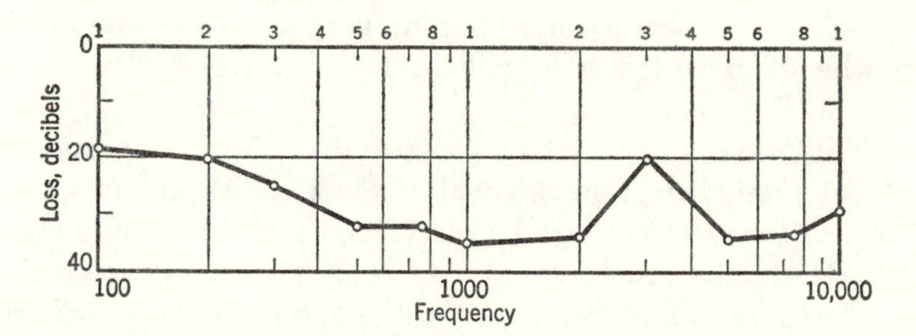

Fig. 21. The loss of sensitivity resulting from removal of the middle ear mechanism in the cat. The differences between the curves of the preceding figure are here expressed in decibels.

tion of frequency. However, we have reason to believe that the average impairment of sensitivity disclosed in the experiments just described is fairly representative of the transformer action of the middle ear.

6

The Nature of the Middle Ear Transformer

At this point we need to consider the particular means by which the middle ear achieves its transformer action. This problem received its first thoroughgoing treatment from Helmholtz (*3,4*), who suggested three processes as combining their effects to this end. These processes are (1) a lever action of the drum membrane, (2) a lever action of the ossicular chain, and (3) a hydraulic action arising from the difference in areas of drum membrane and stapedial footplate. We shall consider these theoretical proposals in detail and attempt to evaluate them in terms of the anatomical and functional evidence.

THE DRUM MEMBRANE AS A CONICAL LEVER

Helmholtz believed that the most important means of pressure transformation in the ear lay in the action of sound on the drum membrane itself. He outlined the idea in his discussion of 1863 and then developed it in considerable descriptive and mathematical detail in his later treatment of 1868. The idea is that the pressures exerted by a sound wave on the gently curved conical surface of the drum membrane are resolved into a greatly amplified force at the apex of the cone.

The principle that Helmholtz had in mind is that of the catenary, as illustrated by a chain suspended between two posts. The tension exerted on the posts greatly exceeds the weight of the chain and it rises rapidly as an attempt is made to pull the chain taut. In fact, the tension would have to rise to infinity in order to make the chain absolutely straight.

Each radial fiber of the drum membrane plays the part of the chain in this example. It is anchored at one end to the edge of the bony tympanic ring and at the other to the tip of the manubrium of the malleus, and its middle portion is relatively free to move in response to sound pressures. In the absence of sound these radial fibers are maintained in a curved form by tensions exerted by the circular fibers of the membrane. The application of sound causes the free middle portions of the radial fibers to move back and forth relative to their equilibrium positions, and this motion,

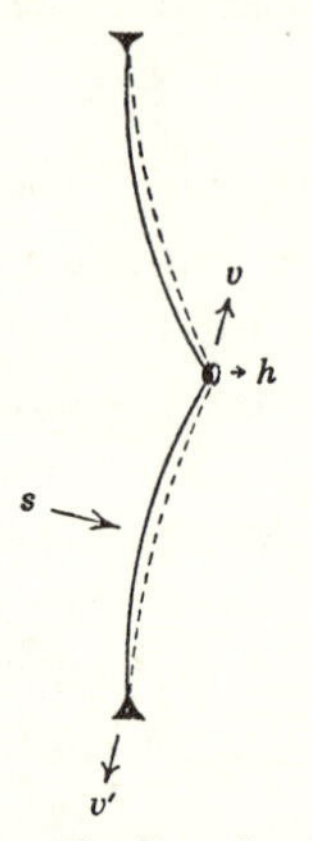

Fig. 22. The catenary principle as applied to the drum membrane. This membrane is shown in cross section, with a sound pressure *s* applied to some of its radial fibers. This local force is transformed into relatively large forces *v*, *v′* at the ends of the fibers, and the force *v* has a horizontal component *h* that tends to move the center of the membrane inward. Helmholtz supposed that the force *h* exceeded the force *s*, but our evidence is to the contrary.

according to the catenary principle as Helmholtz believed, is reduced in amplitude but increased in force at the tip of the manubrium. Thereby is effected an amplification of the force applied to the ossicular chain and the inner ear.

Helmholtz did not attempt any quantitative estimation of this amplification, however. His mathematical treatment of the problem is curiously redundant, so much so as to suggest that though he was convinced of the validity of his conception he was uncertain about its specific processes. He worked out three separate mathematical formulations of the problem and offered them consecutively without any indication of their relations to one another. Actually, these formulations are inconsistent, and two of them if worked out fully do not justify the assumption of any lever action at all. Esser (2) recently offered a mathematical treatment of the problem on the basis of assumptions much like those of Helmholtz and concluded that the drum membrane could provide lever ratios anywhere from one-third to infinity (i.e., giving anything from a pressure reduction to an indefinitely large increase) depending on the properties ascribed to the radial fibers. The lever ratio rises and finally approaches infinity as these fibers are regarded as tense and inextensible. It is clear that this formulation of the problem allows a wide area of uncertainty. Also, because the properties of the tissues are determinative, the problem can only be solved empirically.

The empirical evidence will be treated under three headings. We shall first consider whether the anatomical conditions support the assumptions that the theory makes regarding the structure and form of the drum membrane. We shall then discuss experiments on the form of motion of the membrane in response to sounds. Finally we shall describe direct attempts to measure the transformer action of the membrane.

ANATOMICAL CONSIDERATIONS

Helmholtz's hypothesis requires that the drum membrane have a particular form and mobility as a result of peculiar properties of its radial and circular fibers. The radial fibers must respond to the forces exerted by sound waves by bending but must not be subject to stretching. The circular fibers must maintain the resting curvature of the membrane by exerting a continuing tension and must be readily extensible to allow the membrane to change its form in response to sounds. These two types of fibers therefore are assumed to have different physical properties, and for this assumption there is no obvious anatomical basis. The individual fibers have a similar appearance on microscopical examination. A further difficulty is that the middle portion of the human tympanic membrane—and just that portion of most importance for this theory—is almost wholly lacking in circular fibers.

More direct information on this problem comes from observations by Békésy (*18*) on cadaver specimens. He observed that a fine wire pressed against the tympanic membrane produced a nearly circular indentation. If the tension varied in different directions the indentation would be elliptical. He also used a special scissors to cut out a U-shaped flap at various places in this membrane and studied the form assumed by this partially severed piece of tissue and also the form of the opening produced. In a membrane in which the tension varies in different directions the variation is revealed in a test of this kind by changes in the shapes of the opening and of the piece cut out. The opening enlarges along the line of the greatest tension, whereas the cut-out piece shrinks in this direction. Békésy observed no such changes in the tympanic membrane. In fact, he saw no evidence of any tension at all; this membrane, in the human cadaver at least, appears as a relatively stiff but unstressed tissue.

Békésy studied also the elastic properties of the membrane. By pushing with a calibrated hair on the edge of the U-shaped flap made as just described and observing the resulting deflection he obtained a measurement of the elasticity as 2.0×10^8 dynes per sq cm, and this value was obtained regardless of the region of the membrane in which the flap was cut. The membrane is thereby shown to be uniform in elasticity.

Helmholtz's hypothesis requires that the drum membrane have a particular shape, which is that of a flat cone with slightly incurving sides. This form is essential because the lever action depends upon the tendency of the sound pressure to alter the curvature of the radial fibers, and the lever ratio depends critically on the amount of the curvature. Yet drum membranes vary greatly in form without any obvious relation to sensitivity. Some are retracted, with severe reduction of their radial curvature, yet acuity remains within normal limits. An extreme example was given by Polvogt. In an ear that he studied in serial sections the retraction was so extreme as to bring the membrane into proximity to the bony promontory of the cochlea, thus greatly reducing the radial curvature, and yet the hearing before the person's death was described as normal. According to Helmholtz's hypothesis the amplifying action approaches infinity as the curvature is reduced, and in this ear it should have been greater than normal.

Further evidence comes from experiments on the effects of altering the air pressure in the middle ear cavity, as considered in detail in Chapter 11. According to Helmholtz's assumptions, a negative pressure by reducing the radial curvature of the drum membrane ought to increase the amplification of sound, whereas a positive pressure by increasing this curvature ought to decrease the amplification. The fact is, however, as will be seen in Figs. 76

and 79, both kinds of pressure have a similar effect, which in general is one of reducing the transmission.

PATTERNS OF MOVEMENT OF THE DRUM MEMBRANE

Numerous attempts have been made to study the actions of the middle ear by direct methods. The most obvious method, which is to observe the parts visually, is handicapped by the small scale of the movements. Even if these movements were as large as the movements of the air particles, which is hardly to be expected, they would be only 0.54 mm in peak amplitude for a 100~ tone presented at the very great pressure of 1000 dynes per sq cm, and this amplitude falls off proportionately as the frequency is raised so that it is only 0.0054 mm for 10,000~ at the same pressure. Dimensions such as these are of course discernible microscopically, but they require relatively high powers of magnification and hence a close approach with the microscope objective. Such an approach is inconvenient for a structure of complex form, and investigators therefore have made use of mechanical methods of enlarging the movement. Politzer (2) in studying the ossicles attached fine levers to them. Helmholtz adopted this method, but others objected to the severe loading that the levers produced. Köhler in studying the movements of the drum membrane attached a small mirror from which a beam of light was reflected. This mirror can be made minute in size so as to have a small mass with respect to the mass of the moving parts. Therefore its interference with the movement is less serious than that of mechanical levers, though probably it is not altogether negligible.

Dahmann's Experiments. Dahmann used the mirror method in an extensive study of drum membrane motion. He placed mirrors at various points on the membranes of cadaver specimens and observed the angular deflections resulting from the application of static or slowly varying pressures. The results obtained from 9 positions are indicated in Fig. 23. The deflections, as shown by the sizes of the black areas, are relatively large for the peripheral positions numbered 7, 5, 3, and 9, are smaller for the medial positions 6, 4, and 2, and are still smaller for the manubrial position 1. It must be kept in mind that these are angular deflections, which do not directly portray the in-and-out amplitudes of displacement of the membrane; rather they represent gradients of those amplitudes. The angular deflections are largest at the extreme periphery because there the amplitude gradient is the steepest: the mirror is resting with one edge on the stationary bony ring and another on the moving membrane and therefore is caused to rotate on its

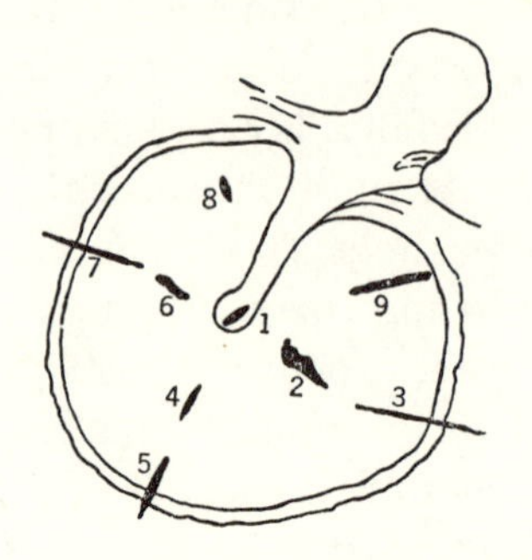

Fig. 23. Dahmann's observations of drum membrane movement. The irregular black spots indicate the relative angular deflections in different regions produced by a sound. From the *Zeitschrift für Hals- Nasen- und Ohrenheilkunde.*

fixed edge. In regions of the membrane where all points are moving with the same amplitude there is no gradient and the mirror shows no angular deflection. Thus if the membrane movement were symmetrical a mirror at its center would give zero deflection no matter how great the amplitude. Evidently, as the image of point 1 in the figure shows, the center undergoes a small angular deflection. The motion here is not symmetrical, but the tip of the manubrium is undergoing a rocking motion.

Apart from the asymmetry just mentioned, Dahmann's results, as far as they go, seem to indicate a diminishing gradient of motion from periphery to center. The form of motion is a flat curve, like *b* of Fig. 24. This is the form of movement taken by a

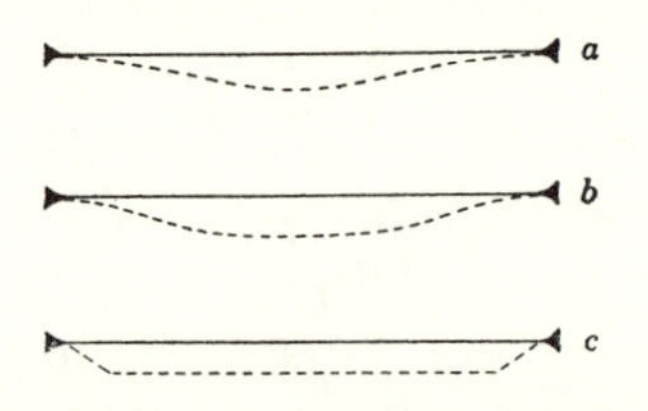

Fig. 24. Three conceptions of the motion of a simple membrane. The solid lines represent the equilibrium position and the dashed lines the position taken in response to a downward force. At *a* is the displacement of a uniformly elastic membrane, at *b* that of one somewhat stiffened in its central region, and at *c* that of one stiffened over almost its whole surface.

simple stretched membrane that is somewhat stiffened in the central area. It is not the form that Helmholtz's hypothesis calls for, which is a double loop, like Fig. 22, with small amplitude at the umbo and maximum amplitudes between this point and the outer edges. Such a motion would have given minimum angular deflections at such points as 4 and 2 in Fig. 23 and larger deflections a little closer to the umbo as at point 6.

Békésy's Experiments. An improved method of determining the form of movement of the drum membrane was used by Békésy (*11*). He used a special instrument, a capacitative probe, with which he was able to measure vibratory amplitudes as small as 10^{-5} mm. The probe was a fine wire placed at some standard distance (usually 0.5 mm) from the vibrating surface. Its end served as one plate of a condenser and the moving surface served as the other plate, and a high-frequency current of 100,000~ was passed through this condenser. With such an arrangement, any movements of the surface toward or away from the probe will change the capacity of the condenser and thereby change the current flow. A vibration of the surface modulates the high-frequency current at its vibration rate. The amplitude of the vibration is measured by amplifying and detecting the modulated current.

With this instrument Békésy measured the amplitude of motion of various parts of the drum membrane in response to a constant stimulus. He found that for all tones up to about 2400~ the whole central portion of the membrane with the manubrium works as a unit. However, this portion does not vibrate simply in and out, but rotates around an axis located at the superior edge of the membrane. Figure 25 shows his sketch of the lines of equal amplitude for a tone of 2000~, and it is evident that the greatest move-

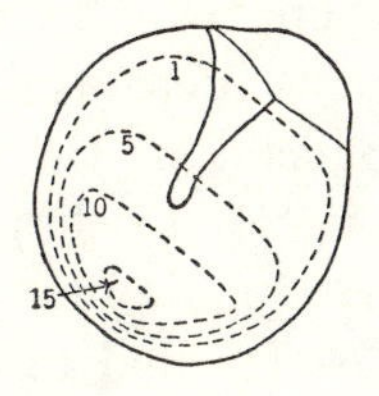

Fig. 25. The form of drum membrane motion according to Békésy (*11*). The broken lines represent contours of equal amplitude, whose relative magnitudes are indicated by the numbers.

ment occurs near the inferior edge of the membrane directly opposite the manubrium. He measured the area of the stiff central portion of the membrane as 55 sq mm in a specimen in which the total area was 85 sq mm.

For tones above 2400~, according to Békésy, the conical part of the membrane seems to lose its stiffness and the manubrium no longer moves in synchronism with it. The manubrial vibration lags behind that of the adjacent parts of the membrane.

For the principal range of tones, up to 2400~, the indicated movement is neither like that of a stretched membrane as shown at *a* in Fig. 24 nor like that assumed by Helmholtz as shown in Fig. 22. Its form, though complicated by the presence of rotation, is equivalent to that shown at *c* in Fig. 24. The results agree with Dahmann's in indicating a rocking motion about an axis, but they do not agree in indicating other regional variations over the surface of the membrane. Dahmann's results are to be explained in part by the experimental errors inherent in the method, but mainly, we think, by his use of pressures of such extraordinary intensity as to modify the characteristics of the membrane. He applied a pressure of ±60 mm of mercury, which in a sinusoidal wave is a root-mean-square pressure of 56,600 dynes per sq cm and a hundred times or more what the ear can safely withstand. Under such a pressure the rigidity of the central portion of the membrane will largely disappear and the membrane will act like a stretched membrane.

DIRECT TEST OF THE HYPOTHESIS

We should have a direct test of the drum-lever hypothesis if we could measure the force produced at the manubrium for a given pressure acting on the membrane. If the manubrial force is found to exceed the product of the membrane pressure and the area of the membrane, then Helmholtz's hypothesis is upheld. Unfortunately, it is difficult to measure forces as small as the ones involved here, and no one has done the experiment.

Helmholtz (*4*) sought to test his hypothesis by an alternative method, in terms of the relative amplitudes of manubrial and drum membrane movements. The applied force (pressure times membrane area) multiplied by the mean amplitude of membrane displacement must be equal to the resulting manubrial force multiplied by its amplitude, and so if the force is increased the amplitude is correspondingly diminished (frictional losses being neglected).

Helmholtz's Observations. Helmholtz used a cadaver specimen and worked the problem backward, that is, he found the membrane movement for a given manubrial displacement. After fitting a manometer tube to the external auditory meatus he filled both tube and meatus with water and observed the volume displacement when the tip of the manubrium was moved inward or outward. From the volume displacement and the area of the drum membrane he calculated the mean deflection of this membrane. He found that when the manubrium was moved 0.036 mm the mean deflection of the drum membrane was about 0.11 mm or three times as much. He then pointed out that because the edges of the membrane are fixed the free middle portion must have been displaced considerably more than three times as much as the manubrium. Thus he considered his hypothesis to be verified, and we must agree if we accept his observations. However, we have repeated his experiment with results that differ widely from his.

A Repetition of the Helmholtz Experiment. We carried out our investigation on the anesthetized cat by two methods. The first method followed the one used by Helmholtz. A rubber tube was sealed into the external auditory meatus and connected to a manometer tube, and this system was filled with water. The manometer tube was a capillary pipette graduated in cubic millimeters and giving measurements by interpolation to 0.1 cubic mm. The auditory bulla was opened to allow an approach to the drum membrane from its medial side. For moving the tip of the manubrium we used a special manipulator as shown in Fig. 26. It consisted of a lever with one arm moved by means of a screw micrometer and bearing on its shorter arm a fine steel rod so shaped as to permit its insertion into the bulla cavity. The lever acted as a reducing lever and gave movements at the manubrium that could be read to 0.0005 mm.

When under these conditions we pushed the tip of the manubrium outward a distance of 0.036 mm, the distance that Helmholtz used, we obtained (in a sample experiment) a volume displacement of 0.31 cubic mm. As the drum membrane measured 35.6 sq mm in this ear, the mean displacement of its surface was

0.0087 mm, or less than one-twelfth of what Helmholtz reported. Instead of a threefold gain in amplitude, as Helmholtz claimed, we obtained a fourfold loss.

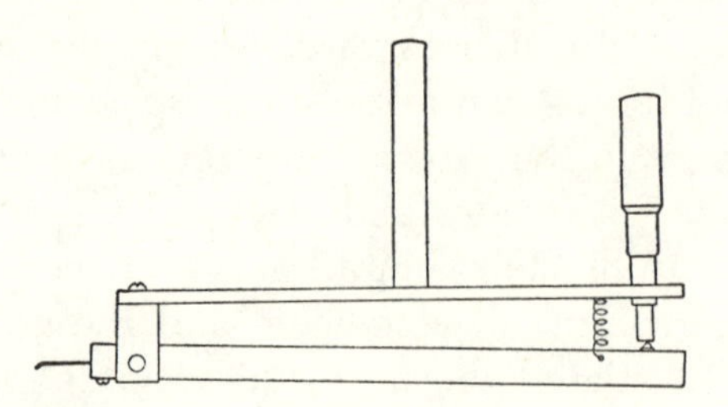

Fig. 26. Manipulator used for moving the manubrium.

Because the loading of one side of the drum membrane with a water column is obviously an abnormal condition that might alter the membrane's behavior we continued our experiments by a modified method. We no longer filled the meatus with water but transmitted the membrane displacements simply by air pressure. We placed the pipette horizontally at the level of the meatus and introduced a droplet of colored fluid into it to act as an indicator. This droplet was observed under a microscope giving 15 times magnification. Under these conditions we obtained results like those given in Fig. 27.

These results show that as the manubrium is moved from its equilibrium position the volume displacement of the drum membrane rises at first linearly and then at a continually decreasing rate. The linear portion of the curve extends to an amplitude of 0.13 mm, at which point the volume displacement is 2.6 cubic mm.

Certain irregularities in the results will be noted near the zero point of the curve. These were found to be due to the fact that as the rod is brought close to the manubrium it is seized by a film of fluid which exerts a somewhat variable negative force on the manubrium, drawing it inward. This effect disappears as soon as a solid contact is established.

An examination of these results shows that a movement of the manubrium of 0.036 mm (the same amount used earlier in our calculations with the first method) gives a volume displacement of 0.80 cubic mm. If we divide this volume displacement by the area of the drum membrane, which in this ear was 37.2 sq mm, we obtain 0.0216 mm as the mean displacement of the membrane. This value is 60 per cent of the amplitude that was imparted to the manubrium. The results varied somewhat in different ears and some gave mean displacements as high as 72 per cent, but in no

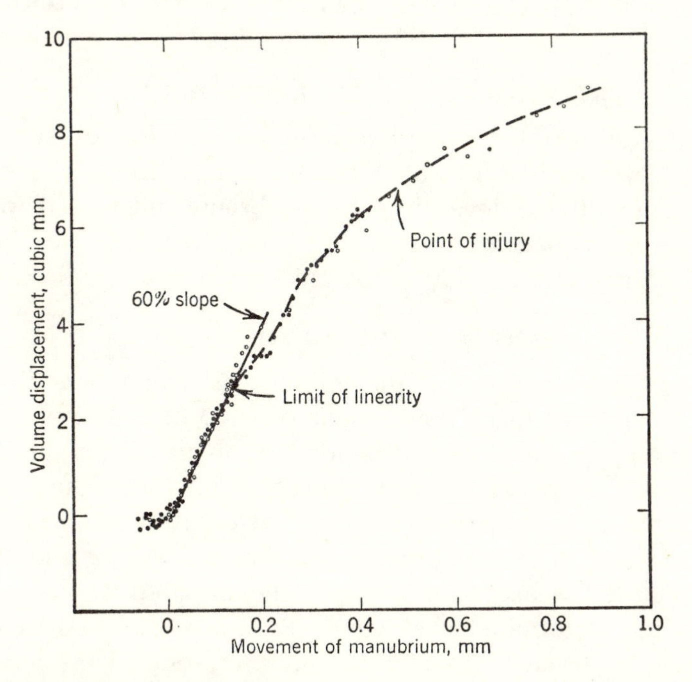

Fig. 27. The volume displacement of the drum membrane of the cat resulting from outward displacements of the manubrium.

instance have we substantiated Helmholtz's conclusion that the drum membrane considered as a whole moves over a greater distance than the manubrium. There is therefore no evidence for the existence of a lever action in the drum membrane as Helmholtz supposed. For the transformer action of the middle ear we must depend upon other principles, as will be dealt with in the next chapter.

7

The Principles of Transformer Action

THE OSSICULAR LEVER HYPOTHESIS

We turn now to Helmholtz's second proposal for the securing of a mechanical advantage in the ear's response to aerial sounds. This is the ossicular lever hypothesis, briefly mentioned in his book *On the Sensations of Tone* in 1863 and then fully developed in 1868 after a careful study of the anatomy of the middle ear.

THE ANATOMICAL BASIS OF THE HYPOTHESIS

The complex structure of the middle ear makes it a difficult matter to discover the physical basis of its action. The parts are irregular in form and are suspended at many points in three dimensions of space, as has been shown in Chapter 1. The suspensory ligaments have tensions that are not easily found, yet these tensions in their relations to one another in both degree and direction establish the axes about which the imposed motions take place and so determine whatever lever action may exist. Helmholtz tried to discover the relative tensions from a consideration of the number and courses of the fibers making up the suspensory ligaments and from visual observations during manipulations of the parts.

By this study Helmholtz arrived at a conception of the motion of the ossicular system that is very complicated indeed. It is based upon three main considerations: (1) the manner of suspension of the malleus, (2) the nature of the joint between malleus and incus, and (3) the manner of anchorage of the incus.

The Suspension of the Malleus. As mentioned earlier, most anatomists have recognized three principal ligaments of the malleus, the large anterior and lateral ligaments and a relatively small superior ligament, in addition to the ligamentary attachments of this ossicle to the tympanic membrane and the incus.

We have already seen (Plate 1) how the manubrium is applied along the superior radius of the drum membrane and how it pulls this membrane inward in the shape of a cone. The attachment is most intimate at the middle of the membrane and grows looser at the superior border. Just above this border the neck of the malleus is firmly anchored by means of the anterior and lateral ligaments to a bony projection that Helmholtz called the spina tympanica major. This spina is a part of the Rivinian recess, a curved margin of bone forming the superior border of the innermost end of the bony meatus. The anchorage is assisted by a thin bony process, the anterior process or processus Folianus, that extends from the malleus into the substance of the anterior ligament and sometimes as far as the anterior bony wall of the middle ear cavity. In children and in mammals below man this process extends much farther anteriorly into the petrotympanic fissure where it is rigidly attached. During adult development in man it is partially absorbed and reduced to a short stump as described. More posteriorly the neck of the malleus is connected to the Rivinian recess also by the lateral ligament, and Helmholtz gave special emphasis to the most posterior bundle of fibers of this ligament, which he described as the strongest and most tense of all and gave the special name of ligamentum mallei posticum. He pointed out that the most anterior fibers of the anterior ligament and these fibers of the ligamentum mallei posticum are attached to the neck of the malleus on nearly opposite sides and exert tension upon it in contrary directions. These fibers therefore form what he regarded as the principal axis of the malleus, and he referred to them collectively as constituting the axial ligament. When sound acts upon the tympanic membrane and moves it back and forth it impels the malleus to rotate about this ligament as an axis. However, Helmholtz said, this is not the only form of motion of the malleus because of restraints afforded by the malleoincudal joint and the anchorage of the incus, now to be described.

The Malleoincudal Articulation. Between the malleus and incus is a saddle-shaped joint, to which Helmholtz attached a special significance. This joint might better be described as a double saddle joint, for the malleus has a depression into which a part of the incus fits, and the incus in turn has a depression into which a part of the malleus fits. Helmholtz saw in the conformation of this joint a kind of cog mechanism. Such a mechanism locks the joint and causes one part to move with the other during rotation in one direction but leaves the parts free during rotation in the other direction. The cog surfaces engage most firmly when inward movement of the manubrium pulls the head of the malleus outward. The incus then is urged to follow the outward movement, and if it is restrained from doing so the articular surfaces, not permitting a simple sliding on one another, are forced to separate. When the head of the malleus moves inward the cogs cease to engage and only the capsular ligament uniting the two surfaces determines the degree of coordinate motion. Helmholtz believed that this ligamentary union was relatively weak and permitted a degree of independent motion of the two ossicles. In his view, this freedom in the connection of the ossicles always introduces changes in the transferred vibrations, and when the driving forces are great, as they are for strong sounds, it introduces outstanding alterations of the wave patterns and thereby aids in the generation of combination tones, as will be discussed later.

The Anchorage of the Incus. The incus is attached at three points: to the malleus as just described, to the head of the stapes, and to the wall of a bony fossa in the posterior floor of the epitympanic recess. The end of the short process of the incus is firmly held in this fossa by the posterior ligament of the incus. The anchoring fibers are short and strong, yet Helmholtz considered them as permitting a limited amount of mobility.

HELMHOLTZ'S CONCEPTION OF OSSICULAR ACTION

We are now ready to consider in detail the form of the ossicular action as Helmholtz envisaged it. An inward motion of the end of the manubrium, as caused by a positive pressure exerted upon the drum membrane, tends to swing the malleus about the ligamentary axis at its neck, but this swinging motion is modified by the restraints provided by the incus. The axis of the malleus lies below the end of the short process of the incus and, as Helmholtz estimated, is about 30° from the plane of the tympanic membrane. Therefore as the manubrium moves inward it tends to move also a little posteriorly and a little upward. Contrariwise, the head of the malleus moves outward and in so doing it tends to move a little anteriorly and downward. The anterior movement of the head of the malleus is largely prevented by the restraint of the incus, but the tension put on the incus causes a partial separation of the malleoincudal joint and pulls the end of the short process of the incus upward in its fossa. Because the anchorage of the incus is below the head of the malleus and internal to it, the restraining force on the head of the malleus not only impels this head posteriorly but also draws it downward. However, the malleus cannot move downward as a whole, for this would cause the anterior process on the anterior side of its neck to strike against the spina tympanica posterior. Instead, the posterior side of the malleus moves downward as on a pivot about the anterior process.

Because the head of the malleus is pulled a little backward its manubrium is impelled forward, and this motion just compensates for the tendency of the end of the manubrium to move backward. At the same time, because the malleus is tipped about the anterior process, the manubrium is moved downward, and this motion practically compensates for the tendency of the end of the manubrium to move upward. Consequently, for a positive pressure on the drum membrane the end of the manubrium moves straight inward and not in an arc. When the pressure on the drum membrane becomes negative all these motions are reversed and the end of the manubrium moves straight outward.

Let us now review the motion as indicated so far. The alternating pressures constituting a sound cause the drum membrane with the manubrium of the malleus to execute simple pistonlike vibrations to and fro. The malleus as a whole rocks on its ligamentary axis and at the same time executes a complicated twist that consists of a rotation and a sliding motion in the vertical plane. The incus swings largely with the malleus, but the looseness of its articulation gives it a moderate degree of independence, so that it does not follow the malleus completely but lags a little during both the inward and outward phases of the vibration.

The motion of the incus is not fully specified in Helmholtz's treatment, but he said that as the body of the incus is drawn outward by an inward movement of the manubrium the end of the long process of the incus is pushed inward. This ossicle therefore rocks on an axis lying between its joint with the malleus and the end of its long process. We cannot know what particular location for this axis might be agreed to by Helmholtz, but our best guess for its location is a line running from the end of the short process of the incus to the axial ligament of the malleus, as shown in Fig. 28.

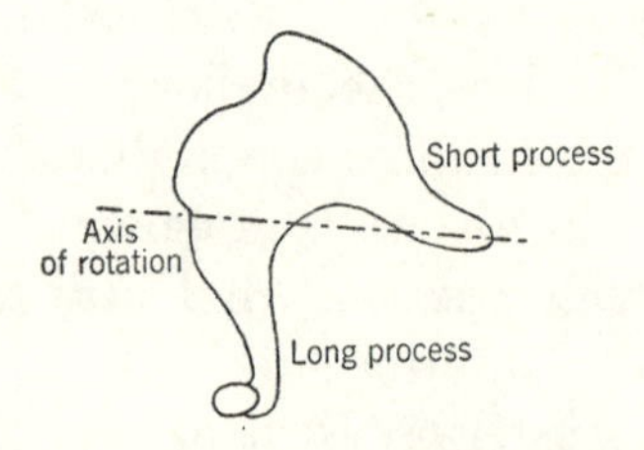

Fig. 28. The axis of rotation of the incus, according to Helmholtz's conception.

The footplate of the stapes is attached to the oval window by an annular ligament that is most tense on its inferior edge and especially strong at the posterior end. Therefore any pressure exerted on the stapes tends to produce a displacement that is a little greater at the superior edge of the footplate than at the inferior edge. Though he recognized this inequality of the displacement, Helmholtz did not fully agree with the suggestions of Henke, Lucae, and Politzer that the stapes only undergoes a rocking motion about a hinge formed by the inferior part of its annular ligament.

The joint between incus and stapes is a delicate one, and the incus can be forcibly drawn away without causing the stapes to follow. This condition, Helmholtz supposed, has the purpose of protecting the stapes against being torn out of its window when undue outward pressure is exerted on the drum membrane.

Now as we follow Helmholtz's treatment further and meet with his description of the ossicular lever action a surprise is in store. We should naturally expect that this action would take its form

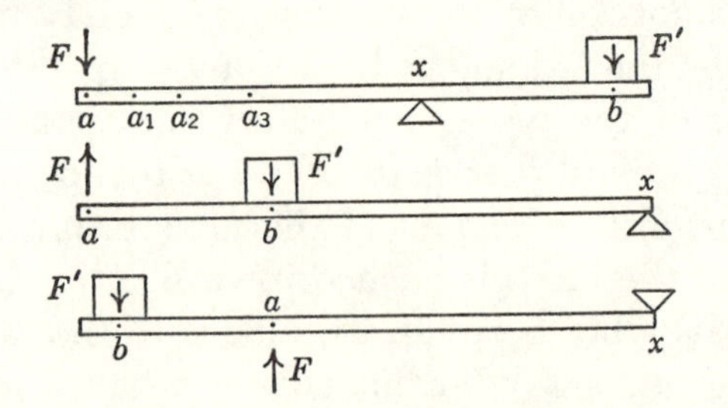

Fig. 29. The three classes of levers. In each the applied force is F, the burden or resisting force is F', the fulcrum is at x, ax is the force arm, and bx is the resistance arm.

from the mechanical relations already disclosed in the middle ear mechanism. From the nature of the mechanical elements and their linkages and the locations of the rotational axes we ought straightaway to determine the type of lever system and from the dimensions obtain a measure of the lever ratio. What has clearly been described is a compound lever system made up of two levers of the first class* connected in series as indicated in Fig. 30. The

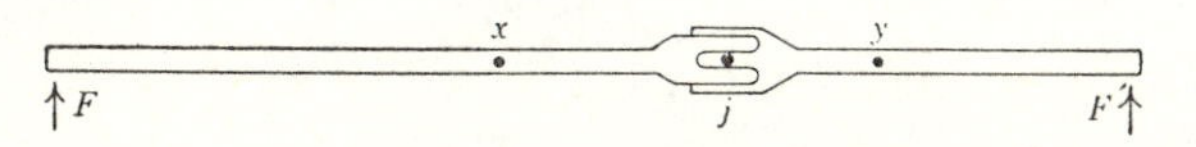

Fig. 30. The form of the ossicular lever according to Helmholtz's main treatment. The force is applied to the manubrium at F and delivered to the stapes at F'. The two levers operate about fulcrums at x and y and are linked at j through a sliding joint (the malleoincudal articulation).

first of these levers Fxj consists of the malleus rotating about its axial ligament x and the second jyF' consists of the incus rotating about its axis y. The first lever amplifies the force F applied to the manubrium and conveys it to the force arm of the second lever through the malleoincudal joint j. In this second lever the force is further transformed (to F') and delivered to the stapes. The ratio of the malleolar lever is easily obtained from Helmholtz's dimensional data. The lever arms are measured as the perpendicular distances between the lines of application of the forces and the rotational axis. These distances for force and resistance arms are 4.5 mm and 2.3 mm, respectively, and their ratio is 1.96 to 1. Corresponding data for the incudal lever are not available from Helmholtz's treatment, but our estimation of the location of the axis gives the force and resistance arms as 1.47 mm and 2.62 mm respectively, and their ratio is 0.56 to 1. The product of these two ratios, which is 1.1 to 1, is the over-all ratio for the system.

What has just been presented is the lever system as it should be derived from Helmholtz's conception of the form of the middle ear apparatus, but it is not the system that Helmholtz described. After his painstaking development of the pattern of forms and connections of the ossicular elements as summarized above, Helmholtz proceeded to offer a picture of the lever action that bears no relation to this pattern. He evidently reverted to a conception

* There are three classes of levers according to the relative locations of force arm, resistance arm, and fulcrum, as pictured in Fig. 29.

that he had formed several years previously. He considered the system as acting as a single lever of the second class as shown in Fig. 31, with its fulcrum at the end of the short process of the

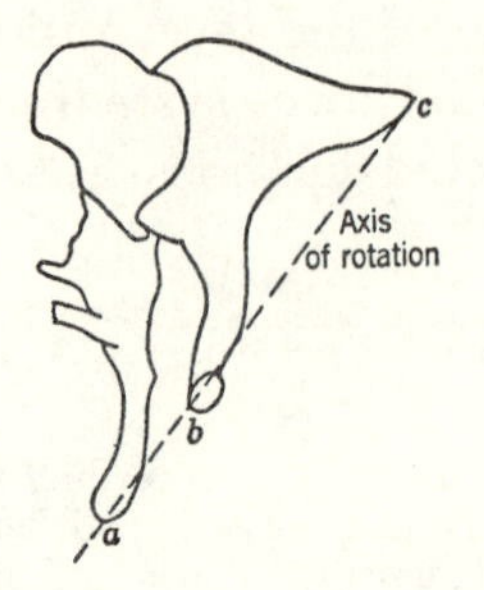

Fig. 31. The ossicular lever system as pictured by Helmholtz. Point *a* is the tip of the manubrium, *b* is the lenticular process of the incus, and *c* is the end of the short process of the incus.

incus *c*, with its force arm measured from the end of the manubrium *a* to this fulcrum, and with its resistance arm measured from the end of the long process of the incus *b* to this same point. The lengths of these two arms are 9.5 mm and 6.3 mm, respectively, and their ratio is about 1.5 to 1. Helmholtz therefore concluded that the force exerted upon the manubrium was increased by 1.5 and the amplitude reduced correspondingly.

This view of the ossicular system ignores the axes heretofore determined and requires a single axis through the end of the short process of the incus. The axial ligament of the malleus, so strongly emphasized in Helmholtz's anatomical description, must exercise no restraint whatsoever. There is no systematic relation between the compound lever system derived from the anatomical considerations and the simple system just described—the second cannot be derived from the first. It is remarkable that this simple lever hypothesis of Helmholtz's has met with general acceptance without any reference to its confused foundation.

EXPERIMENTAL MEASUREMENTS OF THE LEVER RATIO

Helmholtz reported that he had carried out some experiments to test his hypothesis. He used Politzer's method of attaching levers in the form of fine glass rods to the ossicles so as to obtain an amplification of their motions, and he measured the amplitude of displacement of the stapes for a given movement of the tip of the manubrium of the malleus. He reported that the stapedial movement was two-thirds that of the manubrium, as expected. On the other hand, when he produced a movement of the stapes and measured the resulting displacement of the manubrium he obtained the same amplitude as that produced in the stapes instead of an increased amplitude. This discrepancy is puzzling, for obviously a lever system that acts in one direction as a reducer of amplitude ought to act in the other direction as an amplifier of amplitude. Helmholtz suggested that the discrepancy might be the result of a decreased tension in the tissues after death, but this explanation is hardly satisfactory.

Dahmann's Observations. A number of investigators have extended these observations. Of particular interest is the work of Dahmann, carried out by a method similar to that already described in his study of movements of the drum membrane. He used cadaver specimens and observed the angular deflections of a beam of light cast upon small mirrors located at various points on the ossicles.

By this means Dahmann investigated the axes of rotation of the parts. A mirror that is on a rotational axis gives no deflection of a light beam along the line of the axis but only in a plane perpendicular to the axis. Dahmann thus located the main vibratory axis of the ossicular system along a line from the point of anchorage of the short process of the incus through the anterior process of the malleus, as shown in Fig. 32. The effective lever arms run perpendicularly from this axis to the point of operation of the forces. The lever arm for the malleus extends to the tip of the

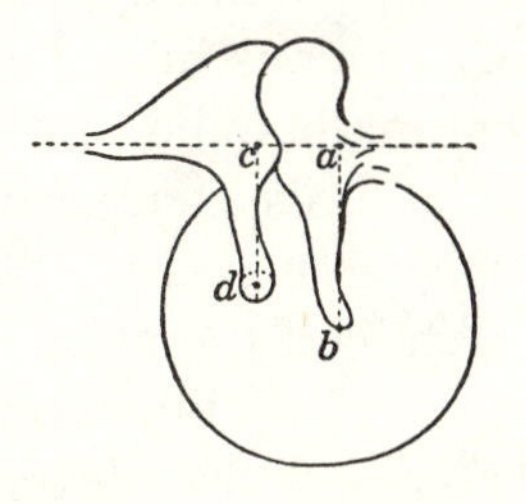

Fig. 32. The ossicular lever system according to Dahmann. The line *ab* is the malleolar lever arm, *cd* is the incudal arm, and the line through *ca* is the rotational axis. From the *Zeitschrift für Hals- Nasen- und Ohrenheilkunde.*

manubrium and the lever arm for the incus extends to the end of its lenticular process.

This conception of the lever mechanism differs considerably from the one portrayed by Helmholtz and also from the one implicit in his functional description. Dahmann therefore obtained a different value for the lever ratio. He found the two lever arms as described to differ in length by 1.31 to 1, thus forming a reducing lever of this magnitude.

Dahmann stimulated usually with periodically altered air pressures of ±60 mm of mercury (56,600 dynes per sq cm) and sometimes with intense whistles and vocal sounds. With these powerful stimuli he found the movements of the ossicular system to be asymmetrical. The outward movement was nearly twice as great as the inward movement. Also, when the ear was intact, the movement grew less as it proceeded along the ossicles, so that the movement of the stapes was only about a third of that of the malleus. However, when the stapes was freed from its connection with the inner ear at the oval window this reduction of motion no longer occurred. The transmitted motion then was about the same as the motion imparted to the malleus. Therefore Dahmann concluded that the reduction of amplitude found with the intact system does not result primarily from the operation of a reducing lever. Rather it represents a yielding of the articulations in the face of the resistance presented by the stapes through its ligamentary attachments to the oval window. He inferred that for small amplitudes of motion and under conditions in which the articular connections of the ossicles are stronger than the fixation of the stapes, the whole ossicular mechanism ought to operate as a single united mass.

Stuhlman's Model Study. Stuhlman, after examination of 9 specimens of human ossicles, constructed an enlarged scale model of the malleus-incus combination and mounted it so that it would rotate about the principal axis that Dahmann had described. Stuhlman then made measurements of motions imparted to the model, from which he obtained the effective lengths of the lever arms. He found that when the joint between malleus and incus was locked the ratio of the manubrial lever arm to the incus lever arm was 1.27 to 1, which is close to the ratio given by Dahmann. When he loosened the coupling between these two ossicles so that they could separate somewhat as one was moved, the axis of rotation was altered in two ways depending on whether the motion was inward or outward. He then obtained a lever ratio of 2 to 1 for inward motion and 1 to 1 for outward motion. Stuhlman accepted this asymmetrical sort of motion as typical of the normal ossicular system, but the basis for this view is obviously insubstantial and we shall later bring forth contrary evidence (page 158).

Fumagalli's Conceptions. Fumagalli made an extensive morphological study of the middle ears of several species of animals, including man. He opposed Helmholtz's description of the lateral and anterior ligaments of the malleus as consisting largely of oppositely directed fibers forming a single ligamentary axis upon which the malleus rotates. Actually, in man and other animals, the fibers of these two ligaments are almost at right angles to one another.

From further observations Fumagalli concluded that the ossicular system has two possible axes of vibratory movement. One of these, which he called the rotational axis, is the same as Dahmann's, running through the anterior process of the malleus and the posterior process of the incus. However, he believed that in most animals this axis is operative only for very great displacements and commonly only for low tones. The second axis, which he called the gravity axis, runs through the anterior process of the malleus and the lenticular process of the incus. It serves in man and most other animals as the functional axis for vibrations of small and moderate amplitude.

Fumagalli located the gravity axis by dissecting free the middle ear mechanism and balancing it on a fine needle or else suspending it by a thread tied to the anterior process of the malleus. In either case, he said, a vertical line extended from the point of support passes through both the anterior process of the manubrium and the posterior process of the incus, and an absence of displacement of the elements shows that the system is in equilibrium. He had no direct evidence for his acceptance of the gravity axis as the one about which the system vibrates, and the only argument that can be offered for his view is that a system whose mass is balanced in this way ought not to be sensitive to shakings of the head in the horizontal plane. Bárány produced evidence that the ear is relatively insensitive to translatory oscillations of the head (i.e., to oscillations limited to the horizontal plane).

Fumagalli's gravity axis is indicated as the line *ef* in Fig. 33. Its force arm runs from the tip of the manubrium *b* to a point on the axis *f* a little distance from the lenticular process of the incus. Its resistance arm runs in a different plane, from point *g* on the

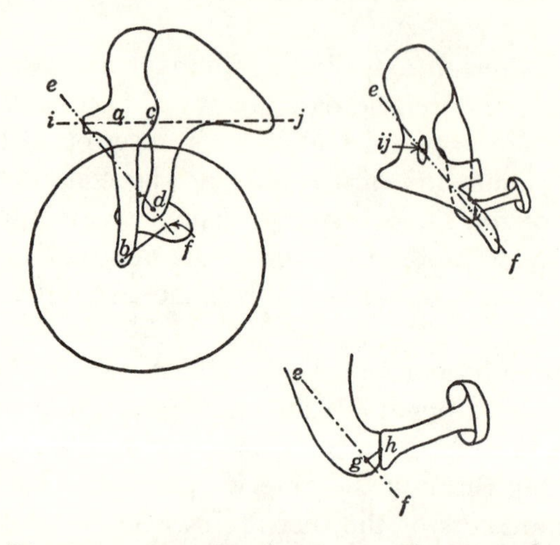

Fig. 33. Fumagalli's gravity axis and its lever arms. The axis is the line *ef*, viewed laterally in the sketch on the left and frontally in the sketches on the right. The force arm is *bf* and the resistance arm is *gh*. The rotational axis is shown as *ij*.

axis to the end of the lenticular process *h*. Fumagalli measured these two lever arms in human cadaver specimens and obtained a lever ratio of 10. Hence he believed that as the motion of the malleus is transmitted to the stapes it is reduced to a tenth of its amplitude and correspondingly multiplied by ten in force.

Rotation about the gravity axis requires that the short process of the incus be sufficiently free in its fossa to execute lateral displacements. Fumagalli believed that such freedom exists in man and most of the other animals that he studied except the guinea pig and rabbit. In these two species he found the ligaments so disposed as to restrain the lateral motion and hence he concluded that in them the only motion that is possible is about the rotational axis.

When in man and most other animals the motion communicated to the short process of the incus exceeds the freedom that this process is allowed in its fossa, the system becomes effectively anchored at this point and the movement changes to one about the rotational axis. This change is gradual as the sound intensity is raised, until above a certain limit the movement is exclusively about the rotational axis. When the system changes from one operating axis to the other its lever ratio changes also, from the 10-fold reduction afforded by the gravity axis to the 1.3-fold reduction afforded by the rotational axis. An important consequence, Fumagalli believed, is a reduction in the effective transfer of vibrations to the inner ear and hence the inner ear is protected against overstimulation.

FURTHER EXPERIMENTAL OBSERVATIONS

The evidence for Fumagalli's gravity axis is far from convincing in itself, and we regard Dahmann's rotational axis as on much firmer anatomical ground. We now turn to further experimental evidence in support of Dahmann's position.

We have previously reported experiments (Wever, Lawrence, and Smith, *1*) in which the cochlear potentials were used in the study of the lever action of the ossicular mechanism of the cat, but the results were not wholly satisfactory because of technical difficulties. We have continued our study by improved methods, with further observations now to be described.

Our procedure consisted of driving the ossicular system at audio frequencies with a mechanical vibrator and observing the cochlear responses, whose magnitudes are proportional to the amplitude of the resulting stapedial motion. This method has a number of advantages over the methods heretofore employed in the study of middle ear function. The ear is stimulated at any desired frequencies and intensities and the results therefore represent the behavior of the mechanism over its normal operating range. There is no need to overstimulate the mechanism or to burden it with levers, mirrors, or other recording devices that restrict and modify its behavior. Most important of all, the observations are carried out in the living animal, and continuous checks can be made as necessary to show whether any changes have occurred in the physiological condition of the ear. There are difficulties in the method, to be sure, but we believe that now they have been suitably dealt with. Our observations have been made in the cat by our usual procedure in which the cochlear potentials are recorded with an electrode on the round window membrane.

To drive the ossicular mechanism we used a vibrator unit consisting of a Rochelle salt crystal of the expander type actuating a steel needle, as described earlier (Chapter 3). The unit was held in a heavy manipulator by which the needle point could be applied to any exposed part of the ossicular system.

The principal difficulty of the procedure lies in the application of the needle. It is necessary to obtain a contact that is sufficiently firm to cause the ossicular structures to follow the driving point faithfully, but at the same time to avoid an excessive pressure that alters the normal ossicular action. Our practice was to lower the point gradually while observing the cochlear potentials and to be guided by their form and magnitude. The first contact is intermittent and generates irregular responses that rise rapidly in magnitude as the pressure is made more firm. From this point on a slight increase in pressure usually causes little or no change, but a further increase has varying effects depending upon the frequency. For the low frequencies the potentials are always reduced by excessive pressure, at first slowly and then more rapidly. For the high frequencies they are usually reduced, but sometimes as the contact pressure mounts they are increased progressively (or increased after a dip so sharp that it might be overlooked if the change of pressure is not gradual enough) and this increase continues to the point of damage to the ear. These high frequencies therefore give greater difficulty of adjustment and the measurements with them are more subject to error than they are with other frequencies. The variations are most prominent in the medium high frequencies where (as will be shown in detail later) the middle ear exhibits rapid changes of resonance characteristics.

To check the fidelity of performance of the vibrator a phonograph pickup cartridge of the crystal type was attached to the case with its pickup needle riding in a tapered hole drilled into the side of the nosepiece. The output of the pickup cartridge was read on a sensitive vacuum tube voltmeter, and by these readings it was found that the instrument maintained linearity of response over a wide range of amplitudes and that its movements were damped only to a negligible extent on bringing the needle to bear on the ossicular mechanism. The use of the vibrator was restricted to its linear range as thus determined.

The experiments were carried out in three steps. The first step was designed simply to ascertain whether a lever action is present in the ossicular mechanism, and consisted of driving this mechanism at various points along the manubrium of the malleus. The manubrium was freely exposed on its exterior side by removing the pinna and dissecting away the walls of the external auditory meatus all the way to the bony tympanic ring. The manubrium is then seen through the transparent drum membrane to which it

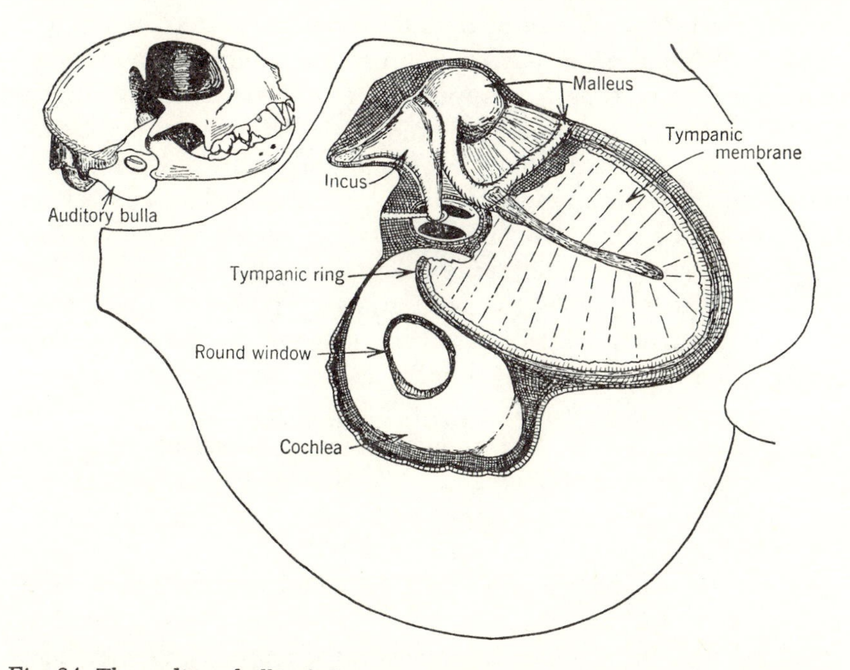

Fig. 34. The auditory bulla of the cat, opened to show the middle ear structures. The tympanic ring and tympanic membrane have been partially cut away to reveal the stapes and the lateral process of the malleus. The stapes is seen in an end view, with its tendon running to the left and its footplate lying in the oval window.

is adherent. The needle was applied from the outside and thus bore upon the manubrium with the drum membrane between.

If the malleus forms one arm of a lever system that is linked to the cochlea, then clearly we must observe changes in cochlear output as we drive this arm at different points along its length, and this will be true regardless of the class of lever that we are dealing with. Consider the upper part of Fig. 29. As we move the point of application of the force F along the arm ax and thereby alter the distance from the fulcrum x, the value of the resultant amplitude must vary inversely. From such observations on any system we can determine whether a lever action is present and obtain a general indication of the direction in which the effective force arm runs.

It perhaps will occur to the reader that if the line of the manubrium represented the arm ax we ought to be able to find the lengths of the arms ax and bx and hence the lever ratio on the basis of the responses observed while stimulating at known points on ax. We should need only to measure the resultant values of F' when a constant force F is applied at any three points a_1, a_2, a_3, and the distances a_1-a_2 and a_2-a_3 are known. Unfortunately, it cannot be assumed that the line of the manubrium follows the direction of this force arm, and indeed our further results show this to be far from the case. We therefore find ourselves limited at this stage of the study to qualitative determinations only.

Series 1. In this series the vibrator was applied to the manubrium at many points from tip to base, and the cochlear potentials were observed as the driving amplitude was kept constant. Figure 35 gives some of the results for a driving frequency of 500~. The driving points followed the somewhat curving course of the manubrium as pictured in the lower part of this figure, and their locations along the line of the manubrium were determined from the calibrated scales of the manipulator.

As seen in this figure, the potentials rose continuously with distance from the manubrial tip until a maximum was reached at a position near the basal end. As shown, the manubrium ends here in a sharply rounded edge, and applications in this region gave only reduced and somewhat unpredictable responses, probably because of the uncertain and varying angles of contact.

The increase of potentials with distance from the tip indicates that the stapes is being driven more and more vigorously. Hence the effective lever arm is being shortened. Near the base of the manubrium the potentials have risen to about 3 times their value at the tip, and thus we know that the lever ratio has been altered by this factor. Beyond this indication, however, the results are not

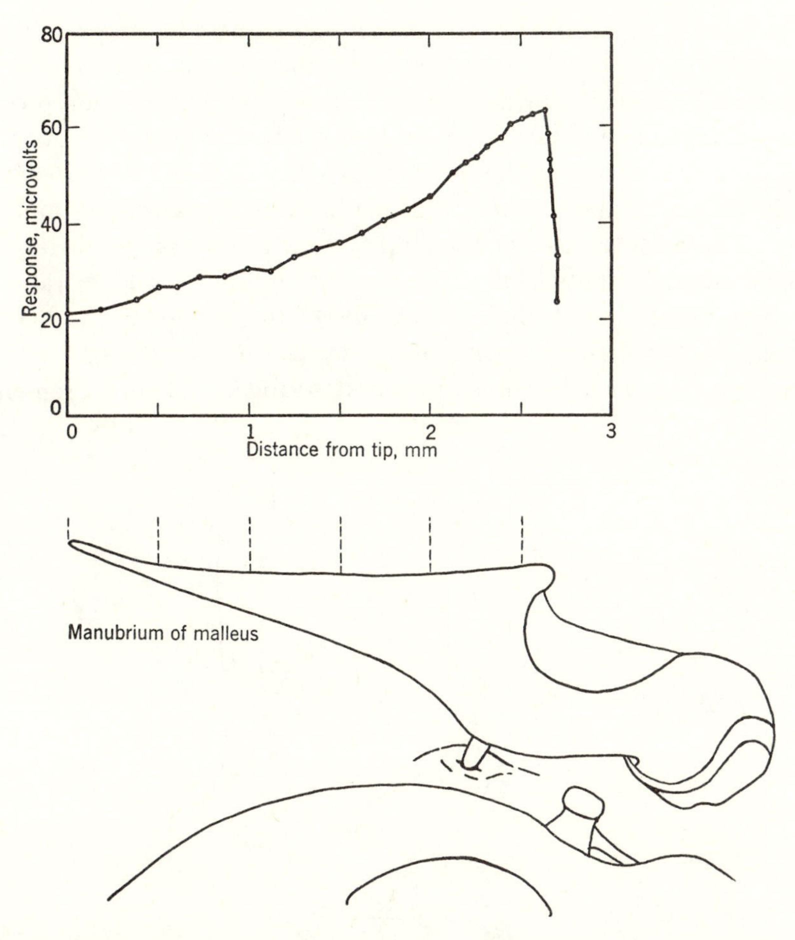

Fig. 35. Responses obtained by driving the malleus at constant amplitude at various points along its manubrium. The sketch below shows the line of driving positions. For a clearer view of the malleus, the incus was removed, but the head of the stapes is shown projecting above the cochlear promontory.

very definite and for more particular information about the system we must extend our method.

Series 2. In the further course of our investigation we exposed the long process of the incus and observed the effects of driving it. We made this exposure by removing a minute segment of the superior margin of the bony tympanic ring and then enlarging the opening superiorly until the long process of the incus and its articulation with the stapes were presented to clear view. If care was taken not to extend the opening so far superiorly as to invade the fossa of the short process of the incus, the exposure could be made without impairment of the ossicular action. This operative approach is an important improvement over our former procedures. It makes it possible to apply the vibrator needle to the end of the long process of the incus where it makes its elbow bend to join the stapes and to orient the needle so that its vibratory motion is in line with the axis of the stapes. This proper angle of driving the stapes is essential for a true measurement of the lever ratio.

The vibrator was applied alternately to the tip of the manubrium of the malleus (point *m* in Fig. 36) and to the end of the incus (point *s*) over the stapes as described, and measurements

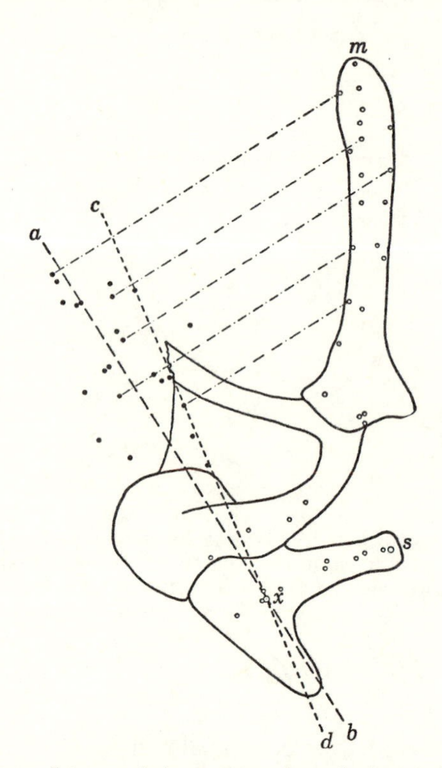

Fig. 36. The malleus and incus of the cat, in a lateral view as in Fig. 34, showing the points of stimulation in Series 2 and 3. The large circle near *s* lies over the center line of the stapes and the one at *x* is the fulcrum of the incus. The line *ab* is the best-fitting tangent to spherical surfaces generated by the ends of lines rotating about the manubrial points. (Some of these lines are represented as dot-dash lines.) When line *ab* is rotated about 34° forward on its axis at *x* it takes a position whose projection on this drawing is *cd*. This line *cd* is the probable axis of rotation of the ossicular lever of the cat.

of cochlear potentials were made for the two positions while maintaining a constant driving amplitude.

The potentials were always greater when driving the incus. This was true regardless of the driving frequency, though the high frequencies gave somewhat variable results for the reasons already mentioned. Figure 37 presents the results for two animals.

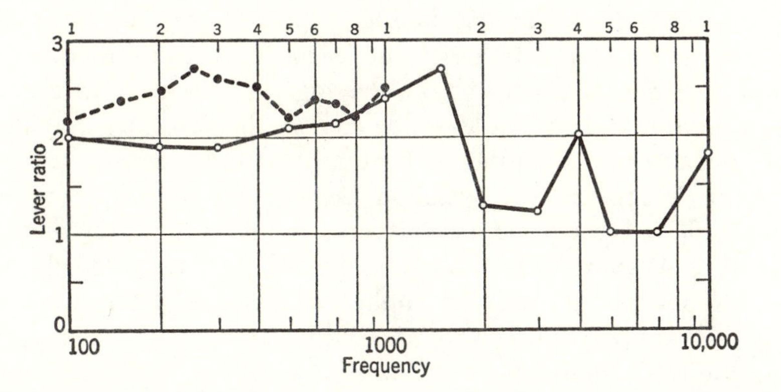

Fig. 37. The ossicular lever ratio as measured in two cat ears, with various driving frequencies.

The ordinate shows the ratios of the potentials obtained under the two conditions, on stimulating at frequencies represented on the abscissa. The solid line shows the results for one animal over the range 100 to 10,000~ and the dashed line shows results for another animal for the low frequencies only. The mean value of the ratio for the first of these animals over the range 100 to 1000~ is 2.07 and for the second animal is 2.39, and these figures may be taken as the most probable lever ratios in these ears. Two other ears when studied similarly gave ratios of 2.8 and 3.0 and an over-all average for the cat is probably about 2.5.

Series 3. Further observations were made on the incus to throw additional light on the vibratory action. Measurements made at various positions on the long process and across the body of the incus gave a characteristic pattern of changes. As we moved along the long process away from the initial position over the stapes the responses increased at first and then as the body of the incus was reached they fell sharply and became highly irregular. When we continued to positions across the body of the incus, on the side opposite the long process, the responses once more became large and regular.

Stimulation on the body of the incus proved to be especially hazardous, as either an undue pressure of application or excessive amplitudes of vibration produced injury to the ear as shown in bleeding from the oval window and an impairment of the cochlear responses.

This pattern of changes in the responses and the hazardous nature of the stimulation become clear from a consideration of what happens in a lever as the driving force is brought close to the fulcrum. In an ideal lever, of course, the resulting amplitude continues to rise and approaches infinity at the fulcrum. After the fulcrum is crossed the function is reversed in form, and the amplitude falls away from infinity at a regular rate as the distance increases. However, this regular behavior depends upon the two conditions that the force be applied at the line of the lever arm and shall act perpendicularly to this line. In practical levers these conditions are met sufficiently well at positions remote from the fulcrum but often are not met when the fulcrum is closely approached. When the lever arm is a beam with considerable thickness the force is usually applied on the surface of the beam and above the true line of the lever arm. Then the force tends to rotate the beam and to displace it laterally. The lateral component of the force is not only useless but tends to disrupt the system.

Our explorations of the region giving these irregularities together with the more quantitative indications from positions on the long process of the incus give the probable locus of the fulcrum of the incus, shown as point x in Fig. 36. The distance from this point to the center line of the stapes (point s) was measured as 1.90 mm in one animal and as 1.93 in another, and these distances therefore are taken as the lengths of the incudal lever arms in these ears.

When the length of the incudal lever arm is known, the effective length of the malleolar lever arm can easily be found from the ratio of the responses observed on driving alternately the end of the incus and the tip of the manubrium. In the first of these ears this ratio was 2.56, which gives a malleolar arm of 4.86 mm, and in the second ear the ratio was 2.80, which gives a malleolar arm of 5.40 mm.

Series 4. We can extend this procedure to give further evidence on the location of the rotational axis. The distance of any manubrial driving point from the rotational axis is equal to the length of the incudal lever arm times the ratio of the responses obtained from driving the incus and driving this manubrial point. By finding these distances for a number of points along the manubrium we can determine the complete course of the axis.

The procedure is further clarified by reference to Fig. 36. In the animal represented here we used 22 driving points scattered over the surface of the manubrium. In the original construction a line was extended from each of these points to the left at the proper scale distance to represent the actual distance from the driving point to the rotational axis. Here, in order not to complicate the drawing unduly, we show only a few of these lines and indicate the end points of the others by the filled circles.

These lines are to be thought of as rotating freely about their origins and thereby generating spherical surfaces, and the rotational axis is determined as the common tangent line along these surfaces. Properly speaking, we should have a tridimensional figure, but an approximate result is obtained by a construction first on one plane and then on another plane perpendicular to this one. On the plane shown in Fig. 36 the results give *ab* as the best-fitting line. We can now think of this line as anchored at point *x* but free elsewhere to move out of the plane of the drawing, yet with any point, such as *a*, keeping a constant distance from the manubrium. Our further treatment has shown that this line ought to be rotated forward so that point *a* comes to lie about 34° above the plane of the drawing. This axis then takes a position whose projection is the line *cd*. This position corresponds closely to the location of the rotational axis as indicated by Dahmann for the human ear and also as described by Fumagalli for a number of other animals. This axis runs through the cranial anchorage of the anterior process of the malleus and the end of the short process of the incus, as shown.

WRIGHTSON'S STAPEDIAL LEVER HYPOTHESIS

A different kind of ossicular lever was suggested by Wrightson as arising from asymmetrical movements of the stapes. Mach and Kessel (3), on the basis of their stroboscopic studies, had described a complex form of motion of the stapedial footplate. They referred to Eysell's observation that the annular ligament provides the greatest anchorage at the two poles, but especially at the posterior pole. They indicated that the normal motion of the head of the stapes for positive pressure on the drum membrane is somewhat forward and upward as well as inward and, contrariwise, for negative pressure it is backward and downward as well as outward. Accordingly, the footplate, though moving as a whole, does not move equally at all points. For positive pressure the upper and anterior edges move strongly toward the vestibule, whereas the lower and posterior edges move but slightly. Indeed, the posterior end is almost completely restrained.

Others have conceived of a somewhat simpler motion in which the whole footplate moves on a hinge provided by the ligament at the posterior end. Fumagalli recently adopted this view. Békésy (9) accepted a form of motion similar to this one—a rotation about a vertical axis—in response to ordinary sounds, but postulated a different mode of motion—about an anteroposterior axis—for intense sounds.

Wrightson based his stapedial lever hypothesis upon this conception of a hinging of the stapes at its posterior end. The stapes constitutes a bent two-armed lever, he said, with the posterior crus forming the force arm and the stapedial footplate forming the resistance arm. The arrangement as he conceived it is shown in Fig. 38. As the manubrium moves inward the malleus and incus rotate about point *b* as a pivot, and the end of the incus carries the head of the stapes along. The stapes rocks on point *d*. Its cen-

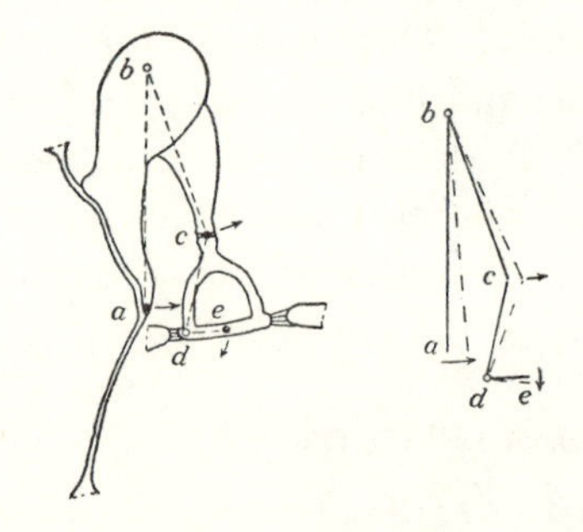

Fig. 38. Wrightson's conception of a stapedial lever. Shown are two bent levers, one the malleoincudal lever *abc* pivoted at *b* and the other the stapedial lever *cde* pivoted at *d*. The stapedial lever ratio is the ratio of *cd* to *de*.

ter *e* then moves with a smaller amplitude but a greater force than its head, as determined by the ratio between the length of the posterior crus *cd* and half the length of the footplate *de*. Wrightson computed this ratio as 2 to 1.

It must be noted that Wrightson's description of the anatomical relations departs considerably from reality, and the drawing of Fig. 38 is correspondingly distorted. The sharp bend at the end of the long process of the incus has been omitted and the stapes is shown as extended along the line of the body of this process. Actually, as may be seen in Plate 1, the stapes lies almost at right angles to the position that this hypothesis requires. It is simply pushed inward by the incus and hence no lever action can arise.

Let us consider further this matter of the local mobility of the stapedial footplate and its effects upon the communicated pressures. If, as we believe, the force exerted by the incus is in line with the axis of the stapes, the effective motion is just as if there were no variations in mobility and the stapes moved as a simple piston. Consider the extreme case just described in which the footplate is regarded as firmly held at its posterior end. A given force applied over the center of area of the footplate then will give three times the volume displacement to the anterior half of the footplate that it gives to the posterior half, but the total displacement will be just the same as if the whole footplate were displaced through the distance moved by its center.

THE HYDRAULIC HYPOTHESIS

As has been pointed out, the difference in areas of drum membrane and stapes footplate gives a transformer action by a sort of hydraulic principle. If we disregard for the moment the lever action of the ossicular chain and consider the drum membrane as directly linked to the stapes, we can say that the total force exerted by the air particles on the drum membrane is communicated to the stapes without loss of amplitude. The pressure at the oval window, which is the force per unit area, is therefore increased by the areal ratio.

Helmholtz estimated the areal ratio for the human ear as lying between 15 and 20, but there have been no systematic measurements. The average area of the drum membrane, based on four measurements, was given earlier as 64.3 sq mm. The average area of the stapes footplate is about 3.2 sq mm. These two figures give an areal ratio of 20.1. This method of comparing the average values for the two structures is not a very good one, however, when the number of ears is small. Because errors of measurement are larger for separate ears than for the ratio of the two areas in a single ear, it is better to make the measurements for drum membrane and stapes footplate in the same ear. We have done this for two human ears—the right and left ears of a single individual—and have obtained areal ratios of 18.2 for one and 19.1 for the other, as reported previously. Fumagalli reported a ratio of 21, evidently obtained in a single ear. Békésy obtained a value of 26.6, again probably in a single ear.

Whittle measured this ratio in 8 guinea pig ears and obtained the following values: 26.1, 26.2, 27.6, 27.8, 27.8, 28.3, 30.4, and 30.9; the mean of these is 28.1. In 4 cat ears we obtained values of 32.7, 33.5, 37.3, and 42.6; and the mean of these is 36.5. Fumagalli reported a ratio of 45 for the dog, 41.6 for the ox, and 17.1 for the horse, all probably obtained in single specimens. It is plain that there are large differences among species and among ears of any one species.

These values represent the anatomical ratios, and we must now consider the fact, as already referred to, that the drum membrane does not work as a whole. Because it is attached to the bony tympanic ring the outermost border is at rest and there is a gradient of motion as we pass toward the center. What we need to determine is the effective area, which is the area that a piston would need to have to displace the same volume when its whole face moves with the amplitude of the center of the drum membrane. This equivalent piston will, of course, be smaller than the drum membrane. How much smaller it must be will depend on the form of the gradient of motion over the drum membrane.

One possibility, entertained by Geffcken, is that the gradient is continuous from edge to center, as it is in a simple flat membrane like the rubber covering of a tambour (see Fig. 24*a*). The effective area then is a little less than a third of the whole area.

It is more reasonable to suppose that the gradient of motion is steep near the outer boundary of the membrane and then quickly reduces to zero. This means that a central portion of the membrane is sufficiently rigid to operate as a unit, with nearly all the flexural motion taking place in a narrow ring near the outer edge (see Fig. 24*c*).

This is the type of motion indicated by Békésy's observations on cadaver specimens, as already mentioned. He described a central cone constituting two-thirds of the total area as moving as a unit. Our observations, already described, on the volume displacements of the drum membrane produced by a given movement imparted to the malleus give information on this problem. These results show variations among individual ears, but in general the effective area of the cat's membrane lies between 60 and 72 per cent of the anatomical area. These data are for static conditions but the same relations can be expected to hold for vibratory stimuli of low frequency, though they may fail for those high frequencies at which the drum membrane loses its effective stiffness and vibrates segmentally.

In these considerations we discover the reason for the drum membrane's conical form. This form adds rigidity to the structure and makes it possible for the central portion to work as a unit, and so adds considerably to the effective area. Loudspeaker diaphragms are given a conical form for the same reason.

THE FINAL TRANSFORMER RATIO

We now have evidence in support of two of the three mechanisms that Helmholtz proposed for the matching of impedances by the ear. These two, the areal and ossicular lever mechanisms, have a combined effect equal to the product of their ratios.

Consider first the situation in the cat, for which we have the most exact measurements. If we take an anatomical ratio between drum membrane and stapes footplate for this animal of 36.5 and reduce it to two-thirds we have an effective areal ratio of 24.3. If we accept an average ossicular lever ratio of 2.5 the final transformer ratio is $24.3 \times 2.5 = 60.7$. The impedance transformation is the square of this value, or 3684. Accordingly, the acoustic resistance of the air, which is 41.5 mechanical ohms per sq cm, is made to appear like 153,000 mechanical ohms per sq cm at the inner ear.

For man our limited data indicate an anatomical ratio between drum membrane and stapes footplate of about 21.0, the average of the five values cited. If again we reduce this ratio to two-thirds we have an effective ratio of 14.0. If we accept Dahmann's figure of 1.31 for the ossicular lever ratio we obtain a combined ratio of $14.0 \times 1.31 = 18.3$. The impedance transformation is then 336, and the acoustic resistance of the air is made to appear like 14,000 mechanical ohms per sq cm. We shall take up in Chapter 17 the question of the adequacy of these transformations.

PART III

The Problem of Distortion

8

The Nature of Distortion in the Ear

We have found that an understanding of the functional significance of the middle ear apparatus comes in a contemplation of its evolutionary origin: this apparatus became a practical necessity as an accessory to the biologically older aquatic ear when certain of the animals left the sea and took up life on land. The effective transfer of aerial energy to the receptor cells then called for a mechanical transformer to give a proper matching of the impedances of aerial and aquatic media. Our discussion further revealed the physical principles upon which this transformer action is based. At the same time we encountered certain difficulties and complications whose detailed consideration was postponed until now.

The difficulties are due to the fact that a mechanical transformer, simple though it is in conception, is not so in actual operation. Ideally such a transformer in matching a tenuous medium to a dense one should simply bring about a multiplication of the acoustic pressure without any further effect: it should not use up any energy or otherwise alter the character of the transmitted vibrations. Unfortunately, no real transformer meets this requirement. Every mechanism has peculiarities of its own that are imposed upon the action. The result is a distortion of the vibratory pattern.

All mechanical transformers have these faults, but they have them in different degrees depending upon their particular construction. It is well known in the design of man-made instruments that there is a good deal of latitude in the choice of physical properties in the obtaining of a given average transformer action, and nearly always a choice can be made that will give good fidelity of performance within some desired range of frequencies. The difficulties of design grow increasingly serious as the instrument is required to operate over a wider frequency range.

These difficulties are illustrated in the development of the mechanical type of phonographic recorder. Edison's original model of this recorder consisted simply of a diaphragm fitted with a sharp stylus that when driven by a sound pricked out a pattern of holes in a sheet of tinfoil. This instrument was remarkable in that it gave a comprehensible reproduction of speech, but it had many shortcomings. It was greatly improved by the addition of a lever between diaphragm and stylus and by the use of a waxed drum as the impressionable surface. The stiffness of the diaphragm and the lever ratio were then varied in trial-and-error fashion to extend the recording to the higher frequencies and thereby permit an acceptable reproduction of music. Still the performance was poor, and nowadays this mechanism has given way to elaborate electromechanical devices in which electronic circuits are incorporated to improve the sensitivity and correct for the bad frequency characteristics. A consideration of these difficulties enhances our respect for Nature's accomplishment in the fabrication of a suitable transformer in the ear by the use of such limited materials as fiber and bone.

THE FORMS OF DISTORTION

Acoustic distortion is of three forms: frequency distortion, phase distortion, and amplitude distortion. Frequency distortion consists of a variation in sensitivity to different frequencies; the system picks up and transmits some tones more efficiently than others, so that the various components of a complex sound are changed in their intensity relations. Phase distortion consists of different degrees of delay in transmission for different tones, so that the various components after passing through the system are altered in their phase relations. Amplitude distortion consists of a variation in the efficiency of transmission according to the intensity of the sound, and usually appears as a progressive reduction in transmission as the intensity level is raised. These three types of distortion are interrelated because they often arise from common causes.

We find in mechanical devices two general causes of distortion. One cause lies in the conditions that determine the response characteristics of the mechanism. Any responsive system inevitably has mass in its moving parts and stiffness in its suspensory system, and friction is generated in its motions. Because certain masses and stiffnesses are present, the system has its own peculiar resonance characteristics: it responds more readily at some vibration rates than at others, thus exhibiting frequency distortion, and at the same time it responds more sluggishly to some frequencies than to others, thus exhibiting phase distortion. If the friction of the moving parts varies with the velocity of the motion, as is usual, the resulting loss of energy will be greater the higher the frequency, which contributes further to the frequency distortion.

The second cause of distortion is nonlinearity in the mechanical action. If the motion that is set up in the recording system is not strictly proportional to the acoustic forces acting on the system the result is amplitude distortion. A common form of nonlinearity exists in the action of the simple lever. If the beam used in the force arm of the lever is too weak for its service and bends progressively more as forces are applied to it the transmitted movements will not bear a constant relation to the applied movements. In general, mechanical elements of this kind work satisfactorily for small forces and then distort above some limit. A spring is another example; its change of form is proportional to the force applied to it up to an elastic limit, and then beyond this limit an extension spring elongates too much whereas a compression spring shortens too little. In the middle ear are lever arms and elastic ligaments that must exhibit these departures from linearity if the forces exerted upon them become excessive.

FREQUENCY DISTORTION

This form of distortion has already been defined as an inequality in the responsiveness of a system to different vibratory frequencies. Some tones are passed readily by the system, others are enfeebled in transmission or practically eliminated. The result is that a complex wave may be much altered in composition, with the relative magnitudes of the components departing greatly from their original values.

The ear like every other acoustic instrument suffers from frequency distortion. The human ear responds to tones from a few cycles to about 24,000~, a wide range relative to that of most instruments, but within this range it varies greatly in sensitivity as has been seen in Fig. 9. The variation shown in this figure is for the ear as a whole and includes neural as well as mechanical characteristics. The mechanical characteristics by themselves are revealed to a limited extent, for frequencies in the vicinity of the resonance point, in the impedance measurements of Fig. 16. These characteristics include the mechanical properties of the inner ear and principally the mass and stiffness of the cochlear fluid.

The over-all sensitivity of the cat's ear, determined by a conditioned response method, is shown in Fig. 39. This ear is similar

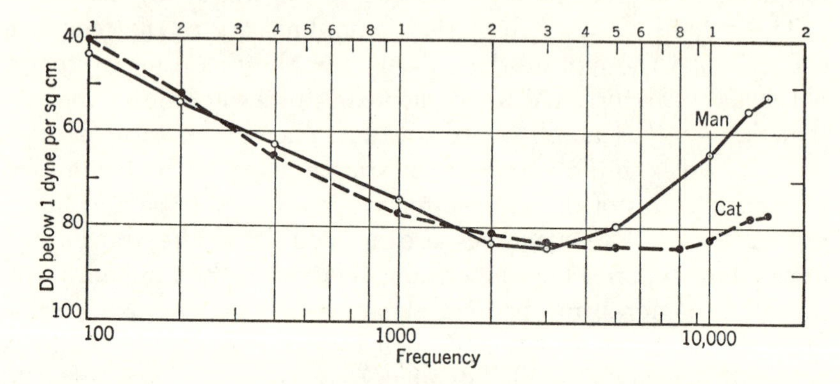

Fig. 39. A comparison of auditory thresholds in man and in the cat. Sound intensity as shown on the ordinate was approximated, but the differences between man and cat are dependable. Data from Dworkin, Katzman, Hutchison, and McCabe.

to man's in sensitivity except for a somewhat greater responsiveness to high tones.

Our observations of cochlear potentials in the cat give information on the middle ear characteristics. Measurements of sensitivity made before and after removal of the middle ear mechanism up to the footplate of the stapes reveal the response characteristics of this portion of the mechanism. These results have been shown in Fig. 21 as the loss of sensitivity due to the removal of the mechanism.

It was not possible to include the footplate of the stapes in these measurements because its presence in the oval window is necessary to retain the cochlear fluid. However, we have obtained an indication of the effects of this part of the ossicular system by another method now to be described.

This method rests upon the assumption that the cochlea would be equally stimulable by way of the oval and round windows if it were not for the presence of the footplate in the oval window. Such an assumption is reasonable in itself in view of the anatomical situation and is supported by experimental evidence to be presented later. The basilar membrane with its sensory structures lies between the two windows and the lengths of the two pathways through the cochlear fluid are not greatly different. There are certain variations in these pathways, to be sure. The oval window pathway leads through the perilymph of the scala vestibuli, across Reissner's membrane, and then through the endolymph of the cochlear duct, whereas the round window pathway traverses only the perilymph of the scala tympani. However, Reissner's membrane is an exceedingly thin and yielding membrane and can have only a slight effect upon the transmission, and the endolymph probably has much the same acoustic properties as the perilymph. Even more important in this relation, as later evidence will show, is the fact that a pressure wave traversing one of these pathways is forced to continue along the other pathway as its only means of egress, and therefore the physical conditions within the cochlea are operative upon both pathways alike. For example, if Reissner's membrane were really a serious barrier to sound it would act as such not only for sounds entering by the oval window but also—and practically to the same extent—for sounds entering by the round window.

Our procedure in this experiment was as follows. We found the stimulus intensity necessary at various frequencies to produce a cochlear response of a standard amount (10 microvolts) when the sounds were delivered by a tube sealed over the oval window with only the stapedial footplate present, and likewise we found the intensity necessary to produce this response when a sound tube was sealed over the round window. On the assumption just made, the differences in these stimulus values represent the effects of the stapedial footplate on the sensitivity. These effects are shown for two different ears in Fig. 40, and as will be noted they are small at all frequencies, never exceeding 9 db. For the most part the sensitivity is a little greater for stimulation by way of the oval window; the average difference over the range of 100 to 10,000~ is 1.6 db for one ear and 4.5 db for the other.

If results like those of Figs. 21 and 40 are combined we obtain a characteristic function for the cat's entire middle ear apparatus,

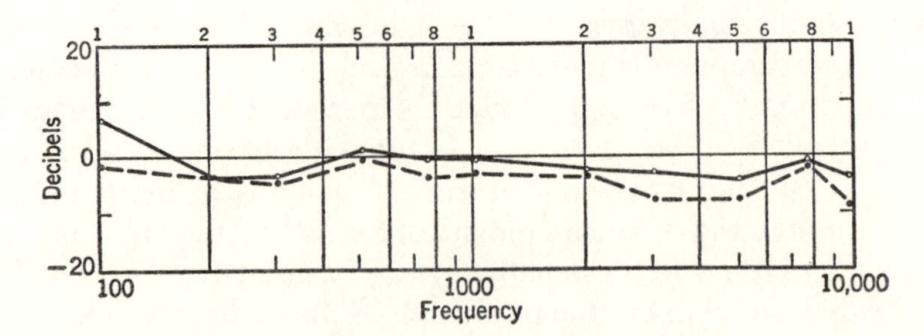

Fig. 40. Sensitivity effects attributed to the footplate of the stapes. The curves show, for two cat ears, the sound intensities required at the oval window relative to those required at the round window in order to produce a standard cochlear response. Thus negative values signify that the oval window route is the more sensitive. Cf. Wever and Lawrence (5).

as shown for two animals in Fig. 41. As will be seen, the advantage to sensitivity afforded by this apparatus increases with frequency up to 1000~ and then is fairly uniform for the higher range except for a sharp variation at 3000~. The average value is 31.7 db for one ear and 27.7 db for the other.

PHASE DISTORTION

The inertia, stiffness, and friction contained in a vibratory system give it peculiarities of temporal behavior in response to sounds, and the behavior varies with the frequency of the sounds. The response wave produced in the system does not usually reach its peak simultaneously with the peak of the imposed waves, but

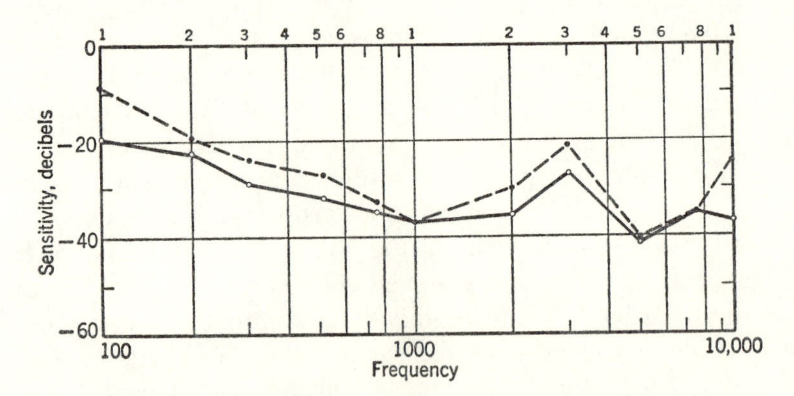

Fig. 41. Sensitivity functions for the middle ear apparatus as a whole, shown for two cat ears. The curves represent differences between the normal ear and an ear theoretically deprived of the whole middle ear apparatus.

is more or less delayed. In a complex sound the component waves are displaced in time relative to one another, usually with marked alterations of wave form.

In general, when a system shows frequency distortion we can expect it to show phase distortion as well. The human ear suffers from frequency distortion as we have seen, and there is no doubt that it suffers from this second kind of distortion also. There is no direct evidence, however.

In the cat's ear this form of distortion can be demonstrated. It is present in the cochlear potentials. The following experiments have revealed in detail the phase changes introduced by the middle ear mechanism (Wever and Lawrence, *4,5*). As in our study of the role of this mechanism in sensitivity, two experimental procedures were necessary, one to measure the phase changes produced by the mechanism up to the stapes and another to find the changes produced by the stapes.

The first procedure was as follows. The auditory bulla was opened in the usual way and an electrode was placed on the round window membrane to record the cochlear potentials. The pinna and the membranous part of the external auditory meatus were removed to give clear access to the drum membrane, and an acoustic probe, as described earlier, was inserted into the remaining meatal cavity with its end about 2 mm from the drum membrane. A sound tube leading from a loudspeaker was placed at the meatal opening and then wax was molded over the region. The sound tube and probe tube thus were enclosed in a small cavity whose inner boundary was the drum membrane. When sound was introduced into this cavity the probe tube could be used to measure its intensity and also to show the relation between its phase and that of the cochlear potentials.

The apparatus for these measurements is shown in Fig. 42, and consists of two channels leading to a dual cathode ray oscillograph. In one channel, the sounds picked up by the probe were converted into electric currents in its microphone and after amplification were led through a phase shifter to a second amplifier contained in one side of the oscillograph. In the second channel, the cochlear potentials recorded from the round window were run through an amplifier to the other side of the oscillograph. This oscillograph employs a dual-beam tube: two electron streams impinge upon the same tube face and give separate waves that may be placed side by side for comparison.

The microphone and cochlear potential signals were applied to the oscillograph so as to give vertical deflections on its tube face while the sweep circuit of the oscillograph gave horizontal deflections that could be synchronized with the frequency of the signals. This sweep was common to both beams so that when the

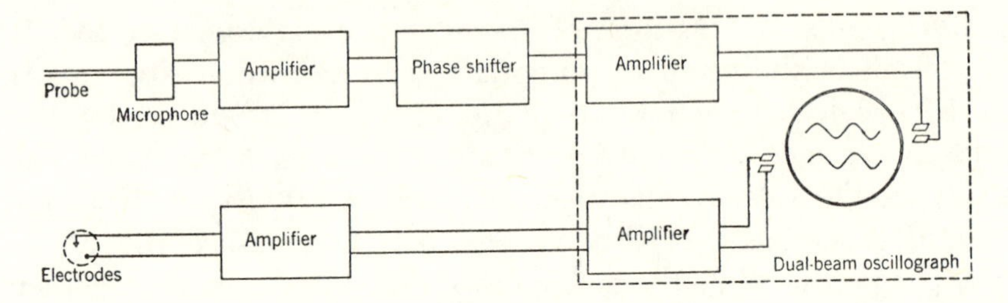

Fig. 42. Apparatus for phase measurements; first arrangement.

two signals were in phase the resulting waves could be made to coincide if suitably adjusted in amplitude and in vertical position on the tube face. When the signals were out of phase the waves were displaced laterally with respect to one another and then could be brought into coincidence only by manipulating the phase shifter in the microphone channel.

A tone was presented to the animal's ear and the phase shifter was adjusted for coincidence of the stimulating and response waves. The reading of the phase shifter was then taken. This reading included the phase characteristics of the ear and also of the apparatus; it had no absolute significance. Such readings were obtained for a great many tones—often as many as 60 to 70—over the range from 100 to 10,000~.

Then the drum membrane was removed and the incudostapedial joint was broken, after which the malleus and incus were removed. The stapes was left in position in the oval window with its tendon still attached. The sound tube remained in its former position at the meatal orifice, but the probe tube was moved close to the stapes to record the sounds impinging upon it. The region was again sealed over with wax to confine the sounds to the oval window location.

Now the ear was stimulated with the same tones as before and new adjustments of the phase shifter were made for coincidence of the stimulating and response waves, and these readings were noted. A comparison of this set of readings with those taken earlier reveals the effect of removal of the principal middle ear structures on the phase of the transmitted sound.

The upper portion of Fig. 43 shows the results obtained on a

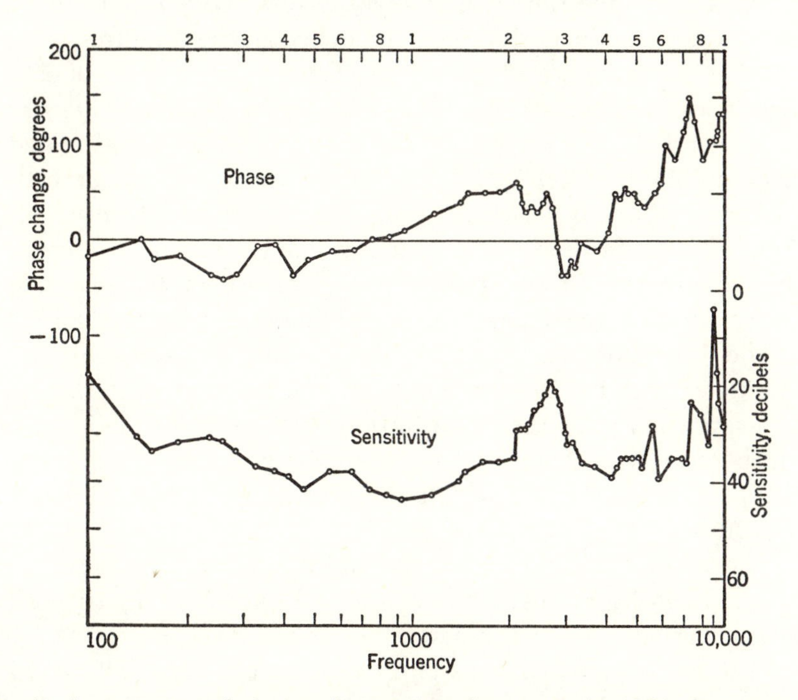

Fig. 43. A comparison of phase and sensitivity changes produced by the principal middle ear structures. Positive degrees represent a retardation of phase. From Wever and Lawrence (*4*) [*Annals of Otology, Rhinology, and Laryngology*].

representative ear. Positive values on the ordinate represent a retardation of phase by the middle ear structures; negative values represent an advancement of phase. We see that over most of the lower range, up to 800~, the middle ear produces a small and fairly uniform advancement of phase, never exceeding 40°. In the region of 800~ the curve crosses the zero line, and here the transmission takes place without phase change. For higher tones the conducted wave is caused to lag in phase, and this lag grows in amount to a maximum at a frequency around 2500~. Thereafter the curve falls sharply and crosses the zero line again at 3000~.

The curve attains a sharp minimum, representing the utmost of phase advancement in this region, and then once more it rises and crosses the zero line. It makes this third crossing around 4000~. In the remaining high-frequency region are the greatest variations. The curve remains above the zero line in the region of phase lag and undergoes a number of rapid fluctuations.

Other ears give phase functions that closely resemble this one in the low frequencies but vary somewhat in the high frequencies. The first upward crossing always occurs in the region of 1000~. The first maximum of phase lag appears in different ears in the range from 2000 to 3000~, and the first downward crossing occurs around 2600 to 3750~. After the minimum is attained, the third crossing, which is upward, occurs in the region from 3800 to 5000~. Thereafter some of the curves, like the one illustrated here, show fluctuations above the line, whereas others, likewise undergoing rapid changes, cross the zero line once more.

The losses of sensitivity resulting from the removal of the middle ear mechanism are plotted in the lower part of Fig. 43 in a positive sense, as the contributions to sensitivity made by this mechanism. We can now compare this sensitivity function with the phase changes for which the middle ear mechanism is responsible.

The sensitivity function for the middle ear contains two regions of relatively high sensitivity separated by a rather sharp peak of low sensitivity. As represented here the lower portions of the curves are regions of maximum sensitivity. The first of these regions is broad; the curve declines slowly to a minimum in the region of 1000~ and then rises a little more rapidly. Note that the location of this minimum corresponds fairly closely with the frequency at which the phase changes from negative to positive. The sharp peak in the sensitivity curve, representing a point of poorest sensitivity, is at 3000~. This is the frequency at which the phase curve falls from positive to negative. The second minimum of the sensitivity curve, representing high sensitivity, lies between 4000 and 5000~, and in this region the phase curve again crosses from negative to positive. For still higher frequencies both the sensitivity and phase functions show numerous variations, but in general the sensitivity is growing poorer and the phase is becoming more and more retarded.

We now turn to our experiments on the phase changes produced by the stapes. Here, as in the study of the effects of this ossicle on sensitivity, it is necessary to make the assumption that when all of the middle ear except the stapes is removed the pathways to the sensory cells of the cochlea by way of the oval and round windows are acoustically different primarily because of the presence of the stapes in the oval window. Then it is only necessary to find the phase differences in the cochlear potentials produced on stimulation by these two pathways.

This experiment was carried out on cats in the following manner. After the drum membrane and the two outer ossicles were removed, two sound tubes were applied, one over the oval window and the other over the round window. The one over the oval window surrounded the stapes but was adjusted to make no contact with it. These tubes led from separate loudspeakers which emitted tones that were of the same frequency but that could be varied independently in other respects. The apparatus arrangement for this purpose is shown in Fig. 44 and consists of

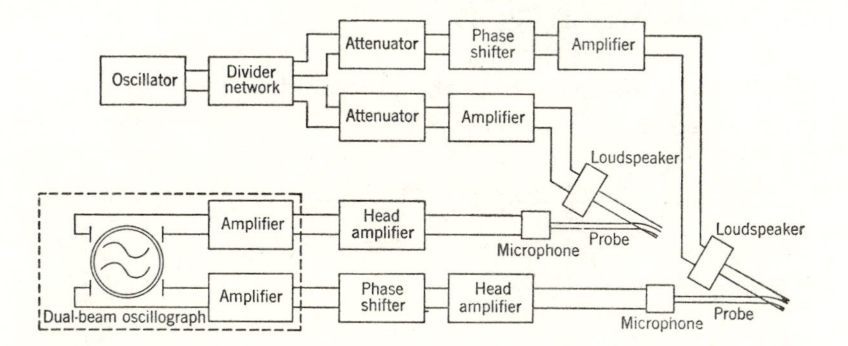

Fig. 44. Apparatus for phase measurements; second arrangement.

a common oscillator working through an isolating network into two channels containing their own attenuators and amplifiers. Also in one of the channels is a phase shifter. By these controls the tones could be given any desired intensities and phase relations. Both sound tubes contained acoustic probes that with their associated apparatus were calibrated to indicate both the intensity and phase of the sounds delivered to the two windows.

The sound tubes were fitted with tubular rubber tips and were applied over the windows with sufficient firmness to prevent any appreciable sound leakage. This isolation of the two sources is essential and was always checked by introducing sound through one tube and recording from both probes.

A needle electrode for the recording of cochlear potentials was placed on the bony capsule of the cochlea as close as possible to the round window.

The further procedure consisted of applying tones simultaneously at the oval and round windows and adjusting the intensity and phase relations so as to obtain a cancellation of cochlear potentials. This was most easily done by first applying the tone at one window and adjusting the intensity for some desired level of response, then applying the tone at the other window with similar adjustments, and finally combining the two stimulations. Then it was only necessary to vary the phase in one channel to reduce the response to zero. The intensities of these two balanced stimuli were obtained from the acoustic probes, and their phase difference was observed also. This procedure was carried out for 60 or more tones from 100 to 15,000~, with results as shown for a representative ear in Fig. 45.

In this figure a positive phase change signifies that to obtain a cancellation of response the tone presented to the oval window had to be advanced in phase relative to the tone at the round

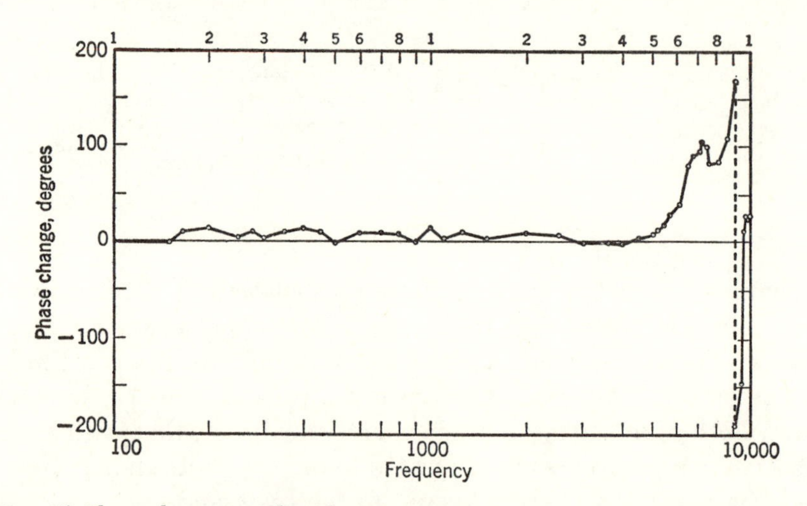

Fig. 45. Phase changes attributed to the stapes. Positive values represent a phase lag in the oval window pathway relative to the round window pathway. From Wever and Lawrence (5) [*Annals of Otology, Rhinology, and Laryngology*].

window. Or, according to our assumption, a positive phase change signifies that the stapes has introduced a phase lag into the transmitted sound.

The curve shows little variation over the lower portion of the frequency range up to 5000~. At no point here does the phase difference exceed 15°. The differences that do appear are consistently positive, however, indicating a phase lag.

Beyond 5000~ the phase lag increases rapidly with frequency and attains a maximum at 9000~. As a lag of phase that exceeds 180° becomes an advance of phase, the function is discontinuous here and suddenly appears as a negative phase. This negative or advancing phase grows less and the curve crosses the zero line at 9400~.

Individual ears showed only moderate variations under these conditions. When, however, various surgical modifications of the stapes were made the curves changed in form at their upper ends. The curve went through its first upward rise at a lower frequency when the stapedial tendon was severed, and executed this rise particularly abruptly when the head and crura were removed and only the footplate remained. During these manipulations the low-frequency portion of the function remained unaltered. Some of the results are shown in Fig. 46. These observations lend weight to

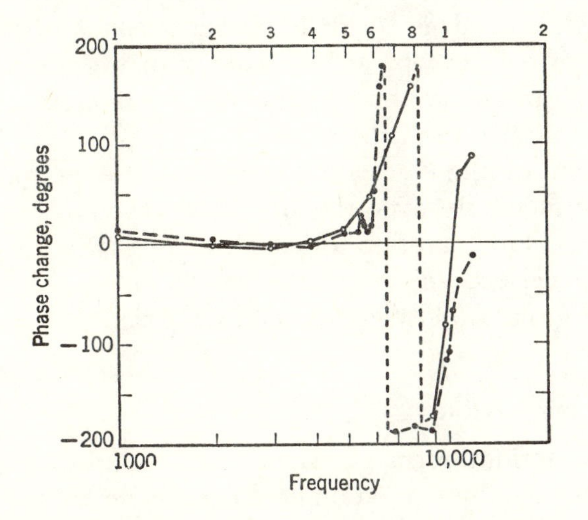

Fig. 46. Variations in the phase functions as a result of manipulations of the stapes. For the solid curve the stapes was intact. For the broken curve its head and crura were removed, leaving only the footplate. From Wever and Lawrence (5) [*Annals of Otology, Rhinology, and Laryngology*].

our assumption that the stapes is responsible for the phase differences found between the two windows.

By combining the results of our two experiments as given in Figs. 43 and 45 we obtain a phase function for the entire middle ear apparatus. This function is represented in Fig. 47. It is neces-

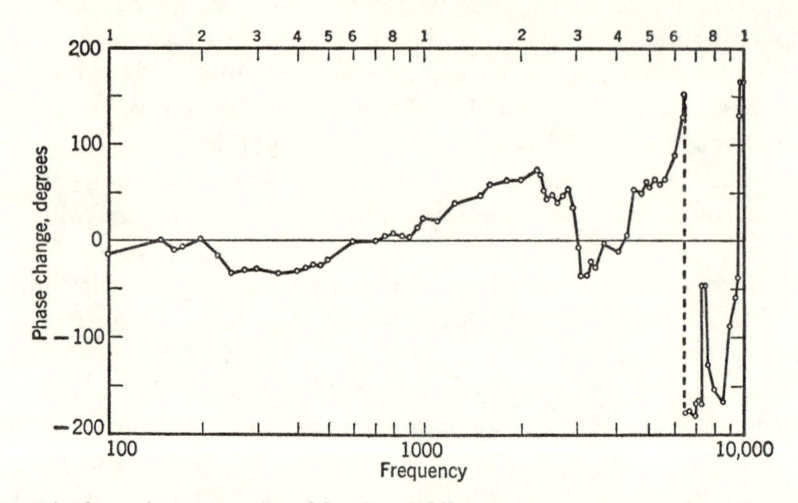

Fig. 47. Phase changes produced by the middle ear apparatus as a whole. Positive degrees represent a retardation of phase.

sary to point out that this curve combines results obtained from two different ears. This is justifiable on account of the general uniformity of ears except at the very high frequencies. At the extreme high-frequency end of the range we must expect considerable variations from the form shown.

The amount of phase variation introduced by the middle ear is surprisingly moderate over the principal portion of its working range. For the low tones, up to 5000~, this variation in the example shown never exceeds 74° and is usually below 40°. As is to be expected, it grows rapidly greater and more variable in the higher frequencies.

AMPLITUDE DISTORTION

The most serious form of distortion is amplitude distortion, arising from a nonlinear relation between the response of a system and the forces acting upon it. Such distortion consists essentially of alterations of wave form, which occur because the changing amplitude in each period of the waves fails to follow the driving force. This distortion can be observed directly if the response is converted into electrical oscillations and led into an oscillograph. More commonly these alterations of wave form are revealed in two other ways: in the failure of the response to reach expected over-all magnitudes, and in the appearance of new frequencies in the wave composition.

A simple example of this form of distortion can be found in the action of a loudspeaker driven by sinusoidal currents, when the diaphragm has been provided with a rigid stop to limit its movements. Moderate currents that do not drive the diaphragm into the stop will then produce sinusoidal waves, but strong currents will produce waves with the peaks cut off. Because the range of amplitude variation is now curtailed the average magnitude of the sound will be reduced below its normal value, and this reduction becomes proportionately more serious as the driving currents are increased. At the same time the sound will be heard to undergo changes of quality; no longer do we hear a simple tone but instead a tone that grows tinny and rough and at extreme levels even noisy in character.

Mathematically, as Fourier first showed, a distorted wave like that just described can be regarded as made up of a series of sine (or cosine) waves of many frequencies in addition to the frequency of the original wave. These frequencies bear a simple integral relation to the fundamental frequency. Thus a 1000~ wave when distorted will be found to contain frequencies of 2000, 3000, 4000, 5000, and so on in addition to the original 1000. Usually in such a harmonic series the magnitudes of the higher components decline rapidly with their position in the series.

When the original wave consists of two primary tones the result of distortion is still more complex. A harmonic series appears for each of the primaries and in addition there are components with frequencies formed by various combinations of the two primary series. These combinations consist of all the possible sums and differences of the original frequencies and their integral multiples. It is simplest to represent these combinations by a formula. If we let h and l stand for the frequencies of the primaries, their combination tone series is represented by $mh \pm nl$, where m and n are given all integral values 1, 2, 3, 4, 5, etc. When m or n is allowed to equal zero this formula expresses the harmonic components of the primaries as well. We designate the order of a component as $m + n - 1$ (Wever, *3*).

Though it was not thought of in this way at the time, the first evidence of distortion in the ear was the observation of combination tones. The formation of a difference tone $(h - l)$ by the interaction of two simple tones was noted by Sorge in 1744 and also by Romieu in 1753. Tartini published similar observations in 1754 and claimed that he had made the discovery in 1714, and his influence was such that the tones came to be known as Tartini's tones.*

Another difference tone in addition to Tartini's was discovered by Thomas Young in 1800, and still others were found a little later perhaps by Scheibler and certainly by Hällström. Hällström in 1832 described four or five combination tones and also attempted a theoretical explanation of them.

The early theory of the origin of combination tones is said to have been suggested by Romieu and was definitely formulated by Lagrange and Young. This is the beat-tone theory, in which the new component is regarded as the result of a fusion of rapid beats between the primaries. This theory arose naturally from the observation that the one combination tone first known, which we now call the first-order difference tone $(h - l)$, has the same frequency as the beats of the primaries. The discovery of other components of different frequencies presented a difficulty for the theory. Hällström tried to meet the difficulty by suggesting that beating could occur not only between the primaries but between any one of them and a combination tone already formed or between two combination tones. In his system, for example, the beating $(h - l)$ produces D_1, the first-order combination tone, and then the beating $(l - D_1)$ produces D_2, a second-order combination tone, and so on.

This type of theory ran into difficulty when an attempt was made to specify the interaction processes and to locate them anatomically. Successive interactions seemed to call for several locations of these processes in the ear, and the lack of any positive evidence for such locations appeared as a weakness in the theory. The theory was further weakened when Helmholtz (*1,2*)

* For a discussion of this discovery, and references, see Jones.

discovered a new kind of combination tone, the summation tone $(h+l)$. Such a tone cannot readily be explained by the fusion of beats.

The Transformation Theory. As a solution of these explanatory difficulties Helmholtz developed his transformation theory, according to which both the difference tones and the summation tones, and the overtones as well, are regarded as the products of nonlinear distortion in the ear. He pointed out that such distortion is characteristic of all acoustic apparatus—of various types of sound generators and sound receivers as well as the ear—when they are driven beyond their proper limits of operation.

A mathematical formulation of the transformation theory is as follows. Let X represent the displacement produced in a vibratory system by the force (or pressure) P. Then as long as the action is linear

$$X = a_0 + aP \tag{1}$$

Here a_0 is a constant representing the equilibrium position of the system; it is usually equal to zero and in any case we shall disregard it henceforth, which means that we shall always measure X from its equilibrium position. The other constant a expresses the sensitivity of the system.

When the action becomes nonlinear this formula no longer holds, and according to the transformation theory the displacement varies to some extent with higher powers of P or, symbolically,

$$X = a_1P + a_2P^2 + a_3P^3 + a_4P^4 + \ldots + a_nP^n \tag{2}$$

The constant a_1 modifying the first power of P is now smaller than a in the linear equation, because some of the exerted force is operating in the more complex ways indicated in the higher terms of this equation. The additional constants a_2, a_3, a_4, etc., have values in accordance with the degree of this transformation, growing larger in relation to a_1 as the distortion increases.

The quantity P represents the pressure at any moment. In a simple wave the pressure varies periodically as a function of time according to the formula

$$P = P_0 \cos \omega t$$

where P_0 is the maximum pressure, attained at the peak of the wave, ω is the angular frequency and is equal to $2\pi f$ where f is the frequency of the vibration, and t is the time measured from any arbitrary starting point.

When we substitute this value of P in the linear equation 1 we get

$$X = aP_0 \cos \omega t$$

which simply means that the displacement undergoes regular sinusoidal variations in time at a frequency equal to f and with peak values equal to aP_0.

When we substitute this value of P in the nonlinear equation 2, however, we obtain a complicated result. Then

$$X = a_1P_0 \cos \omega t + a_2P_0^2 \cos^2 \omega t + a_3P_0^3 \cos^3 \omega t \\ + a_4P_0^4 \cos^4 \omega t + \ldots + a_nP^n \cos^n \omega t \qquad (3)$$

The first term of this series, as in equation 1, represents a periodic displacement at the fundamental frequency f. The second term is a little difficult to interpret as it stands because the form of a squared cosine wave is not immediately obvious. However, by reference to a table of trigonometric relations we discover that in general

$$\cos^2 \theta = \tfrac{1}{2} \cos 2\theta + \tfrac{1}{2}$$

or, in words, the squared cosine of an angle is the same as the cosine of twice the angle modified by constants. In our equation $\theta = \omega t$, hence

$$a_2P_0^2 \cos^2 \omega t = \tfrac{1}{2}\, a_2P_0^2 \cos 2\omega t + \tfrac{1}{2}\, a_2P_0^2$$

Our squared cosine wave is therefore equivalent to a simple cosine wave of double frequency (for $2\omega = 2\pi \times 2f$) and reduced amplitude ($\tfrac{1}{2}\, a_2P_0^2$) together with a constant ($\tfrac{1}{2}\, a_2P_0^2$) that shifts the whole wave relative to the equilibrium position.

A similar treatment of other terms of equation 3, a process that is laborious but not at all difficult, will show that the term involving $\cos^3 \omega t$ introduces a component whose frequency is 3 times that of the fundamental, the term in $\cos^4 \omega t$ introduces one with a frequency 4 times that of the fundamental, and so on. These higher terms also add constants and contribute to the magnitude of components already present. Hence we see that in equation 3 we have a mathematical expression in harmony with the fact that a simple wave when distorted yields a complex wave that in the most general case contains the original frequency and all its integral multiples.

When two driving frequencies are present at the same time the situation is still more complicated. The driving force P is then the sum of two cosine functions of different frequencies that we shall call h and l, or

$$P = P_h \cos \omega_h t + P_l \cos \omega_l t$$

where $\omega_h = 2\pi h$ and $\omega_l = 2\pi l$ represent the two angular frequencies and P_h, P_l represent their magnitudes.

A substitution of this value of P in equation 2 and an expression of the resulting equation in equivalent trigonometric forms then will give a long series of terms representing the two primary frequencies and their multiples and all combinations of these frequencies. This derivation gives all the components needed to explain the facts as observed.

Experimental Evidence; The Exploring-Tone Method. The demonstration of combination tones as a product of interaction in the ear came earlier than the detection of aural overtones because the pioneer observers had to use sources of sound that were rich in overtones to begin with, and these physical overtones obscured the subjective ones. When pure tones became available, first in the carefully designed tuning fork, it was easy to show that overtone components also originate in the ear. A person with training in musical perception easily identifies the first or second overtone along with the fundamental.

The higher overtones and most of the combination tones give difficulty to simple observation, but are readily revealed by the exploring-tone method. In this method an additional tone is introduced and varied in frequency so as to approach closely the frequency of an expected component. Beats are then heard at a rate equal to the difference in frequency between the exploring tone and the subjective tone. For example, if a component of 2000~ is expected we present an exploring tone of, say, 2003~ and look for a beating at the rate of 3 per second. It is found to be helpful to make the loudness of the exploring tone equal to that of the component being studied, for then the beats are most prominent. By this method Wegel and Lane on stimulating with tones of 1200 and 700~ were able to show the presence in the ear of 16 to 17 components, including 5 overtones, 5 or 6 difference tones, and 6 summation tones, in addition to the two primaries. Graham and Cotton obtained further indications of the complexity of the distortion pattern. Graham's results, reported by Fletcher (*1*), were obtained by the exploring-tone method and showed the presence of as many as 11 overtones on stimulation with a 800~ tone. Cotton reported evidence for the existence of 19 combination tones within the interval of an octave.

The exploring-tone method can be used also to give indications of the magnitudes of these subjective tones. Graham's results, just referred to, included such measurements. Békésy (*6*) reported measurements on the first few overtones produced by stimuli in the range from 200 to 3000~ and gave detailed results on the first two overtones of 800~. He gave results also for a difference tone of 300~ produced by stimulating the ear with 2000 and 2300~. Moe made measurements on the first 5 partials of tones of 690 and 950~ when they were presented singly and on 19 of the combination tones produced when these two tones were presented together.

Distortion as Shown in the Cochlear Potentials. The ear's distortion is readily revealed in the cochlear potentials. The simplest evidence is found in the intensity functions as presented in Fig. 48. The response rises as a linear function of sound pressure from the lowest values measurable by present methods, about 0.1 microvolts, to levels that vary systematically with frequency but for low tones attain about 300 microvolts, after which the function is nonlinear. The curve continues to rise as the sound intensity is increased, but at a progressively slower rate, until finally a maximum is reached beyond which an increase of intensity causes a reduction of response. The sound intensities at which this overloading begins to appear vary considerably for different conditions, namely, the species of animal, the individual ear, the frequency of the tone, and the technical arrangements, but in general lie between 1 and 10 dynes per sq cm for cats and guinea pigs,

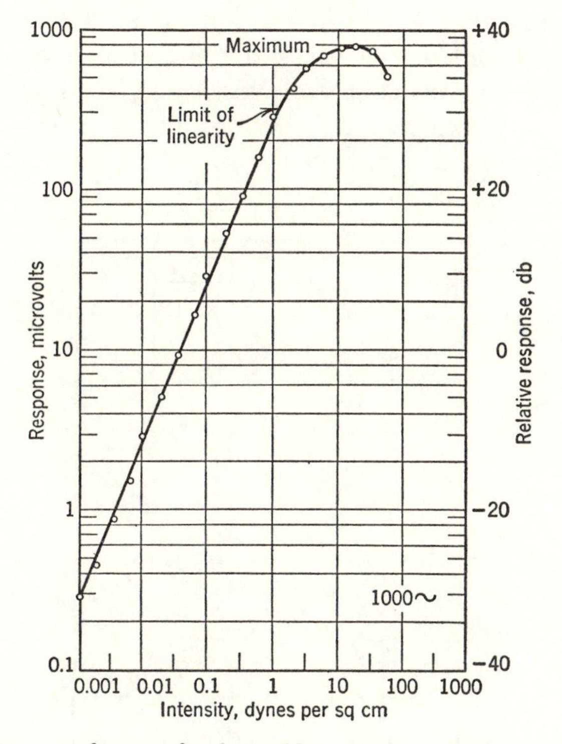

Fig. 48. The intensity function for the cochlear response, observed in the cat. From Wever, *Theory of Hearing*, 1949, John Wiley and Sons.

the animals in which the most extensive measurements have been made.

An examination of the wave form of the cochlear potentials in response to a sinusoidal stimulus always shows a simple sine form at levels below the entrance of overloading. At higher levels the form is no longer sinusoidal, but is distorted as shown in Fig. 49.

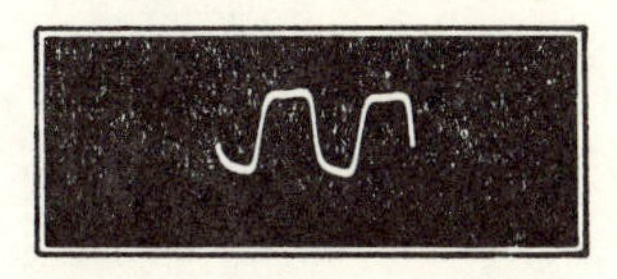

Fig. 49. Wave form of potentials produced in the cat's ear by an intense tone of 1000~.

Series 1; The Harmonic Pattern. A wave analysis of the cochlear responses reveals the distortion pattern in detail. In this analysis a stimulus tone that has been carefully filtered to remove harmonics is delivered to the ear and the resulting potentials are run into an electric wave analyzer, which acts as a highly selective voltmeter. In our experiment, whose results are shown in part in Fig. 50, it was found necessary to use both electric filtration for

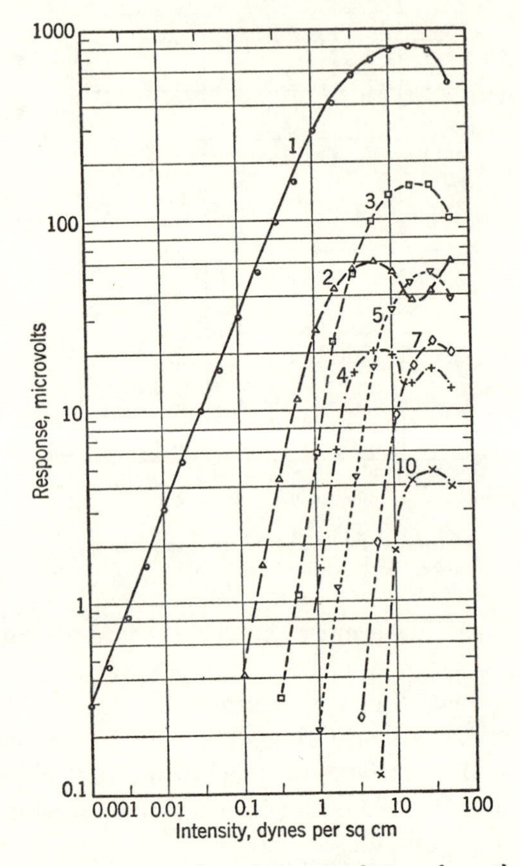

Fig. 50. The harmonic pattern resulting from stimulation of a cat's ear with a pure tone of 1000~. Each curve bears a number indicating its order in the harmonic series; thus the curve marked "2" represents the component of 2000~. From Wever and Bray (5).

the oscillatory currents driving the loudspeaker and acoustic filtration for the generated sounds in order to attain the necessary degree of purity. We stimulated with 1000~ and with the wave analyzer we measured successively the magnitudes of all components present from the fundamental to the 10th harmonic, and sometimes as far as the 16th. We did this at each of 20 intensities of the 1000~ stimulus, with results as shown in the figure. To simplify the figure the curves for some of the high-frequency components have been omitted.

As the figure shows, each harmonic, like the fundamental, has a curve that at first is straight as represented in this double-logarithmic plot, and then bends over at its upper end. The slopes of these higher functions are steeper than the slope of the fundamental, and tend to grow increasingly steeper as we go up the series. Because a straight line on this type of plot represents a power function, we see that the harmonics are all power functions of sound pressure in the early part of their courses and then assume more complicated relations at higher levels.

In general, the higher orders of harmonics are progressively smaller in magnitude. This relation of magnitudes is clear at moderate levels of stimulation but at high levels it becomes partially obscured by the changing slopes and the tendency for the odd harmonics to attain relatively large maximums.

If by the root-mean-square method we sum up the magnitudes of all the components present at each intensity the result gives a curve that only at the upper end is distinguishable from that of the fundamental, and even here it has much the same form. The bending of the total response curve at its highest levels signifies that in overloading there is an actual loss of efficiency of the ear and not merely a conversion of energy into higher partials.

Series 2; the Combination Tone Pattern. When two pure tones are led to the ear and raised to a moderately high level of intensity the cochlear potentials show a complex pattern that on analysis yields a veritable profusion of combination tones. Without any attempt to exhaust the possibilities we have recorded as many as 40 components, including overtones and combination tones, as a result of stimulating the guinea pig's ear with two pure tones.

When tones of 1000 and 2800~ were presented we recorded the components shown in Table 1, plus an overtone of 7000~ and a combination tone of 6200~ that would require another row and column for portrayal. The arrangement of the table shows the relations of the components to the primaries; thus the 800~ component in column 2 and row 4 has the composition $(h - 2l)$. The numbers in brackets in the last column represent components

TABLE 1. COMBINATION TONES

The arrangement shows the relation of the components to the primaries and their multiples. The components indicated in brackets were not found in the cochlear potentials.

	$h = 2800$	$2h = 5600$	$3h = 8400$
$l = 1000$	1800 3800	4600 6600	7400 9400
$2l = 2000$	800 4800	3600 7600	6400 10,400
$3l = 3000$	200 5800	2600 8600	5400 [11,400]
$4l = 4000$	1200 6800	1600 9600	4400 [12,400]
$5l = 5000$	2200 7800	600 10,600	3400 [13,400]
$6l = 6000$	3200 8800	400 11,600	2400 [14,400]

From Wever, Bray, and Lawrence (3) [*Journal of Experimental Psychology*].

looked for but not found; all these are summation tones of rather high frequencies. All these components belong to the series $(mh \pm nl)$, and no others—no subharmonics, meantones, intertones, or any other frequencies not belonging to this series—could be found in spite of a diligent search for them. The results therefore are in full agreement with the transformation theory as to the composition of the distorted wave.

The highest order of component in the table is 2400~ = $(3h - 6l)$ and is of the 8th order, and the 6200~ component omitted from the table is of this order also. However, it is easy to obtain components of still higher orders when the primaries are selected so that the components will have frequencies within the animal's range of good sensitivity. Thus, with 1000 and 1300~ we obtained 1800~, which is given both as $(7l - 4h)$ of the 10th order and as $(6h - 6l)$ of the 11th order. With 100 and 10,000~ we obtained 8000~, which is $(h - 20l)$ and of the 20th order.

The production of combination tones does not seem to be limited by the frequency separation of the primary tones. These components appear for primaries of closely adjacent frequency and also for primaries near the two ends of the frequency scale. Thus with 1000 and 1025~ we obtained a difference tone $(h-l)$ of 25~ and with 100 and 10,000~ we obtained the whole series of difference tones $(h-l)$, $(h-2l)$, $(h-3l)$, etc., up to $(h-20l)$, and the series of summation tones $(h+l)$, $(h+2l)$, $(h+3l)$, etc., up to $(h+15l)$, a total of 35 combination tones.

We have carried out a quantitative study of many of the lower orders of combination tones, and have found a systematic relation between their magnitudes and the intensities of their stimuli.

When one of the two primaries h and l used to generate the tone $(h-l)$ is kept constant and the other primary is varied in intensity the magnitude of $(h-l)$ rises at first linearly with the primary that is varied and then bends over at high intensities. This relation can be seen in Fig. 51, where h is varied and l is kept constant for any given function. When l is given larger values the curve obtained by varying h has a more extensive linear course and attains greater magnitudes before bending over. Almost exactly the same forms of functions are obtained by keeping h constant and varying l. The largest value of this difference tone therefore is attained when both of the primaries are strong and about equally so. Within the limits imposed by overloading, the magnitude of the first-order difference tone is proportional to the intensities of both its primaries or, symbolically,

$$X_{(h-l)} \propto P_h P_l \qquad \text{or} \qquad X_{(h-l)} = k_1 P_h P_l$$

where P_h, P_l are the sound pressures of the primary tones h and l and k_1 is a constant representing both the degree of distortion and the sensitivity of the ear.

When similarly we measure the first-order summation tone $(h+l)$ we obtain results that are like those for $(h-l)$. We find here also that the magnitude of this component is a linear function of the intensities of its primaries up to the level of overloading.

On the other hand, when we study combination tones of different composition we obtain different results. Thus the second-order difference tone $(2h-l)$ rises at a relatively rapid rate as h is varied and l is kept constant. This rate approximates the square of the sound pressure; as actually measured the slopes of

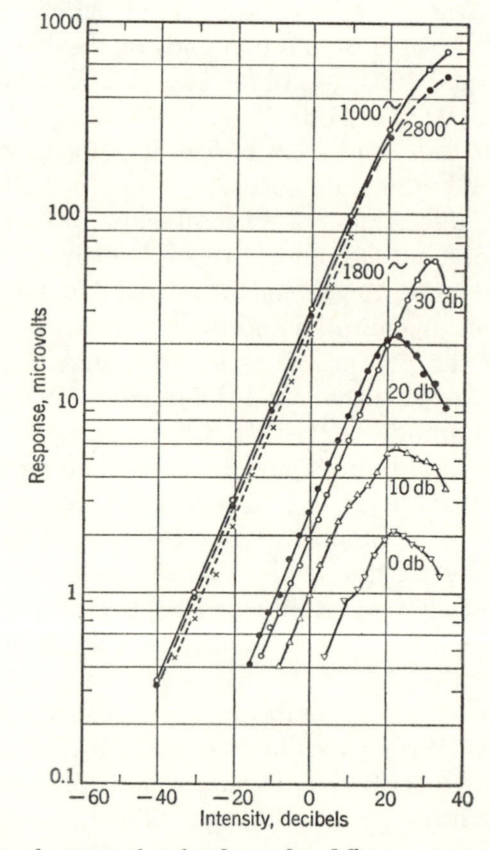

Fig. 51. Intensity functions for the first-order difference tone $(h - l) = 1800$~, produced by the primaries 2800 and 1000~. On the left are curves for the primaries and also for a stimulus tone of 1800~. On the right are the difference tone curves, plotted with the 2800~ primary as abscissa and the 1000~ tone as parameter. Both the abscissa and parameter values are given in decibels relative to 1 dyne per sq cm. From Wever, Bray, and Lawrence (3) [*Journal of Experimental Psychology*].

the curves along their straight portions varied from 1.82 to 1.98. A graph of $(2h - l)$ when h is kept constant and l is varied, however, gives results of the same form as those of Fig. 51; this difference tone increases only linearly with l. Therefore we express its magnitude in general as

$$X_{(2h-l)} \propto P_h^2 P_l$$

The corresponding summation tone $(2h + l)$ was found to have the same form of dependence upon its stimuli.

Measurements were carried out in a similar way on other components like $(3h - l)$, $(h + 3l)$, $(2h - 3l)$, and $(3h - 3l)$. For the component $(2h - 3l)$, for example, the slope of the function when h was varied was measured as 1.81 and the slope when l was varied was measured as 3.13. We see that the slopes depart somewhat from integral values. This was generally true for the higher orders of components, and most often the departure was downward. When we averaged the values of slopes obtained for a large number of components of the form $(mh \pm nl)$ for values of m or n of 1 and 2, the means came out as 1.02 and 2.06 respectively. When m or n was 3 the mean value of the slope was 2.83, and higher values of m or n gave slopes even smaller than expected from theory. We ascribe this decline in the slope for higher orders of components to errors in measuring the small amounts of potential present and especially to the entrance of overloading. Within this limitation, however, it appears that the functional variation of a combination tone can be predicted from its composition with respect to the primaries. Or, in general,

$$X_{(mh \pm nl)} \propto P_h^m P_l^n$$

This empirical formula is vastly more simple than the formulas developed in the mathematical expression of the transformation theory. As already mentioned, these formulas show every component to be contributed to by alternate terms in the power series beyond the one in which the component first appears. If all these terms had an appreciable effect upon the magnitude of the component the slope would obviously be steeper than that produced by the first of these terms alone. Hence we conclude that the constants in the power series must fall off continually in magnitude, or in other words that the series is convergent, and sufficiently so that for any component we can neglect all terms beyond the earliest one that produces it. Yet this convergence cannot be too rapid; it must still permit the appearance of components of very high orders, at least as high as $(h - 20l)$ of the 20th order which requires an appreciable variation with P^{21}. We can more readily satisfy these two conflicting requirements if we assume that the decline in the magnitude of the constants in equation 2 is rapid for the first few terms and more gradual thereafter. A further possibility is that the contributions made to a given component by the higher terms of the power series have random phase relations so that they tend to cancel one another. The transformation theory as stated heretofore gives no indication of such varying phase relations but it could easily be reformulated in order to do so.

In other respects the observations are fully in accord with the transformation theory. Especially significant is the fact that summation tones and difference tones of the same form, like $(h + l)$ and $(h - l)$, bear the same relations to their stimuli. They evidently arise by the same process, namely, the distortion involving P^2, as the theory states. This evidence finally disposes of the argument offered by proponents of the beat-tone theory that the summation tone is actually another sort of component. Thus Preyer contended that what is called the summation tone $(h + l)$ was the result of an interaction between the second harmonic of the higher tone $(2h)$ and the difference tone $(h - l)$, that is, was constituted as $[2h - (h - l)]$. Algebraically this expression reduces to $(h + l)$ and gives the proper frequency. However, such a component ought to vary as P_h^3, whereas actually it is found to vary as P_h.

The same consideration disposes of the contention that the summation tone is a difference tone of higher order. It is true that the proper frequency can be obtained as a difference tone, though often it is necessary to go to a very high order to do so. Thus, for example, the frequency of 3800 arising from the primaries 2800 and 1000~ can be obtained as a difference tone of the composition $(6h - 13l)$. But if the 3800~ component arises in this way it should vary as the 6th power of the 2800~ tone and as the 13th power of the 1000~ tone, whereas actually it varies linearly with each of these primaries. Another argument of interest was presented by Peterson some time ago. It is that mistuning one of the primaries ought to have a different effect on the frequency of the component depending upon its origin. If in this example the 3800~ component is of the composition $(h + l)$ it ought to be raised 1~ when the 1000~ primary is increased to 1001~, whereas if it is of the composition $(6h - 13l)$ it ought to be reduced by 13~. Observation readily proves that it is increased by 1~.

9

The Locus of Distortion in the Ear

THE HELMHOLTZ THEORY OF AURAL DISTORTION

HELMHOLTZ located the distortion process in the middle ear, and especially in the actions of the drum membrane and of the joint between malleus and incus. He opposed the view expressed by Riemann that for the purposes of hearing the character of external sounds must be faithfully represented to the inner receptor processes. Riemann had argued that the middle ear apparatus is a delicate and precise amplifying mechanism that conveys the pattern of the external pressure changes with perfect accuracy and without any alteration in the intensity relations among its components. In other words, he assumed a linear relation between the external stimulus and its actions upon the ear. Helmholtz maintained on the contrary, as we shall see in more detail presently, that every tone is subjected to distortion in its passage through the middle ear. For a suitable representation of the world of sound he saw no need for the strict correspondence that Riemann had postulated. For every tone to possess a certain perceptual identity it is only necessary, he said, that it have the same effect upon the ear every time it is presented. He accepted a nonlinear relationship because through it he was able to explain the production of overtones and combination tones.

The reason for Helmholtz's insistence upon the middle ear as the seat of nonlinearity is to be found in his general theoretical position. In a simple resonance theory such as his, which holds to a spatial analysis of complex sounds in the cochlea by the action of specific elements, it is essential that the distortion by which combination tones arise shall take place before this spatial analysis has occurred. The generation of these tones requires nonlinear action at some site where the two primary tones are simultaneously present. This site cannot be the cochlea if, as this theory maintains, an analysis occurs there by which the primary tones are relegated to different places and thus are operating independently. Hence the site must be the middle ear.

In this theory the overtones also are best explained as arising in the middle ear through the same distortion process. It is possible, however, by the use of a special assumption, to grant these tones a separate origin in the inner ear. The necessary assumption is that overtone frequencies are generated through nonlinearity in the primary cochlear resonator—the one that is responding to the stimulating tone—and that these new frequencies are then radiated throughout the cochlea and are able secondarily to set in action the other resonators tuned to them. By these secondary excitations the overtones come to be perceived and at the same time to possess the proper specific qualities.

It is obviously simpler to hold that the two products of distortion, both overtones and combination tones, have their origin in the same processes and at the same site. The transformation theory provides a secure analytical basis for this view. Therefore the adherents of the simple place theory have generally considered the middle ear as the common site. With this as their site of origin the overtones and combination tones can be conducted on to the cochlea and have their own regional actions there just like any other tones. Helmholtz and most other place theorists have taken a stand regarding the seat of distortion that is consistent with their general position and requires the fewest assumptions. On the other hand, Békésy (*6*) assumed two separate sites. On the basis of his own experimental observations he argued that combination tones have their origin in the middle ear whereas overtones have their origin in the cochlea.

As we return to our examination of Helmholtz's theory, we see that though his assumption of middle ear distortion seemed to solve one of his theoretical difficulties it left another difficulty in its wake. On close scrutiny his conception of distortion as always present for every tone proves to be a serious complication. As is well known, his cochlear theory rests upon two fundamental tenets, the principle of specific energies of nerves and Ohm's law of tonal analysis. By the specific energies principle, every tone of recognizable pitch stimulates a specific resonator in the cochlea and this action in turn is represented to the higher centers in the specific nerve fibers. By Ohm's law, the ear always perceives sinusoidal vibrations as simple and resolves all other forms of vibration into a series of sinusoidal components that are separately perceived. If every tone is first subjected to distortion in the middle ear the simple representation indicated by the specific energies principle can no longer hold. What is transmitted to the cochlea is complex and must excite a whole series of resonators, one for every component revealed in a Fourier analysis of the complex wave. Consider, now, what must happen when the ear is presented with a musical chord containing the tone just mentioned and other tones of the frequencies of the overtones. In the cochlea this chord must excite exactly the same series of resonators as are excited by the single tone. The only difference between the second cochlear pattern and the first will be in the intensity relations among the responding elements. Yet Helmholtz's theory requires that we shall perceive the first pattern as simple and as representing only one pure tone and that we shall perceive the second pattern as complex and as made up of a tone and its series of overtones. This is hardly a specific energies theory, but a theory in which pitch perception depends upon a pattern of relative intensities. Such a theory can be maintained, and indeed is eminently reasonable, but it is not the theory presented by Helmholtz in his other discussions on this subject or the theory that has continued to be held and defended in his name.

No one can doubt that the middle ear mechanism and the mechanism of the inner ear as well will introduce nonlinear distortion into the transmitted sounds if these mechanisms are driven at sufficiently high intensities. Every responsive system of whatever sort will do so. Our real problem therefore is to discover the levels at which distortion enters for the different parts of these mechanisms and to understand the nature of the processes. We need to know what parts first suffer distortion as the intensity is raised and continue to suffer it the most seriously, for the actions of these parts will dominate the whole operation. Especially, as will presently be made clear, we need to evaluate the performance of the middle ear in relation to the mechanical and electromechanical processes of the inner ear.

THE HELMHOLTZ THEORY IN DETAIL

Helmholtz directed attention to two particular sites for the ear's distortion processes, one in the drum membrane and the other in the joint between malleus and incus.

Distortion in the Drum Membrane. Helmholtz's belief that the drum membrane is one of the principal sources of distortion grew out of his particular conception of the form of this membrane and its manner of operation as a lever mechanism. As we have seen, he supposed that a critical balance exists in the relative tensions of the radial and circular fibers of the membrane, a balance that serves to maintain the radial fibers in a gently curved form as shown in Fig. 22. The lever action depends upon this curvature and the alternating changes in it caused by the varying pressures of a sound wave. In his opinion, an increase in pressure diminishes the existing curvature and exerts a greatly amplified force upon the malleus, whereas a decrease in pressure enhances the curvature and exerts a force that is amplified to a smaller extent.

When we examine this theory we see that it requires that all sound transmission be nonlinear. The positive variations of a sound wave are represented with greater amplification than the negative variations, and the greater the amount of these variations the greater the degree of asymmetry introduced into the wave. A sine wave will be transformed into a wave with peaked positive halves and blunted negative halves, and such a wave is rich in harmonics.

We now turn to the experimental evidence on this question. Dahmann, in a study mentioned earlier, used the deflections from a mirror placed on the head of the malleus to measure the effects of slow changes of air pressure applied to the drum membrane through the external auditory meatus. The observations were made in the cadaver, with pressure variations of ±60 mm of mercury (±56,600 dynes per sq cm, rms value, or ±80,000 dynes per sq cm, maximum value). He found that the outward movements of drum membrane and manubrium resulting from negative pressures exceeded the inward movements caused by positive pressures by a ratio of 5 to 3. He concluded that this difference is due to the restraint imposed by the ossicles, which is greater for inward than for outward displacements. A difference such as this, if manifested under ordinary conditions, during stimulation with sounds, would constitute nonlinear distortion. However, it is to be noted that Dahmann's pressures were enormous, exceeding by far any that the ear can endure without damage. His evidence does not show that distortion of this nature would arise with normal stimuli.

Kobrak (3) repeated these observations on the cadaver by the same method with fuller details. The results are shown in Fig. 52,

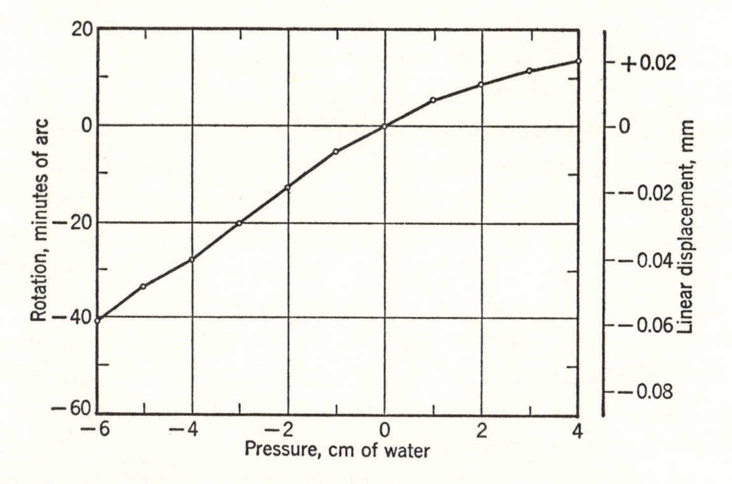

Fig. 52. The movements of the drum membrane in response to static air pressures according to Kobrak (3). The measurements were of angular rotation, from which the linear displacements of the umbo were calculated.

expressed both as angular rotations of the malleus as observed and as amplitudes of displacement of the umbo as calculated from the anatomical dimensions and an assumed location of the axis of rotation. As shown, these measurements were made over a range of pressures from +4 to −6 cm of water (+3920 to −5880 dynes per sq cm). Kobrak, like Dahmann, obtained greater outward than inward excursions over the main course of the measurements, but it is again to be noted that the pressure range extends well beyond the normal physiological limits. Between +1 and −1 cm of water (±980 dynes per sq cm) the function is linear; within these limits, which include the functional range for sounds, no distortion is indicated.

Of interest in this connection are the data already given in Fig. 27 on the volume displacements of the drum membrane of the living cat as produced by outward deflections of the end of the manubrium. As that figure shows, the membrane displacements at first increase linearly with the movement of the manubrium until a volume displacement of 2.6 cubic mm is reached for an outward deflection of 0.13 mm. Thereafter the curve ceases to be linear and bends progressively: the volume displacement rises more and more slowly as the manubrial deflection is further increased.

The deflection of 0.13 mm represents the limit of linear action, but the membrane will withstand considerably greater displacements without obvious damage. We carried out the following experiment to test the practical limit for the ear as a whole. We measured the cochlear potentials produced by an aerial sound of constant magnitude at the beginning of the experiment and again after each of a series of trials in which the manubrium was displaced by progressively larger amounts. In each trial the needle was applied to the manubrium and the membrane was pushed outward a desired amount; then all contact was removed, the tone was turned on, and the cochlear potentials were read. At first, when the displacements were small, the response was promptly and completely restored to its initial value after the contact was removed. Then as the displacements grew greater the response still returned to normal but only after a little time. This recovery period increased progressively with the magnitude of the displacements until it amounted to several seconds. A microscopical examination of the drum membrane at this stage revealed a prominent wrinkling of its surface. As the amplitude of displacement exceeded about 0.3 mm this wrinkling had grown to a serious distortion of form. After a displacement of 0.48 mm was imposed the recovery was incomplete; the responses rose over a period of a minute but reached only a third of their initial magnitude. The loss of 10 db represents a permanent injury to the ear.

We have not located the site of this injury with certainty. There was no visible impairment of the drum membrane. Moreover, its volume changes in response to displacements of the manubrium remained the same as before. Therefore we do not believe that the injury occurred in this membrane. Because in some instances a slight hemorrhage was noticed about the footplate of the stapes after this treatment we suspect that the injury was to its annular ligament. We have added to the graph of Fig. 27 an indication of this practical limit of manubrial displacement.

The range of displacements studied by Dahmann are well beyond the limit of linearity for the cat, but Kobrak's extreme is within this limit. It appears that the drum membrane of the living cat gives a much better performance than a dead human membrane.

Let us consider the cat's function a little further. Let us assume, as a maximum possible value, that the drum membrane on stimulation by sounds will move with the full amplitude of the air particles. Let us assume also that the static function shown in Fig. 27 will be maintained for the alternating forces produced by sounds. Then the limit of linearity for the drum membrane would be reached for a 100~ tone at a root-mean-square pressure of 240 dynes per sq cm and for a 1000~ tone at a root-mean-square pressure of 2400 dynes per sq cm. Experience proves, however, that the ear as a whole will not sustain these pressures without distortion or indeed without damage. For such stimulation a serious degree of distortion is shown in the electrical potentials of the cochlea, and when the stimulation is maintained for a few minutes a later histological examination reveals injury to the hair cells of the organ of Corti. Indeed, our studies show that the inner ear structures can be almost totally destroyed by overstimulation without any sign of damage to the middle ear. It appears therefore that the drum membrane has an elastic limit well beyond anything demanded of it in its normal response to sounds. We turn now to the second portion of Helmholtz's theory.

Distortion in the Malleoincudal Articulation. Even more important for distortion, according to Helmholtz, is the form of the connection between malleus and incus. This joint is a loose one, he said, and permits a good deal of independent motion of the two ossicles. The incus does not directly follow the vibratory motions of the malleus, but lags and twists away in a peculiar fashion, as described in Chapter 7. As the head of the malleus moves outward its cog tooth catches on the incus and impels it to follow, but the restraining forces acting on the incus prevent its following completely and as a result the joint surfaces separate slightly. Therefore when the head of the malleus reaches its most outward position the incus is lagging, and just as the malleus begins its inward excursion it takes up this slack and delivers a resounding slap to the incus. The malleus then carries the incus before it, to a degree depending upon the capsular connections, which are rather weak. Because of their weakness the joint surfaces separate a little at this phase of the motion also, especially for strong sounds.

These variations in the movements of malleus and incus constitute nonlinear distortion, and Helmholtz believed that he had direct evidence of such distortion in the performance of his own ears. He reported that when a strongly vibrating tuning fork of a low frequency was held to the ear he heard a jarring or buzzing sound, rich in overtones. This alteration of the simple fork tone he attributed to the impact of the malleus on the incus during each vibration as just described. However, this buzzing effect can be accounted for otherwise. It is only indicative of the presence of distortion somewhere in the responsive system and not necessarily in the middle ear.

Dahmann in his experiments with powerful stimuli observed the relative motions between the ossicles that Helmholtz had spoken of. He reported that the amplitude of motion of the malleus was not fully transmitted to the incus, but some of this amplitude was lost in the resiliency of the joint. However, on theoretical grounds he supposed that when the stimuli were more moderate, and the displacements called for in the ossicles were not so great as to produce stresses in their articulations that exceeded the limits of yielding of the footplate in the oval window, the malleus and incus would vibrate together as a single mass. This, we will recall, is the view taken earlier by Eduard Weber and Riemann.

Frey provided anatomical support for this single-mass concept in a careful study of the histological nature of the malleoincudal joint. He reported that in general a true joint does not exist here. In most mammals the cartilagenous surfaces of the ossicles are partly replaced with what he called an intervening substance, a complex of fiber and bone. The result is an ankylosis of the joint preventing any noticeable amount of mobility.

Fumagalli carried these anatomical studies further, with similar results. He found that in man the joint as first formed is fully movable, with its surfaces covered with articular cartilage, but in early childhood a gradual transformation begins by which the joint is finally ankylosed and made immobile to ordinary forces.

He observed a similar progressive ankylosis in all other species studied except the rabbit and guinea pig. In these two species the joint is immobile even at the outset. It is described as a synchondrosis, which is a joint in which the parts are united by cartilage that gradually calcifies and turns to bone. Though Fumagalli regarded these joints as immobile to ordinary forces he considered the possibility of yielding in the face of forces of an injurious nature. It is also significant that the articular ligaments of these joints consist of yellow elastic fiber rather than the relatively weak and flexible white fiber found in joints that are freely movable.

The following experiments carried out in the cat in connection with our study of the ossicular lever action are pertinent to this question of the character of the malleoincudal joint, for if this joint is a flexible one the effective lever ratio will not be constant but will progressively diminish as more intense sounds are applied. We measured the magnitudes of cochlear potentials produced by driving the ossicular system with a mechanical vibrator applied alternately at two points, at the tip of the manubrium of the malleus and at the head of the stapes, and with wide variations of the vibratory amplitude. Results obtained with a vibratory frequency of 1000~ are shown in Fig. 53. Note that the two curves take a parallel course over their main extents, up to the region of overloading, which signifies that over this main course they differ by a constant ratio. At any given vibratory amplitude the driving of the stapes produced the larger responses. For example, at −15 db on the abscissa scale the driving of the stapes produced a response of 70 microvolts and that of the malleus produced a response of 24 microvolts, a difference of 2.92-fold, which is the ossicular lever ratio.

This ratio is constantly maintained, within the error of measurement, over the course of the functions up to about −5 db where overloading sets in. At their upper ends the two curves bend in the same fashion and attain the same maximum value. It is clear that the overloading does not arise in the malleoincudal articulation but somewhere beyond, for if it arose in this articulation it would be shown in the malleus curve but not in the other. These results show no evidence of an alteration of the efficiency of transmission by a change of lever ratio at any level of intensity within the practical range.

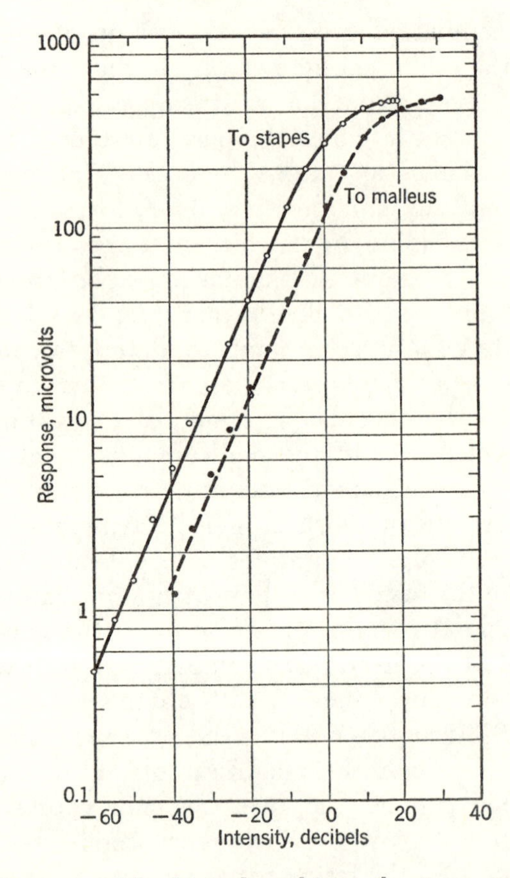

Fig. 53. Cochlear potentials obtained on driving the stapes and the malleus with a mechanical vibrator at a frequency of 1000~.

Distortion in the Incudostapedial Articulation. We turn now to the incudostapedial joint. It is generally agreed that great mobility is present here. This connection is a compound diarthrosis, a ball-and-socket joint that allows rotation in all directions. Therefore the stapes, when acted upon by forces from the incus, is free to execute whatever forms of movement its ligamentary attachments permit. We have already seen that Mach and Kessel believed that the stapedial footplate moves in the oval window as on a hinge formed by the inferior-posterior border of its annular ligament, whereas Helmholtz thought that the motion was mainly in and out of the window like a piston. Others, however, from the observation that the annular ligament is narrowest at its posterior end just below the foot of the short crus, regarded this end as forming the principal rotational axis. Dahmann described also a second axis running lengthwise through the stapedial footplate and thought that the stapes might rotate around it to a limited extent. Békésy (9), as we have seen, accepted both these axes and believed each to be operative under suitable conditions of

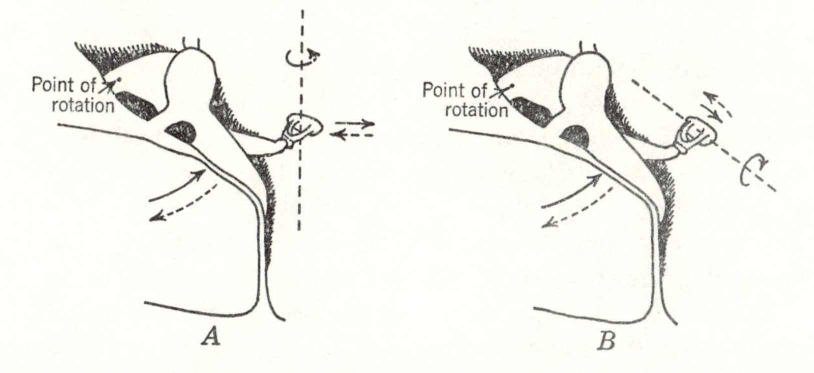

Fig. 54. Békésy's conception of stapedial motion. For moderate sounds the stapes is said to rotate about a vertical axis as at *A*, whereas for intense sounds it rotates about an anteroposterior axis as at *B*. The point of rotation of the peripheral part of the ossicular system is shown also. From Békésy (*10*) [*Acta Oto-Laryngologica*].

sound pressure. His observations showed that for moderate pressures the stapes rotates around its vertical axis at the posterior end, but as the pressure is raised the form of movement changes rather suddenly and the stapes rotates around its long axis. The first form of movement causes an effective in-and-out displacement of the cochlear fluid whereas the second form merely causes the fluid just behind the footplate to surge back and forth between upper and lower halves of the oval window region without any effect upon the interior of the cochlea. This change in the mode of movement Békésy believed to be made possible by a certain degree of freedom in the main anchorage of the ossicular chain at the end of the long process of the incus. The function of this change is protective, he said; excessive sounds are not permitted to communicate their full effects to the cochlear endings. Such a change in the mode of movement would introduce nonlinear distortion at high levels.

We now consider experiments that apply more generally to the question of the fidelity of operation of middle ear structures.

FURTHER EXPERIMENTAL CONSIDERATIONS

Clinical Evidence. Dennert objected to Helmholtz's theory of middle ear distortion on the basis of his observation that many persons without drum membranes, and some without the malleus and incus as well, are still able to hear the same combination tones that normal persons hear. Bingham examined a person who had lost the drum membrane and the two larger ossicles in both ears and yet was able to identify two difference tones. Lewis and Reger tested three persons with middle ear defects of this kind and found them able to hear various subjective tones and to match them properly with objective frequencies. These authors expressed doubt that the drum membrane and ossicles play the major role, if any at all, in the production of combination tones.

This clinical evidence proves that the middle ear structures peripheral to the stapes are not the only source of distortion in the ear. However, most of the persons with middle ear defects had considerable difficulty in hearing the combination tones, and probably more difficulty than normal persons. It could still be argued that the middle ear is the usual source of distortion, though other structures also contribute and in the absence of the middle ear give a similar result. Thus Schaefer suggested that when the middle ear was absent a noticeable amount of distortion might arise within the labyrinth, perhaps in the movements of the cochlear fluids.

Békésy's Experiments. An extensive series of experiments on this problem was carried out by Békésy (*6*). He first tried to discover by objective measurements whether overtones are present in the external auditory meatus when the ear is stimulated with a pure tone. He pointed out that if overtones are generated in the movements of the drum membrane they ought to be reflected into the neighboring air and be identifiable there. He used the auscultation method, with a tube running from the stimulated ear to the ear of another person who served as listener. An acoustic filter was interposed between the ends of this tube to eliminate the fundamental frequency while permitting the passage of any overtones. For a stimulating tone of 500~ at an intensity of 10 dynes per sq cm no overtones could be detected at the listening point. Yet to the stimulated ear the overtones were prominent. A somewhat similar method was used to ascertain whether a difference tone could be picked up from the vicinity of the drum membrane while stimulating with two pure tones, and the results showed that such a tone if present at all must be very weak. These observations therefore indicate that the drum membrane is not responsible for the distortion pattern, at least for intensities up to about 10 dynes per sq cm, which was the greatest intensity used here.

Békésy continued his investigation by other methods to test the possibility of distortion in other parts of the middle ear. He observed that the introduction of a steady air pressure into the external auditory meatus had no effect upon the overtone pattern provided that the stimulus intensity was raised to compensate for the weakening of the fundamental tone that was produced by the air pressure. On the other hand, such pressure changes were found to cause noticeable variations in the loudness of difference tones. Likewise, contractions of the middle ear muscles were found to produce these variations. Békésy therefore concluded that though the middle ear is not responsible for overtones it does play a part in the production of difference tones.

Békésy also found that if the ear is strongly fatigued with a tone of a certain frequency (say 2130~) and then is stimulated with two primary tones of adjacent frequencies (say 2000 and 2260~) the loudness of their difference tone (of 260~) rises at once to its full magnitude and steadily maintains this magnitude over several seconds, while the primaries, which have suffered from the fatigue, are heard at first only faintly and then at rapidly increasing strength. He concluded that the difference tone is purely of mechanical origin because it is unaffected by the fatiguing process which he attributed to nervous elements.

From all this evidence Békésy finally concluded that the difference tones have their origin in the middle ear, whereas the overtones arise in the cochlea. The cochlear distortion, he suggested, is a local rectifier action associated with the fluid motion produced by the sounds. He argued that this harmonic distortion is specific to each tone, arising in the local areas of the cochlea to which that tone is relegated and not in some common distortion process. He took this position because he failed to observe a modulation of the intensity of a subjective overtone of 1000~ by a low tone of 100~. The 1000~ overtone was made to beat with an exploring tone of 1004~ and then the 100~ tone was introduced, with no noticeable effect upon the purity and smoothness of the beating overtone. On the other hand, an overtone that was produced in a distorting electric circuit was found to be strongly modulated on the introduction of a 100~ tone.

Cochlear Potential Studies. To study this problem further the following four experiments were carried out on animals by means of the electrical potentials of the cochlea.

Series 1. The first experiment (Wever, Bray, and Lawrence, *1*) consisted of measurements of the harmonic pattern resulting from stimulation of the ear with a pure tone under two conditions, with the ear intact (except for the usual exposure of the round window) and after removal of the middle ear structures peripheral to the stapes. The experiments were carried out on guinea pigs under deep anesthesia. Under the first or "normal" condition the stimulation was with loudspeaker tones that were carefully filtered to eliminate objective harmonics. Under the second condition the stimulation was with a mechanical vibrator of the crystal type as described earlier. The needle point of this vibrator was applied to the head of the stapes, with care to keep its direction of vibration in line with the axis of the stapes and to maintain the stapes in its normal position. A firm pressure on the head of the stapes was required to secure the necessary negative motion, but once this pressure was obtained (as determined by observing the regular nature of the responses) any further increase had little or no effect until the pressure was carried too far and a point was reached at which the ear was injured.

The stimulation was carried out with three frequencies, 300, 1000, and 3000~, at intensities from the lowest that gave measurable responses to the highest that could be used without overloading the stimulation apparatus, a range of about 80 db. With a wave analyzer we measured the first 5 harmonics (the fundamental and 4 overtones) in the electrical responses for each stimulating frequency under the two conditions described. Because of the time-consuming nature of the measurements it was not usually feasible to carry out the readings for these two conditions on the same ear or even in the same animal. Our results showed, however, that individual differences in the distortion patterns are small, and therefore it is proper to make a comparison of the two methods as applied to different ears.

Some of the results of this experiment are given in Figs. 55 and 56. Figure 55 shows the functions obtained for an aerial tone

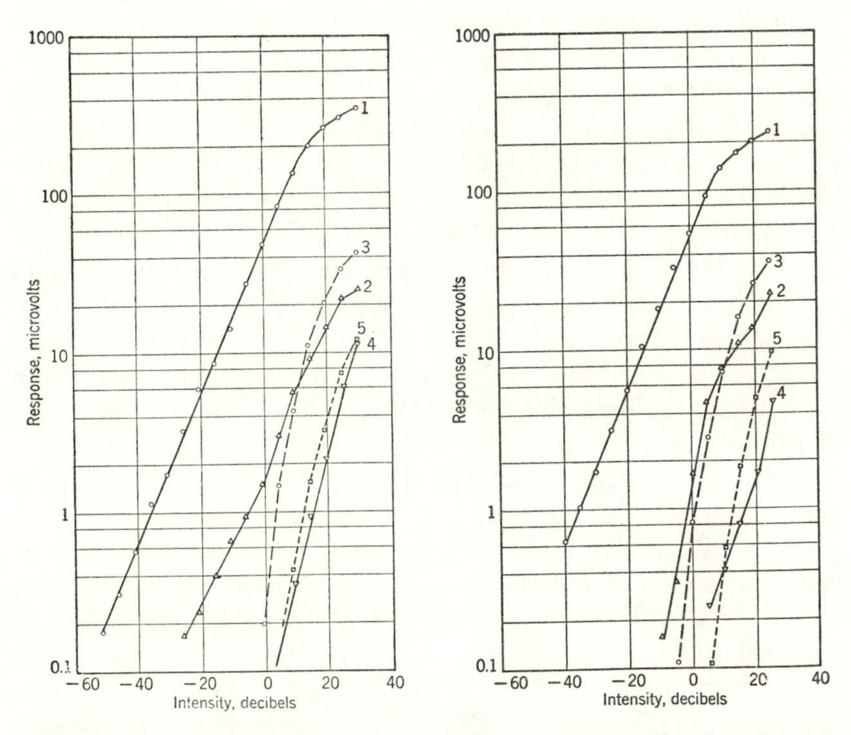

Figs. 55 and 56. On the left, analysis of cochlear responses produced in the intact guinea pig ear by a pure aerial tone of 1000~. The number on each curve shows the harmonic that it represents. On the right, analysis of responses produced by mechanical vibrations of 1000~ applied directly to the stapes after removal of the peripheral parts of the middle ear. From Wever, Bray, and Lawrence (*1*) [*Journal of the Acoustical Society of America*].

of 1000~ when the middle ear was intact and Fig. 56 shows the functions for the same tone produced by the mechanical vibrator when the drum membrane, malleus, and incus were absent. As the stimulus intensity is raised from a low level the second harmonic appears first and then soon is joined by the third harmonic, and finally by others of higher order. The two sets of functions in Figs. 55 and 56 are closely similar, varying no more from one another than two sets taken from different ears by the same method. The removal of the middle ear structures up to the stapes has caused no significant alteration of the harmonic pattern. It follows that this pattern does not arise in these structures but in other structures farther along in the line of transmission to the cochlea.

Series 2. A second experiment (Wever, Bray, and Lawrence, 2) dealt similarly with the problem of the place of origin of the combination tones. Again guinea pigs were stimulated by two methods, by aerial sounds from a loudspeaker and by the mechanical motions of a crystal vibrator applied to the stapes after removal of the other middle ear structures, but in this instance the sound consisted of two tones delivered simultaneously from two oscillators. Several pairs of frequencies were used, though most of the measurements were made with 1000 and 2800~. The two primary tones were first presented singly and adjusted in intensity to give some arbitrary amount of response and then they were combined and thereafter raised and lowered by equal steps of 5 db.

Measurements were made by means of a wave analyzer on 8 components of the cochlear response wave in addition to the primaries themselves. These components included the two combination tones of the first order $(h-l)$ and $(h+l)$, the four of the second order $(h-2l)$, $(h+2l)$, $(2h-l)$, and $(2h+l)$, and two of the third order $(2h-2l)$ and $(2h+2l)$. Often other components were measured up to the fifth order and at times still others up to the eleventh order. As in the preceding experiment the two sets of measurements, on the normal ear and on the operated ear, were usually carried out in different animals but in two instances both sets were made successively on the same ear.

Representative results of this experiment are given in Figs. 57 to 60. Figures 57 and 58 show results for aerial stimulation of the intact middle ear, with the curves for difference and summation tones plotted separately to avoid confusion. Similarly, Figs. 59 and 60 show results for mechanical stimulation of the stapes after removal of the drum membrane, malleus, and incus. The curves are plotted with reference to an intensity (zero db) that for present purposes may be considered as arbitrary; actually it is a

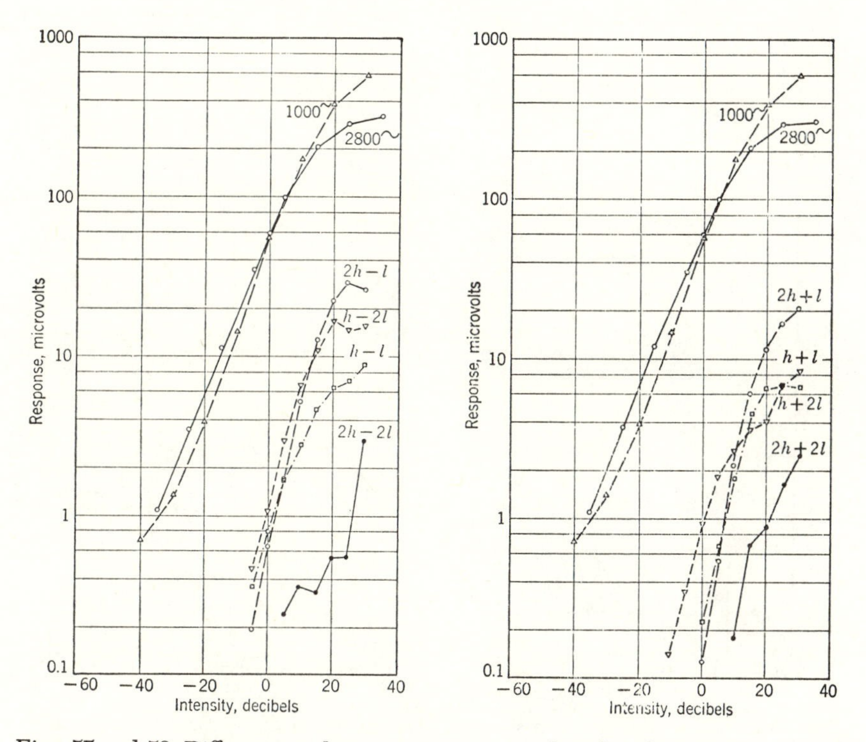

Figs. 57 and 58. Difference and summation tones produced in the intact guinea pig ear by aerial stimulation with 1000 and 2800~. In each figure, functions are shown for the primaries as well as for the combination tones as indicated.

stimulus value that gives the same magnitude of response as ordinarily obtained from an aerial tone of 1000~ at 1 dyne per sq cm. The abscissa therefore represents both the 1000 and 2800~ primaries at the same effective level but not (usually) at the same physical level.

As shown, the combination tone patterns have a fairly constant form. The first-order components $(h - l)$ and $(h + l)$ become noticeable when the primaries reach about —10 db and then these components rise rapidly until they bend around +20 to +30 db. They are soon joined by the components of higher orders, which likewise rise and show bending. For the most part these functions are fairly uniform, but some irregularities appear, especially for the components of highest orders that are present only in minute amounts. The combination tone functions in general are somewhat less stable than the functions for the primaries or for their

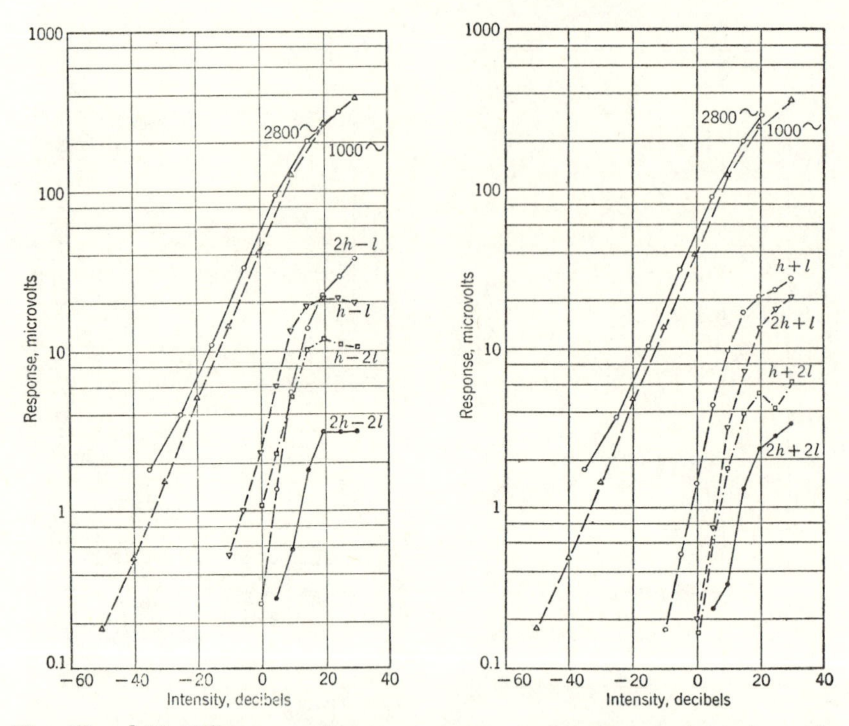

Figs. 59 and 60. Difference and summation tones produced by mechanical stimulation of the stapes. Compare with the two preceding figures. These four figures from Wever, Bray, and Lawrence (2) [*Journal of Experimental Psychology*].

harmonics, and repeated measurements often give changes in the curves, especially in their upper portions where they undergo rapid bending.

From these results we were able to conclude that the combination tones arise mainly, if not wholly, in structures beyond the stapes. Because of the presence of the variations just mentioned we could not be as positive on this point as we were in regard to the overtones and thus could not rule out altogether any participation by the middle ear.

Series 3. A third experiment (Wever, Bray, and Lawrence, *6*) used the cochlear response method to study the effects upon distortion of a change in the air pressure in the middle ear. It will be recalled that Békésy carried out this experiment on human ears and found that a steady air pressure had no effect upon overtones when the loss of sensitivity to the stimulating tone was compensated for, but such pressure did produce noticeable variations in difference tones. Our tests were carried out on cats under deep anesthesia and with curarization to eliminate reflexes of the middle ear muscles.

Measurements were made of the harmonic pattern during stimulation with 1000~ both with a normal pressure in the middle ear and with an increased pressure of 20 mm of mercury. This pattern was much the same under the two conditions, though as expected the stimulus intensity had to be increased greatly during the introduction of air pressure to compensate for its effects upon the fundamental tone. There were only minor variations in the relative magnitudes of the components, and the maximum values attained were similar.

Measurements were made also of the combination tone patterns in the presence of middle ear pressure. The ear was stimulated simultaneously with 1000 and 2800~ at various intensities and measurements were made of several of the more prominent combination tones. Some of the results for the second-order difference tone ($2h - l$) are given in Fig. 61. Functions are given for normal conditions and after the introduction of a middle ear pressure of 20 mm of mercury. The curves in the presence of pressure are displaced to the right, indicating an impairment of sensitivity, but their form is the same as that of the normal curves. The other combination tones showed a like constancy of pattern in the presence of middle ear pressure.

These observations lend further weight to the evidence already presented on the locus of distortion. If this distortion occurred in the middle ear we should expect it to be modified by a mechanical change as profound as a change in middle ear pressure. The experiment next to be described provides final evidence on this problem.

Series 4. The method of this experiment consists essentially of a determination of the distortion occurring along the pathway of transmission through the ear by means of an acoustic probe so placed as to record the sound waves produced in the vicinity of the round window as a result of fluid movements inside the cochlea. Such fluid movements and the sound waves that they cause at the round window must reflect any distortions occurring in the middle ear structures as well as any mechanical distortions within the cochlea. By a comparison of the pattern of distortion revealed in these sounds with the one found in the cochlear potentials we discover the importance of these purely mechanical effects in relation to other effects that may appear farther along

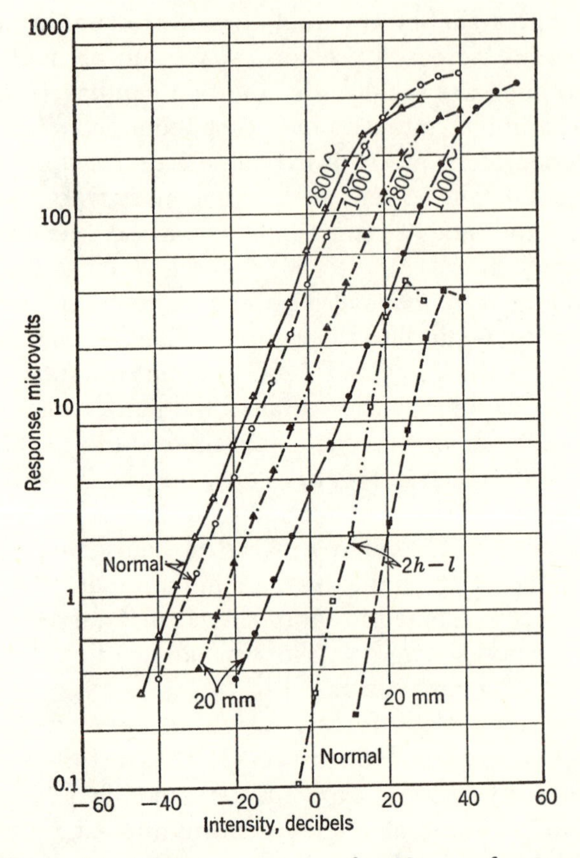

Fig. 61. Effects of a middle ear pressure of +20 mm of mercury on cochlear response functions for primary tones of 1000 and 2800~ and for their difference tone $(2h - l) = 4600$~. From Wever, Bray, and Lawrence (6) [*Journal of the Acoustical Society of America*].

in the sensory process, as in the electromechanical actions by which the cochlear potentials arise.

These experiments were carried out in cats with a stimulating tone of 1000~ and with particular care to eliminate objective overtones. A series of acoustic filters was introduced along the conduit leading from the loudspeaker to the ear and these filters were adjusted for minimum transmission of second and third harmonics. A "dummy ear," a chamber of appropriate volume and dissipation, was used for optimum adjustments of the filters and for final measurements of the distortion remaining in the stimulating sounds.

A tight connection was made between the ear and the conduit from the loudspeaker by means of a connector tube whose end was securely tied into the external auditory meatus. This connector tube, as described earlier, carried a probe tube from a condenser microphone. This probe tube served to pick up the sound at the surface of the drum membrane and to record its intensity and wave form at that position.

A second tube was placed over the round window and firmly sealed there with wax. This tube also carried an acoustic probe whose end was near the round window membrane. The outer end of this tube was closed so as to form an air cavity of 1.5 cubic cm into which the round window opened. This arrangement served to isolate the round window from the stimulating sound, and tests showed that the degree of isolation was always in excess of 20 db. The air cavity prevented any appreciable restraint being imposed upon the vibratory movements of the round window membrane or the fluids behind it. After the tube was placed over the round window the bulla was closed with wax, thus restoring it to a condition approximating the normal.

The probe tube in the meatus recorded the stimulating sounds, but only after a possible modification caused by the ear's presence. As has already been pointed out, any distortion produced by the drum membrane ought to be reflected into the air in its vicinity and therefore will be recorded by the meatus probe to an extent depending upon the coupling between the drum membrane and this air. Moreover, any distortion appearing farther along in the transmission system ought also to be reflected to this location to an extent depending in addition upon the degree of coupling between these parts and the drum membrane. The indications are that these couplings are close, and especially so at 1000~ which is a resonance frequency for the cat's middle ear mechanism.

These observations were carried out for a range of intensities extending far beyond that ordinarily used in auditory stimulation. The stimulating tone was first presented at a moderate level, at which no overtones were measurable either in the sounds or in the cochlear response, and the intensity was raised by 5 or 10 db steps, with analyses of both entering and emitted sounds and of cochlear responses at each step, until the limit of safe performance of the loudspeaker was reached. This limit was usually about 3000 dynes per sq cm, but sometimes exceeded 10,000 dynes per sq cm.

A preliminary picture of the distortion arising under these conditions is easily obtained from an examination of the wave forms of the cochlear potentials and of the sounds entering and leaving the ear. It must be borne in mind, however, that only the higher degrees of distortion are revealed in such a visual study. Under the conditions of this experiment, when the early distortion is dominated by the second harmonic, the wave changes become barely perceptible when the distortion reaches about 10 per cent and they become obvious for a distortion of about 20 per cent.

Figure 62 is a display of some of the oscillographic records in relation to the cochlear response function. The cochlear response waves appear sinusoidal until a level of 120 microvolts is reached, for a sound of 4.4 dynes per sq cm, at which point a slight amount of distortion is just discernible. The distortion increases with intensity of stimulation until it is obvious at 245 microvolts and above. The sound waves as measured at the drum membrane are sinusoidal in appearance throughout the series, up to the maximum used. The sound waves as measured at the round window remain sinusoidal also until after the response curve has passed well beyond its maximum. In this ear, when the response had fallen to a level of 300 microvolts on stimulating at 475 dynes per sq cm, these waves began to show distortion, and they showed it more prominently as the intensity was raised further. Not all ears showed this distortion at the round window; indeed the majority of those studied in this way continued to show only simple waves to the very limit of the intensities available.

Results obtained on the cochlear potentials, such as those represented in Fig. 62, are stable and easily repeatable, provided that in the region of the maximum of the response curve the stimuli are applied only momentarily—only for a second or two as required for a measurement—and periods for recovery are allowed after any slight impairments of response are noticed. However, the use of still higher intensities, which carry the response curve far

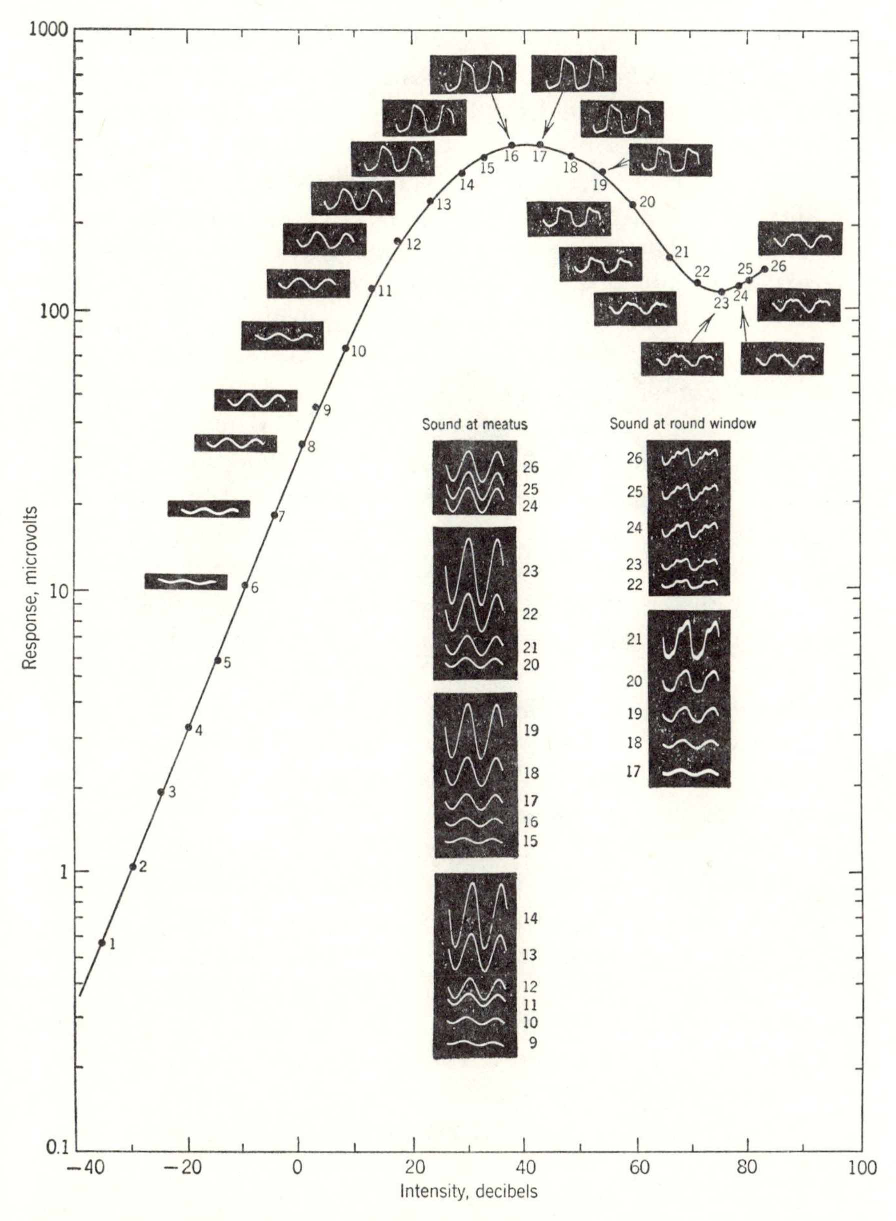

Fig. 62. The intensity function for the cochlear responses of the cat's ear, with a display of the wave forms of this response at different levels. The intensity scale is in decibels relative to 1 dyne per sq cm. The points of measurement as shown along the curve are numbered, and from 6 on each has adjacent to it an oscillogram of the cochlear response obtained. For these waves the amplification of the oscillograph was constant up to point 9 and then was reduced and held constant thereafter for points 10 to 26. In two columns below are oscillograms for some of the sounds entering the ear at the meatus and also for sounds picked up from the round window. In recording these sounds the amplification of the oscillograph was constant for each block of pictures but was varied between blocks.

beyond its maximum, cause serious injury and bring about progressive impairment of the responses.

For finer details of the distortion pattern we turn to our analyses of cochlear and microphone outputs made by means of an electrical wave analyzer. Because of large variations in the absolute magnitudes of the measured quantities it is convenient to express the data in relative terms, as percentages of the fundamental frequency. It is convenient also to obtain a single figure for the overtone distortion by combining the results for the first four overtones. This was done by calculating the root-mean-square sum of their observed voltages and then expressing this sum as a percentage of the fundamental voltage.

As already mentioned, the amount of distortion that was measurable in the sounds at the drum membrane and at the round window varied considerably in the different ears. In most of the ears studied we were unable with the equipment available to detect any distortion at these points. Certain ears evidently are particularly susceptible to distortion and we shall discuss the results from one such ear in detail. Results from such an ear have already been shown in Fig. 62 and are now given in a more quantitative form in Fig. 63.

Consider first the nature of the stimulating tone as seen in the "dummy ear" measurements. This tone contains only 0.014 of 1 per cent of harmonics at 10 dynes per sq cm, and then the amount rises progressively to 4.3 per cent at 3120 dynes per sq cm. This curve runs well below all the others in the figure. It is 32.0 db below the nearest curve (that for sound at the drum membrane) at 100 dynes per sq cm and though it rises more steeply than that other curve it is still 11 db below it at 3120 dynes per sq cm. At this last point the distortion in the stimulus might have an effect of raising or lowering the readings at the drum membrane by as much as 4 per cent, but elsewhere its effect is altogether negligible.

Distortion in the sound at the drum membrane was first measurable at a stimulating intensity of 1 dyne per sq cm, where it was 0.16 per cent of the fundamental intensity, and this distortion rose progressively as the intensity was raised. It attained a maximum of 23 per cent at 1000 dynes per sq cm and then fell off somewhat at 3120 dynes per sq cm. No measurements were possible at 0.1 dynes per sq cm, but we know from the sensitivity of our apparatus that the harmonic content did not exceed 0.014 per cent as indicated by the extension of the function to this point.

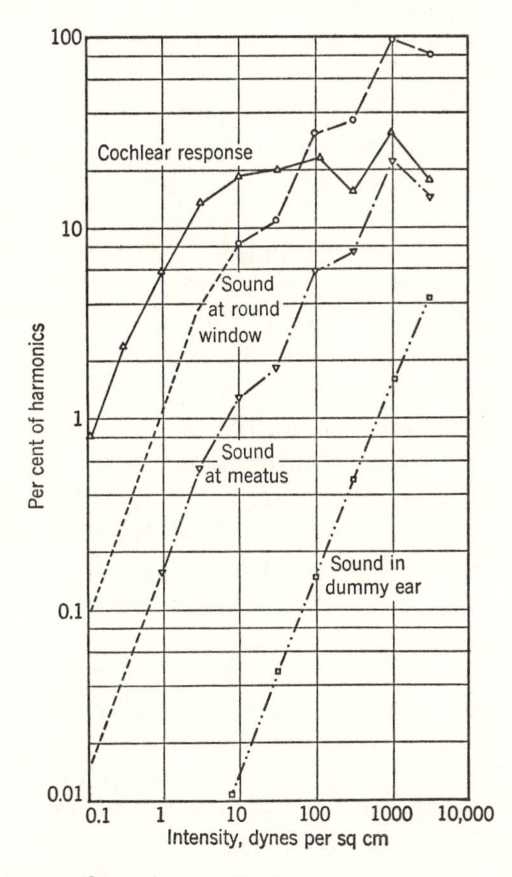

Fig. 63. Distortion in the cat's ear. The four curves represent the percentage of harmonics observed in the cochlear response, in the sound at the round window, in the sound in the meatus adjacent to the drum membrane, and in the sound applied to a dummy ear.

This extension represents a maximum value of the distortion, and the true value may well be much lower.

Our ability to measure the distortion appearing at the round window is limited by the fact that the sounds at this point have been subjected to considerable attenuation in passing through the ear. The amount of this attenuation varies a good deal in different ears, but always exceeds 20 db. In this ear it was 32 db. Therefore the harmonics at this position became measurable only after the stimulating sound was raised to 10 dynes per sq cm. For intensities above this level, note the closely parallel courses of the two curves for sounds entering and leaving the ear. The round window curve has been extrapolated downward for reasons that will be brought out presently.

Note finally the curve for harmonics in the cochlear response. These harmonics were first measurable for a stimulating intensity of 0.1 dynes per sq cm, for which they amounted to 0.8 per cent, and they rose with intensity and then leveled off around 20 per cent at 30 dynes per sq cm. Up to this leveling region the cochlear response curve runs well above the other curves, but around 70 dynes per sq cm this curve is crossed by the round window curve. For stronger stimuli the response curve varies up and down in a curious manner in the region of 20 per cent as shown.

In considering these results for the more intense stimuli it must be borne in mind that the measurements were carried out in the face of progressive impairments of the ear. Also it should be remembered that these observations are for an ear that showed more than the average amount of distortion. The results therefore are to be taken as representing an upper limit of distortion under the conditions.

The measurements were made in ascending steps, so that the complications resulting from injury to the ear are less important at first and become increasingly serious at the upper levels. By careful procedure this injury can be limited to a few decibels. We know from earlier studies (Smith, 2; Wever and Smith) of the effects of overstimulation that an injury produced by 1000~ affects a large portion of the organ of Corti and impairs the sensitivity almost equally for all tones. Hence the relative picture as represented even under these abnormal conditions can reasonably be accepted as indicative of the nature of the ear's distortion processes.

Our measurement of a small percentage of distortion at the drum membrane obviously means that we are recording here a composite of the incident wave and of another wave reflected from the ear that has encountered distortion processes somewhere in its path. The nature of this distortion is not clearly revealed in these measurements because we do not know the relative magnitudes of incident and reflected waves. However, if we can make the assumption that the reflected wave is transmitted to our recording position at the drum membrane with the same composition as found at the round window we can work the problem backward and obtain the relative magnitudes of the two waves. Our calculations on this basis show that the reflected wave bears a ratio to the incident wave of 0.13 to 0.18, depending on the phase relation existing between these two. We also find that the attenuation in the pathway of conduction from the meatus to the site of the distortion in the cochlea is about 8 db.

The assumption just made, that the reflected wave at the drum membrane represents the same distortion pattern as the sounds recorded at the round window, is supported by the parallel courses taken by the drum membrane and round window curves. If we could remove the diluting effect of the incident wave from the pattern recorded at the drum membrane this drum membrane curve ought to have the same form as the round window curve. These considerations justify our extrapolation of the round window curve as shown in Fig. 63.

On the basis of this extrapolation we can say that the harmonic content of the cochlear response exceeds that of the sounds in the cochlea over the whole range of the measurements up to the leveling of the response function. This fact signifies that the cochlear potentials or the processes by which they are generated have passed through a distortion process in addition to that to which the sounds traversing the cochlea have been subjected. If these two distortion processes, the mechanical and the electromechanical, combine their effects as a vector sum we can easily show that the electromechanical process suffers more than six times as much distortion as the other. Stated differently, the presence of the mechanical distortion adds at most only 0.3 db to the total distortion at high levels in a particularly susceptible ear. Under practical conditions, when the ear is below the region of severe overloading, this mechanical distortion is altogether negligible.

Similar experiments were carried out with two stimulating tones of 1000 and 2800~ applied simultaneously to the ear, with measurements of the resulting combination tones. These results showed the same relations between the distortion patterns in the cochlear response, in the sounds at the drum membrane, and in the sounds at the round window as described in detail for the overtones. We conclude that the combination tones have the same places of origin as the overtones, and likewise that such tones of mechanical origin are negligible in relation to those attributable to the electromechanical processes.

These results support the conclusion already reached from our earlier experiments that the ear's distortion is produced in the cochlea and not in the middle ear. We can now carry this statement farther and say that the distortion is largely in the final sensory processes, or at any rate is not in such mechanical processes of the cochlea as are closely coupled to the cochlear fluid. We can thus discount the frequent suggestions that distortion may occur in the stapedial movements, in the gross movements of the basilar membrane, or in the cochlear fluid itself, for distortion in any of these would be disclosed in the sounds passing out of the round window. These experiments support the view already expressed that the middle ear mechanism carries out its function of sound transmission with great fidelity.

TWO STAGES OF SENSORY DISTORTION

Let us now consider further the distortion seen in the electrical responses. It has already been suggested, from other, indirect evidence, that there are two stages of the distortion that is recorded in the sensory activities, one shown in the early appearance of overtones and combination tones and another responsible for the limitation in the total output of the cochlea that we call overloading. The evidence comes from a study of the phenomenon of interference (Wever, Bray, and Lawrence, *4*).

Interference is found in the cochlear potentials as a reduction in the magnitude of responses to one tone due to the introduction of a second tone. To observe the phenomenon one tone is presented to the ear and the resulting potentials are measured with a wave analyzer, used here as a selective voltmeter. If now a second (interfering) tone is introduced and raised to a sufficient intensity the potentials produced by the first tone are diminished in magnitude. The effect is absent at moderate intensities of the interfering tone and then sets in rather suddenly as the intensity is raised. As the intensity is raised further the interference increases rapidly until it begins to level off at intensities near the ear's practical limit of operation, as shown in Fig. 64. This interference bears an intimate relation to the transformation process by which overtones and combination tones are generated, and indeed

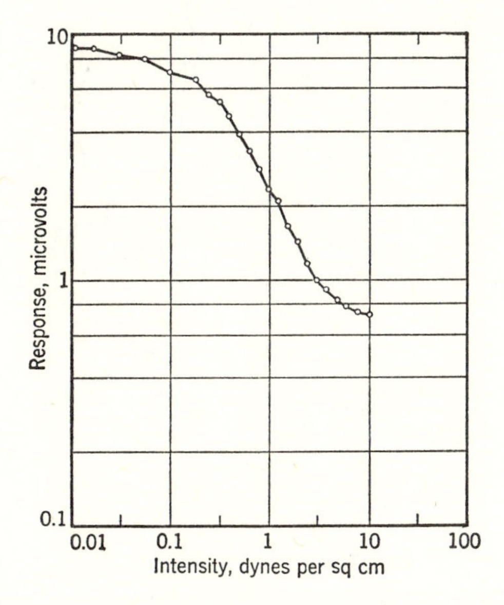

Fig. 64. The interference phenomenon. The curve shows the varying magnitudes of cochlear response to a tone of 1000~ when acting in the presence of a 3000~ tone whose intensity was varied as represented on the abscissa. From Wever, Bray, and Lawrence (*4*) [*Journal of the Acoustical Society of America*].

may be merely one aspect of that process. In its early stages at least the loss of response observed in the tone interfered with can be accounted for as a diversion of energy into new frequencies, the overtones and combination tones.

The degree of interference caused by a given tone bears a close relation to the ear's sensitivity to that tone, when the sensitivity is measured in terms of the cochlear potentials. A number of interfering tones of different frequencies but with the same sound pressure will not produce the same amount of interference upon some chosen tone, but, generally speaking, those interfering tones to which the ear responds with the larger potentials will produce the greater interference.

This fact is clearly shown in Fig. 65 by the curve marked "1 dyne." For this curve a number of tones were adjusted to an absolute intensity of 1 dyne per sq cm and then were made to interfere upon a constant tone of 1015~. The loss of response suffered by the 1015~ tone is shown in decibels on the ordinate. This loss is far from uniform, and in fact its variations bear a striking resemblance to the sensitivity curve.

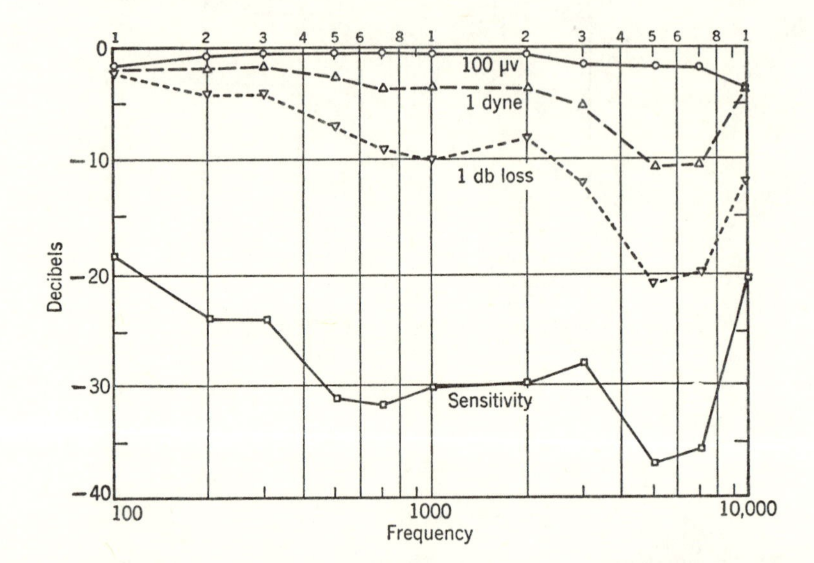

Fig. 65. Interference curves for a guinea pig ear, together with a sensitivity curve. The interference curves were obtained with a tone of 1015~, adjusted to give a response of 10 microvolts by itself, and then acted upon by various tones according to three procedures as described in the text. The sensitivity curve shows the sound intensity, in decibels relative to 1 dyne per sq cm, required at various frequencies to produce a standard response of 10 microvolts. From Wever, Bray, and Lawrence (*4*) [*Journal of the Acoustical Society of America*].

Similar results are seen in the curve marked "1 db loss," which was obtained by adjusting the interfering tones to whatever intensities were necessary to produce a loss of 1 db in the response to the 1015~ tone. These intensities are represented on the ordinate in decibels with reference to 1 dyne per sq cm.

A more careful study shows, however, that the relation between interference and sensitivity is not a perfect one. Consider the following test. Let the intensities of the interfering tones be chosen so that each tone when presented by itself produces some constant magnitude of response; then the sensitivity will have been compensated for. Now let us present these tones in conjunction with some other tone in which interfering effects are observed; if interference simply depends upon sensitivity the resulting interference will be the same for all the tones. This was done in obtaining the uppermost curve of Fig. 65. All the interfering tones were first presented alone and adjusted in intensity to give a response of 100 microvolts, and then their effects upon the 1015~ tone were ascertained. The curve shows in decibels the resulting loss in the response to this tone. It is evident that the interference is indeed made more uniform, but the uniformity is not perfect. Over the middle range of frequencies the interference is practically constant, but it is relatively greater for the very low and the very high frequencies.

From these observations it appears that the sensitivity at the site of generation of the cochlear potentials differs somewhat from the sensitivity that is operating at the locus of the interference process; evidently some process intervenes that discriminates against the extreme frequencies.

More specific information on the locus of interference is obtained in the study of the relation between interference and cochlear injury (Wever and Lawrence, *1*). When the ear is injured by overstimulation with sounds the cochlear potentials are impaired and a later histological examination shows that the hair cells of the organ of Corti have been disrupted over a wide area of the basilar membrane. The study of interference before and after such injury proves that the interference takes place at the site of the injury, that is, at the hair cells. The proof is derived from the forms of alteration of the interference functions after the injury. If interference took place in some process peripheral to the injury it would not be affected by the injury, though its effects in the cochlear potentials would have to be viewed through the injury supervening. Its functional relations to the stimuli, however, would remain unchanged. This is not the case; after injury the intensity of the interfering tone necessary to produce a given amount of interference is considerably greater than before. If on the other hand interference took place in some process beyond the site of the injury it should be unaltered if the injury is compensated for by raising the stimulus intensity to restore the former level of response. This is not the case either; an interfering tone at a level that produces the same response as before causes a much greater amount of interference. It follows by a process of elimination that the interference occurs at the site of the injury.

Overloading occurs beyond the process of interference, because when interference is present the overloading is reduced or prevented altogether. It is supposed that transformation and interference represent an early sensory process and overloading represents a late sensory process. Hence we locate two stages of distortion in the sensory activities, one earlier, connected with transformation and interference, and another later, associated with overloading.

A specific suggestion is that the early process occurs in the cuticular rods of the Deiters cells in the course of their communication of basilar membrane movements to the hair cells. These rods may undergo a certain amount of bending when the forces acting upon them exceed some limit and thereafter they fail to move the hair cells as much as they should. The late process may occur within the hair cell itself in intimate relation to the action by which the cochlear potentials are generated (Wever, 5).

If we accept the suggestion that the first process is a mechanical one and occurs in the transfer of movement to the hair cells by the rods of Deiters, we must consider why its distorting effects are not communicated to the cochlear fluids and observed in the sounds recorded from the round window. It is necessary to suppose that there is only a small energy transfer from the hair cells and rods of Deiters back to the cochlear fluid. This is reasonable in view of the comparatively slight mass of these structures and the limited stiffness of the rods of Deiters. Indeed, it is necessary to make the assumption in general of a loose coupling between the cochlear fluid and parts of the basilar membrane and organ of Corti if we are to suppose, as most auditory theories do nowadays, that the different parts of this sensory structure respond with some degree of individuality.

PART IV

Further Characteristics of the Middle Ear

10

The Tympanic Muscles

As we have seen, the middle ear apparatus is provided with two muscles, the tensor tympani muscle, which is inserted on the manubrium of the malleus, and the stapedius muscle, inserted on the neck of the stapes. Both muscles are of the pennate type, consisting of many relatively short fibers arranged in parallel, as shown in Plates 5, 6, and 7. This arrangement gives great tension and slight displacement. The tension is the resultant of the tensions of all the fibers, taking account of the angles at which they work, whereas the displacement is that of an individual fiber, or perhaps more exactly that permitted by the shortest fibers.

The body of the tensor tympani muscle in man lies in a slender canal that runs just above the Eustachian tube and is separated from that tube by a thin partition of bone and fibrous tissue. The muscle fibers have their anterior ends anchored to the walls of the canal and run posteriorly to insert into the tensor tendon. This tendon enters a narrow bony channel that leads it around a rather sharp bend whereupon it emerges into the tympanic cavity and runs to its point of insertion on the upper part of the manubrium of the malleus. The bony channel acts as a pulley to change the direction of action of the muscle from anterior to anteromedial. Measurements of this muscle made in four specimens gave an average length of 25 mm and a cross-sectional area of 5.85 sq mm.

When the tensor tympani muscle contracts it pulls on the malleus in an inward and forward direction and thus it exerts a tension nearly at right angles to the course of the ossicular chain as a whole, as shown in Plate 2.

The stapedius muscle in man occupies a bony canal that lies in a nearly vertical position posterior to the tympanic cavity and alongside the canal that transmits the facial nerve. Its fibers are attached below to the bony walls of the canal and above to a minute tendon. This tendon bends almost to a horizontal position as it leaves the canal through an opening in a little elevation of the canal wall (called the pyramidal eminence) and attaches to a point on the posterior aspect of the neck of the stapes. Measurements in three specimens gave an average length of this muscle of 6.3 mm and a cross section of 4.9 sq mm.

The stapedius muscle pulls upon the stapes in a posterior direction, about at right angles to the main axis of this ossicle and so, like the tensor tympani, it exerts a tension lateral to the course of the ossicular chain. However, this direction of tension is almost opposite in direction to that exerted by the tensor tympani, as may be seen in Plate 2.

The tensor tympani muscle is supplied by a branch from the mandibular division of the trigeminal or fifth cranial nerve, whereas the stapedius muscle is supplied by a branch of the facial or seventh cranial nerve (Politzer, *1*).

NORMAL STIMULATION BY SOUNDS

The tympanic muscles are normally stimulated by sounds and thus their contractions represent an acoustic reflex. That some kind of action occurs in response to sounds is easily observed by anyone in his own ears. A sudden, sharp, and unexpected sound produces kinesthetic and pressure sensations referred to the ears, sensations that ordinarily are ignored but which become obvious when attention is directed to them. The action also produces a kind of rushing noise that is doubtless due to the muscle tremor that is transmitted to the ossicular mechanism. Further subjective evidence is found in a modification of the loudness and quality of a continuing sound at the moment that a reflex contraction of the muscles is elicited. However, the particular nature of these changes is difficult to determine because of the complexity of the perceptual situation.

Direct evidence that the tympanic muscles contract in response to sounds has come from animal experiments. Hensen in 1878 made the first observations on tensor tympani contractions in dogs by opening the middle ear cavity and inserting a needle into the tendon. On stimulation with sounds he was able to see a movement of the needle. Many have repeated his experiments with similar results. When the contractions are large they can be seen directly with the unaided eye or, more easily, by the use of

Plate 5. Photomicrograph of a section of a normal tensor tympani muscle in an adult ear. From the collection of the Lempert Institute of Otology, by courtesy of Dr. Julius Lempert.

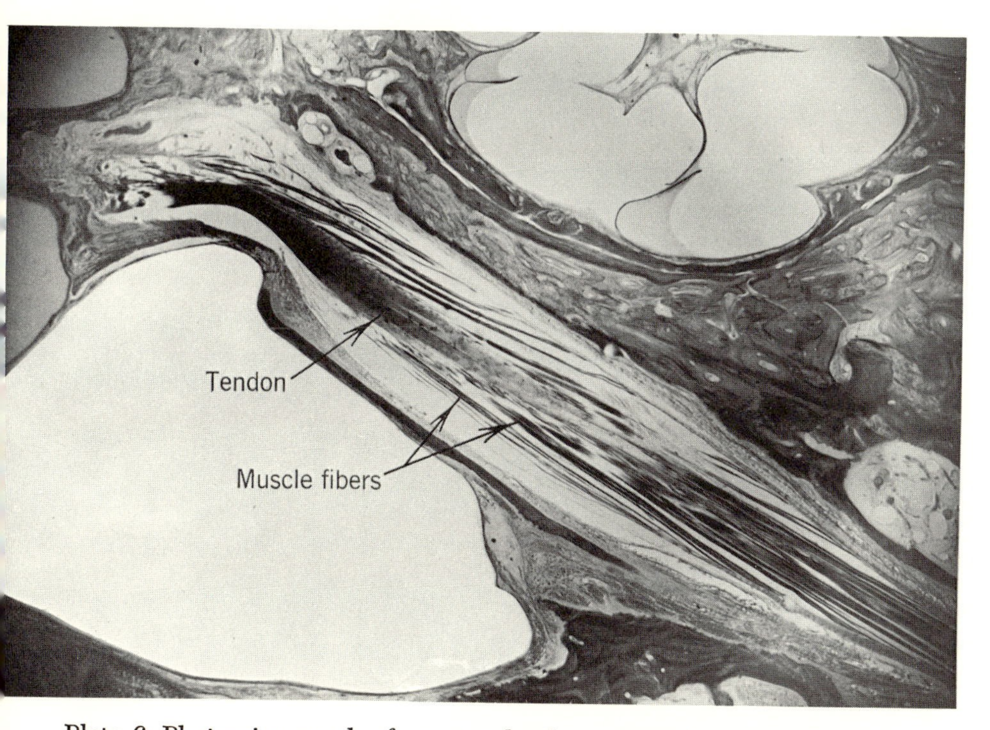

Plate 6. Photomicrograph of an atrophied or undeveloped tensor tympani muscle in a child. Because the fibers are few in number their pennate arrangement and relation to the tendon are more easily seen than in the normal ear. From the collection of the Washington University School of Medicine, transmitted by Dr. Dorothy Wolff.

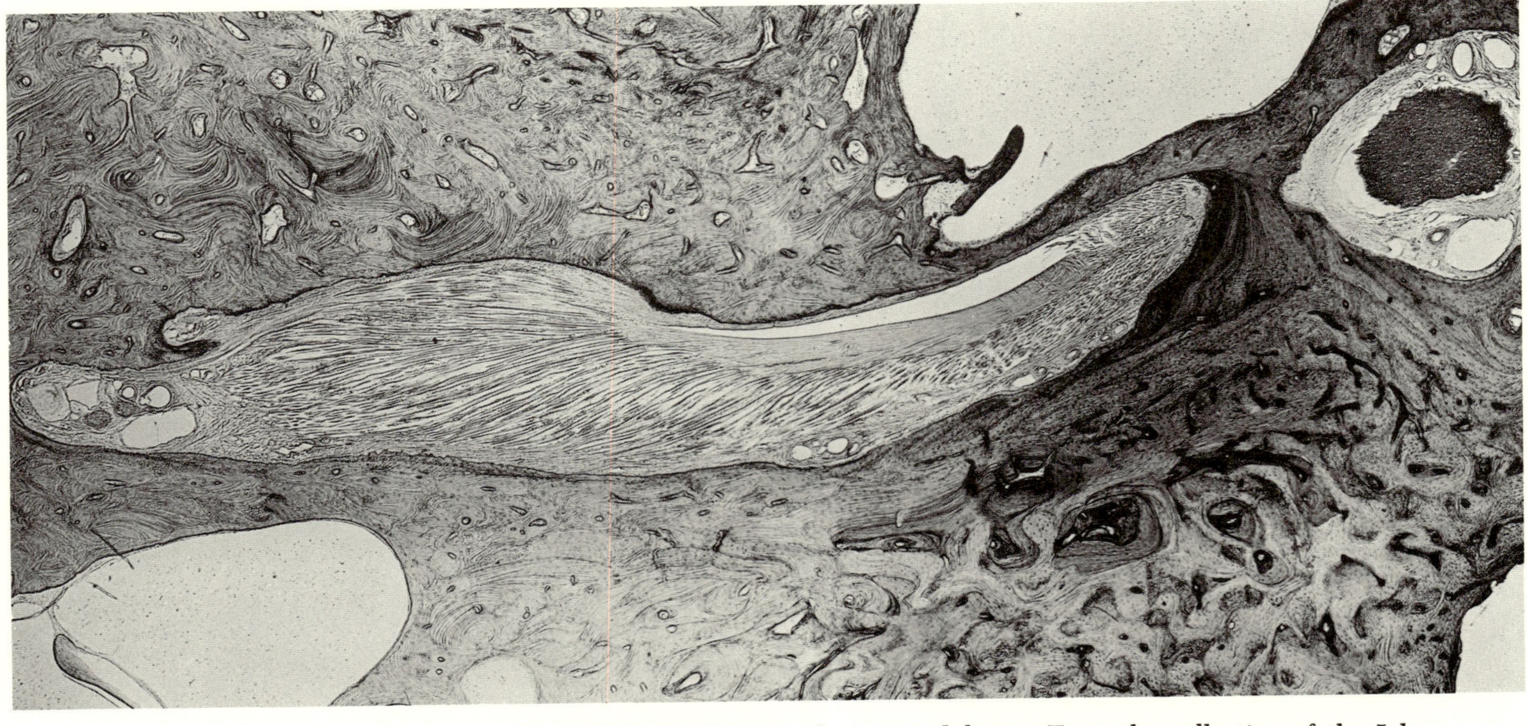

Plate 7. Photomicrograph of a section of the stapedius muscle in an adult ear. From the collection of the Johns Hopkins Medical School, by courtesy of Dr. Stacy R. Guild.

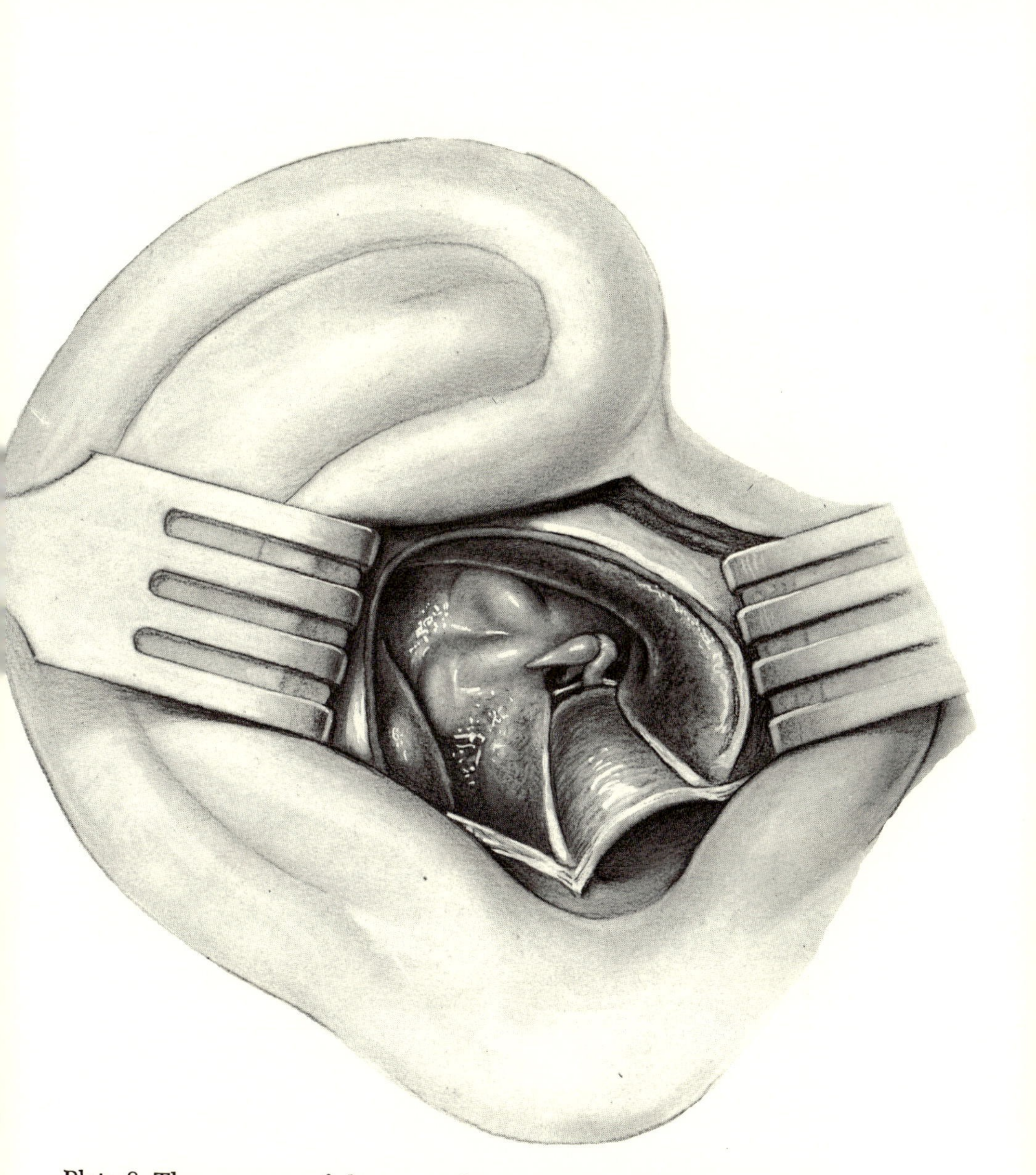

Plate 8. The exposure of the ear in the Lempert fenestration operation. Three bulges are seen, representing the lateral bony walls of the three semicircular canals. The incus and head of the malleus are clearly in view, but the stapes is deep in shadow, in the fossula leading to the oval window. Running behind the malleus is the chorda tympani nerve, which serves the sense of taste. The posterior bony wall of the external auditory meatus has been removed, exposing the skin of the meatus which will be used to form a flap to cover the canal region after the fenestra is made. Drawing by Alfred Feinberg, under direction of Dr. Julius Lempert.

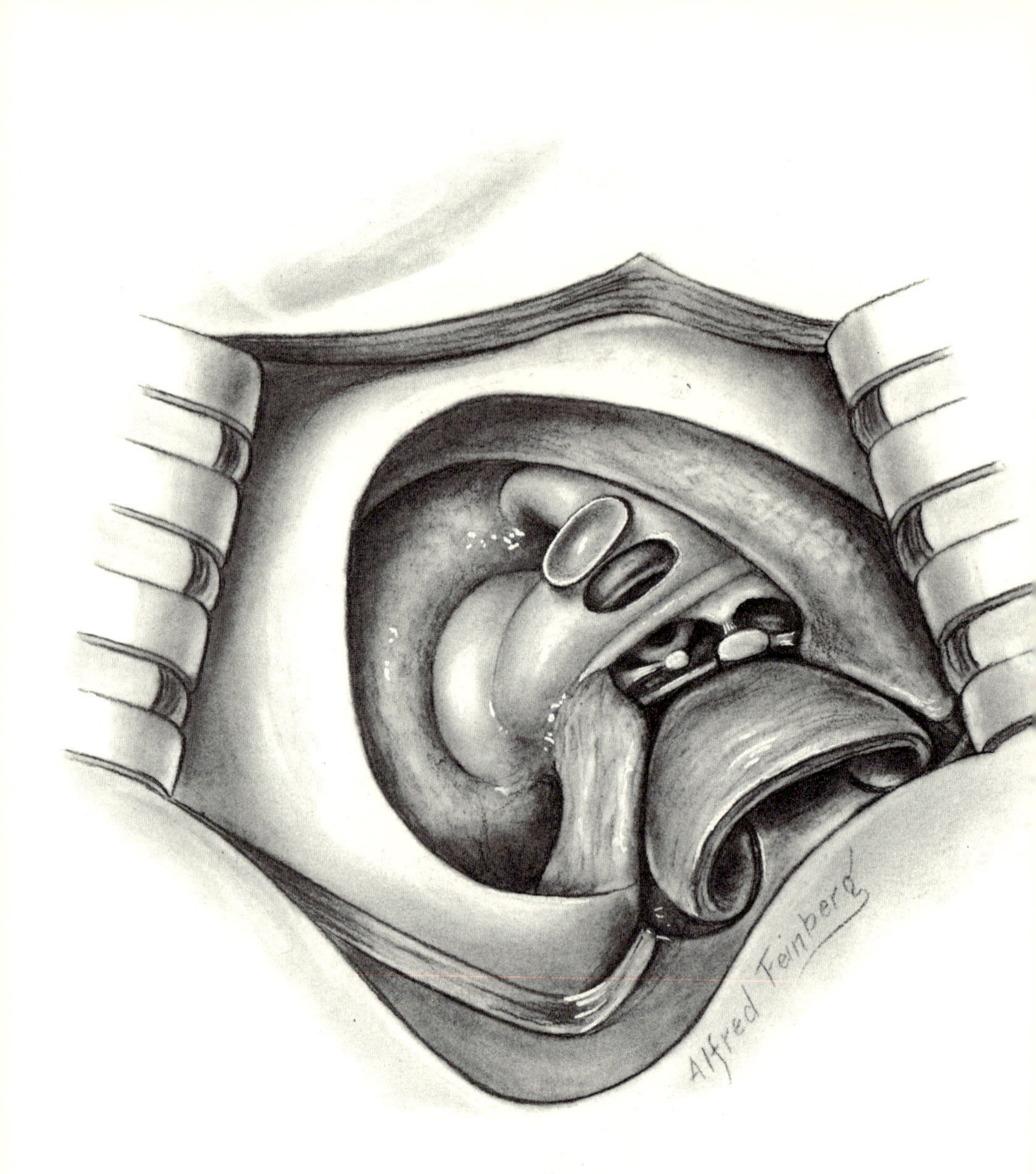

Plate 9. Fenestration of the external semicircular canal in the Lempert operation. A bony cupola has been removed, exposing the perilymph space and the membranous canal within. Also seen in this drawing are the facial nerve below and to the right of the fenestra, the stapes in its window and with its tendon intact, the handle of the malleus from which the head has been severed, and the skin flap now curled out of the way. The incus has been removed. From Lempert (*6*) [*Proceedings of the Royal Society of Medicine*].

a microscope. Less study of this kind has been carried out on the stapedius muscle. Kato in work on cats and rabbits obtained clear evidence of stapedial movements on tying a thread to the tendon and connecting it to a recording tambour. The stapedial movements were seen also by Luescher (*1*), Potter, and others in human ears in which there was a large perforation of the drum membrane. When sounds were presented the movements of the tendon or of the stapes were visible through an ear microscope.

The reflex effects of sounds may also be observed indirectly in the movements of the malleus or of the drum membrane, in changes of air pressure produced in the middle ear cavity or in the external auditory meatus, in changes of fluid pressure in the labyrinth, and in changes of the impedance of the ear. Bockendahl, working on dogs and cats, noted changes in the tension of the drum membrane on tonal stimulation, and found that the degree of tension increased with the tonal intensity. Waar, on the other hand, failed to observe the drum membrane movements in man even for loud sounds when precautions were taken to prevent general head movements. Hammerschlag obtained movements of both the malleus and the drum membrane in dogs and cats under special conditions. Kato found that the observations were made easier by strewing a metallic powder over the drum membrane. Mangold (*1*) reported a diminution of pressure in the external auditory meatus, evidently as a result of an inward movement of the drum membrane. Tsukamoto sealed a capillary tube into the wall of the cochlea of rabbits to serve as a manometer to indicate changes of labyrinthine volume. On stimulating with the sound of a whistle he recorded complex changes of volume, consisting of a sharp rise, a rapid decline, then a slow rise to above normal, and finally a slow decline to normal. These changes did not appear after the tendons of both muscles were severed. The volume changes were of the order of 0.02 cubic mm. Changes of a corresponding degree of complexity were observed by Wiggers in the electrical potentials of the cochlea, and these changes he regarded as due to pressure variations in the labyrinthine fluid caused by the muscle contractions. One of his records is reproduced in Fig. 66. Metz in 1951 introduced the method of measuring the changes in the impedance of the ear while eliciting the acoustic reflex.

The stimulus frequencies found effective in producing the

muscle reflexes vary greatly in different experiments according to the animals used and the particular conditions, but altogether they cover the greater part of the normal auditory scale. Hensen (2) obtained tensor responses in dogs and cats over a range from 200 to 6000~. Kato reported a lower limit around 128~ for rabbits and 200~ for cats, and an upper limit in both these species

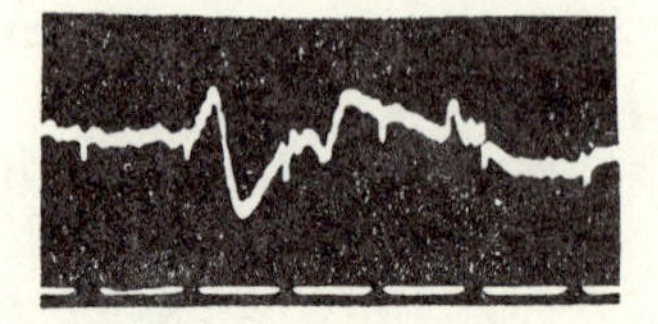

Fig. 66. Changes in electrical potentials at the round window caused by spontaneous contractions of the tympanic muscles. The interrupted line below marks quarter seconds. The largest downward deflection has a magnitude of 700 microvolts. From Wiggers [*American Journal of Physiology*].

of 50,000~. However, because of uncertainties of calibration of the Galton whistle, which was used as the high-frequency source, this upper figure must be regarded with skepticism. The same is true of Philip's report of responses up to 45,000~, also obtained with the Galton whistle. Luescher (*1*) in his observations of stapedius contractions in man obtained an effective range from 90 to 14,000~, again with the Galton whistle as the source for the high tones. Lorente de Nó and Harris observed responses of both muscles in the rabbit to tones over a wide range, at least up to 10,000~.

Bockendahl noted that the tensor contractions increase progressively with the intensity of the stimulating sound and that the muscle continues in contraction as the stimulation is maintained. He also remarked that high tones were more effective than low tones. More often the middle tones or the medium high tones are reported as the most effective in eliciting the reflex. Thus Kobrak (*1*) reported that the most effective frequencies were in the region of c^4 (perhaps 2048~). However, it must be borne in mind that in these experiments there was no absolute calibration of intensity and it is probable that the whistles that were used as sound sources were most efficient in this region of frequency. Kato found that for a given tone the stapedius muscle contracts at a much lower intensity than the tensor tympani muscle does. For faint tones, therefore, this muscle acts alone, and then at a higher intensity it is joined by the tensor tympani muscle. Lorente de Nó and Harris, working on rabbits, verified this relationship for tones below 3000~ and perhaps above 10,000~ but in the range from 3000 to 10,000~ they found the two muscles to have about equal reflex sensitivity. These relations are shown in Fig. 67, where the

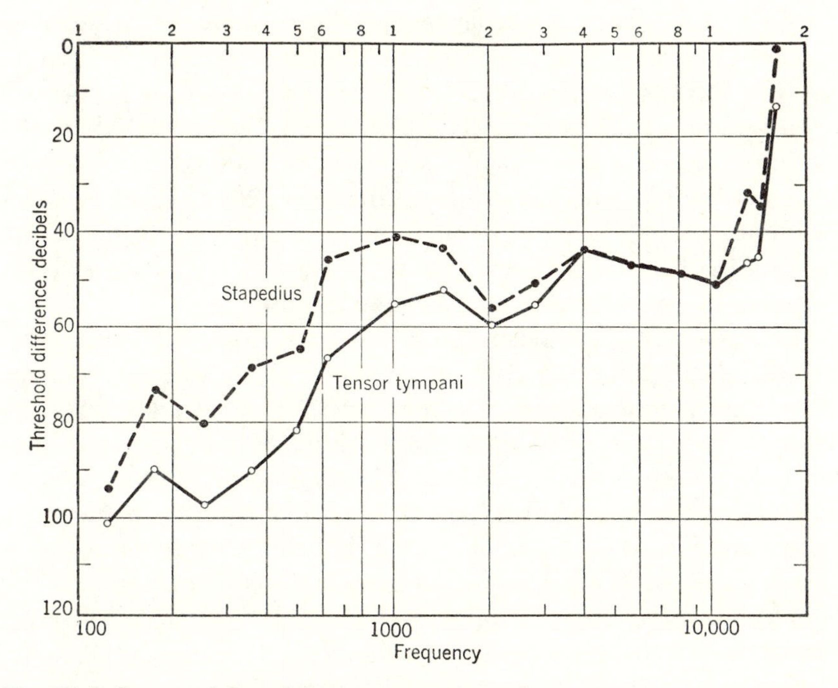

Fig. 67. Reflex excitability of the tensor tympani and stapedius muscles in response to sounds. The ordinate values represent differences between reflex thresholds in the rabbit and tonal thresholds in man. For the low tones the stapedius reflex is the more sensitive. From Lorente de Nó and Harris.

thresholds of the two muscle reflexes are represented relative to the absolute threshold of tonal perception in man. It is evident that for the greater part of the auditory range the muscle reflexes occur only when the stimulating tones rise to a high level of intensity.

The same is true in man, as Jepsen found by using the impedance method. He determined the acoustic threshold for the combined action of the two muscles, and also measured the threshold for tonal perception in the same ears. The average results obtained from 98 normal ears showed a reflex sensitivity about 80 db poorer than the tonal sensitivity, with no significant variation with frequency over the range of 250 to 4000~.

Hammerschlag, in work on dogs and cats, extended the observations made by Pollak and proved that the tensor tympani reflex is bilateral. He obtained muscle movements on the right side when the sound stimulus was conducted to the left ear, and continued to do so even when the cochlea and auditory nerve were destroyed on the right side. The movements still remained after destruction of the left trigeminal nerve and the left (or right) cerebral hemisphere. The movements disappeared, however, when the auditory nerve was cut on the left side. Then it was not possible to elicit the reflex by electrical stimulation of the cut end of either auditory nerve. It should be added that the reflexes to sound are absent in deaf animals (Pollak) and cease after the ear is severely injured by overstimulation by sound (Kato).

Hammerschlag's experiments proved that the tensor reflex takes a path through the cochlear nucleus of one side to both the ipselateral and contralateral trigeminal nuclei. The reflex does not require the presence of the temporal lobes or indeed of any part of the cerebrum. Hammerschlag (2) made a number of sections in regions of the medulla oblongata in an effort to trace the reflex pathway in further detail. He found that the reflex remained after he had severed all the fibers taking a dorsal course from the cochlear nuclei to ipselateral and contralateral regions of the medulla. He then concluded that the nerve fibers essential to the reflex run ventrally from the cochlear nuclei of one side through the trapezoid body, and thence (by an undisclosed route) to the two trigeminal nuclei. It must be added, however, that Hammerschlag's evidence does not preclude the possibility that this reflex could be carried also by the dorsal pathways, for his experiments did not include a sectioning of the trapezoid body itself.

Luescher (*1*) in his studies on man found the stapedial reflex to be as easily stimulated by applying the sound to the opposite ear as to the ear on the side on which the action was being observed. Lindsay, Kobrak, and Perlman verified this observation.

OTHER MEANS OF STIMULATION

Contractions of the tympanic muscles are not produced exclusively by sounds. It has long been accepted that some persons can contract one or perhaps both of these muscles voluntarily. Fabricius ab Aquapendente in 1600 said that he could do so, and there have been many similar claims since that time. The evidence of course is subjective only, and some, like Schapringer, have raised doubts whether the activity is actually that of the tympanic muscles. Schapringer was able at will to obtain a feeling of pressure and a strong muscle noise in one or both ears, but he accepted the suggestion of Politzer that the effect was due to an action of the tensor veli palatini. This muscle is one of the muscles of the palate that is attached to the lateral wall of the Eustachian tube. It is supposed that contractions of this muscle will distort the form of the Eustachian tube and alter the air pressure in the middle ear cavity. The alteration of air pressure is heard as a clicking sound. Contractions of this muscle occur regularly in swallowing and sometimes during yawning, and produce an easily audible sound. Often the clicking sounds produced voluntarily can be heard by another person who places his ear opposite the ear of the one making the sounds. A better method is to connect the two ears by a rubber tube. The two observers now describe the sound in much the same way, as a brief, sharp click. It seems likely that this sound is indeed caused by the tensor veli palatini, whereas the more prolonged, rushing type of sound that some persons report is caused by one or both of the tympanic muscles. When this rushing sound is present there are prominent changes in the loudness and quality of continuing sounds from external sources. Low tones seem to be most impaired in loudness, whereas high tones are affected but little, and sometimes are said to be intensified. Therefore a complex sound changes in timbre, with the high-frequency components becoming more prominent.

The contractions of the tympanic muscles may also be elicited by various noxious stimuli applied in the region of the face. Kato observed contractions in cats and rabbits in response to rather mild mechanical stimulations such as stroking or pinching the skin of the pinna or of the external auditory meatus or blowing on the pinna, and by more extreme measures such as pushing a needle into the oval or round window, cutting the eighth nerve, injuring the acoustic regions of the medulla, or electrically stimulating the facial nerve. If these measures cause too much neural damage, however, the contractions occur but once and then cease. The tympanic muscle contractions therefore are found to be a part of the general facial complex, and occur in response to most stimuli that cause pinna movements, facial twitches, blinking, and whisker movements in these animals. The effects are outstanding in an animal that has been mildly poisoned with strychnine.

Wiggers reported spontaneous contractions of the tympanic muscles in guinea pigs that were lightly anesthetized with a mixture of diallyl-barbituric acid and ethyl carbamate and operated upon so as to expose the middle ear and cochlea for the recording of cochlear potentials. The contractions occurred in coincidence with respiratory movements, and it seems likely that here is another instance of noxious stimulation produced incidentally by these movements.

LATENCIES OF THE REFLEXES

Only a few measurements have been made of the latent times of the tympanic reflexes as elicited by sounds. Hensen (2) measured these times for the tensor tympani contractions in two dogs as 0.073 and 0.092 seconds. Kato obtained values of 0.04 seconds for the tensor tympani and 0.02 seconds for the stapedius contractions in the cat, but regarded his results as only approximate. He found that the latencies varied with the frequency and intensity of the stimulating sounds, with low tones giving longer times and high intensities giving shorter times. Kobrak (*1*) reported a range of 0.07 to 0.14 seconds and an average of 0.11 seconds for the tensor tympani and a range of 0.031 to 0.15 seconds and an average of 0.07 seconds for the stapedius response in the rabbit. Tsukamoto obtained a latency of 0.29 seconds for the tensor tympani and 0.13 seconds for the stapedius response in the rabbit. Perlman and Case by recording action currents from an electrode on the stapedius tendon in four human subjects with large perforations of the drum membrane obtained an average latency of 0.01 seconds for stapedius contractions in response to a loud tone of 1000~. The variability of these data is obvious, and is due to the many uncontrolled conditions of the experiments and often to the relatively crude methods of measurement. The results are in agreement in indicating a significantly shorter latency for the stapedius reflex; the means of the values, for what they are worth, are 0.15 seconds for the tensor tympani and 0.06 seconds for the stapedius muscle. Mangold (*2*) obtained voluntary reaction times for the muscles in man of 0.17 to 0.18 seconds. Metz, using the impedance method in man, observed a latency for the combined muscle action that became progressively shorter as the stimulus intensity was increased. For a tone of 1000~ the latency was about 0.2 seconds near the reflex threshold and became about 0.04 seconds when the intensity was increased by 30 or 40 db.

The temporal pattern of contraction of these muscles has been but little studied. Bockendahl said that the tensor tympani muscle remained in contraction when a tone was continued. Luescher (*1*) in his observations on the stapedius muscle in man reported that continuous weak sounds gave only a momentary contraction at the beginning of the sound, moderate sounds gave several successive contractions, and strong sounds gave steady contractions that lasted throughout the presentation of the sound up to 10 or 15 seconds. Köhler found that the responses grew smaller on the rapid repetition of trials. Mangold reported a rapid fatigue in attempts to sustain a voluntary contraction of the muscles.

Metz in his studies on human ears by the impedance method found that the contractions rapidly attained a maximum and then usually remained constant over many seconds as the stimulating tone was maintained. Then, while the tone was still sounding, a slow relaxation set in. This process is one of reflex adaptation and not of muscular fatigue, for if the tone is stopped and then immediately presented again the response appears as before in full magnitude. Also, after the adaptation has occurred to a continuous tone, the contraction reappears if another tone is introduced.

FUNCTIONS OF THE TYMPANIC MUSCLES

Since the early discovery of the tympanic muscles there has been much theorizing as to their possible functions in hearing. Four general types of theories have been proposed. These are (1) the intensity-control theory, (2) the frequency-selection theory, (3) the fixation theory, and (4) the labyrinthine pressure theory. In this theorizing the major consideration has been given to the tensor tympani muscle. The stapedius muscle has usually been regarded as cooperating with the tensor in the general action. At times, however, the two muscles have been regarded as antagonists (e.g., by Toynbee).

The Intensity-Control Theory. The intensity-control theory supposes that the muscles by their contractions are able to alter the efficiency of transmission of sound by the middle ear mechanism. The theory in its most complete form asserts that this alteration can be in either direction: the efficiency can be either increased or decreased. The early formulation of this theory is attributed to Molinettus; according to Morgagni, who wrote in 1764, he claimed that the tensor tympani muscle serves to exclude strong and uncomfortable noises and to amplify weak ones. Haller, a contemporary of Morgagni's, expressed this view in the following words: "The tensor, working through the malleus, adapts the tympanic membrane for the hearing of weak sounds, and the other muscle of the malleus, if it exists, moderates the sounds that are too strong by pulling the malleus away from the incus and thereby interrupting the propagation of the sonorous vibrations."

This theory in a more limited form, known as the protection theory, supposes that the efficiency can only be decreased. According to this view the muscle acts so as to safeguard the inner ear against overstimulation by strong sounds. There were many early approaches to this theory, but the formulation that had most influence upon subsequent thinking was Johannes Müller's, given in 1838. Müller suggested that the tensor acts reflexly to loud sounds to induce a "deadening or muffling of the ears."

The Frequency-Selection Theory. The frequency-selection theory holds that the muscle contractions may be varied so as to cause certain tones to be favored in transmission. DuVerney in 1683 clearly expressed this notion; the drum membrane, he said, is put into the proper state of tension to receive particular sounds. This theory has usually been called the accommodation theory in a fancied analogy with the process of accommodation in the eye. Mach in 1863, following this always dangerous line of reasoning by analogy, concluded that the ear fixes and follows tones just as the eye does, and picks out particular components in a complex sound by adjusting the resonance frequency of the ear to them. Increased muscular tension should adapt the ear for high tones and reduced tension should adapt it for low tones. Later Mach became doubtful about his theory when his experiments failed to give it any certain support (Mach, *2*; Mach and Kessel).

The Fixation Theory. The fixation theory states that the tympanic muscles assist the suspensory ligaments in maintaining the ossicular mechanism in its proper position and in a state of tension suitable for the reception of sounds. Valsalva in 1707 was perhaps the first to give a clear statement of this theory. He declared that initially in the hearing of a sound the tympanic membrane is not suitably disposed to respond to the vibrations, for it hangs loose and relaxed, and then "the major muscle of the malleus" (the tensor tympani) is strongly contracted to cause this membrane to be most readily moved by the incoming sounds. Magnus took the simpler position that the tensor muscle is continually in a state of tonus to give rigidity to the ossicular mechanism, but regarded the muscle as too feeble to have any outstanding effects. Helmholtz (*4*) pointed out that the two tympanic muscles are so placed that a moderate tension can produce large effects upon the ossicular system. He considered the muscles as functioning primarily in preventing excessive motion of the ossicular chain lateral to its operating axis. However, he admitted a certain variability in the muscular tension according to the needs of the ear.

The Labyrinthine Pressure Theory. The labyrinthine pressure theory states that the effect of the muscular contraction is to raise the pressure of the labyrinthine fluid and in so doing to alter the mechanical performance of the ear. Lucae believed that tensor contractions by increasing the intralabyrinthine pressure gave an improvement in the conduction of low tones. Zimmermann argued that the ossicular mechanism is not concerned in sound transmission but rather serves for communicating the tensions of the tympanic muscles to the labyrinthine fluid. These tensions, he supposed, raise the fluid pressure and restrain the movements of the basilar membrane, especially for low tones. We shall return to a consideration of these theories after reviewing the experimental evidence.

EXPERIMENTAL EVIDENCE

Several experimental studies have been made to test the above theories. Politzer (*1,2*), working on freshly killed dogs, was able to produce a contraction of the tensor tympani muscle by electrical stimulation of the trigeminal nerve. If this was done while observing the movements of the malleus in response to a tone of 512~ the movements were sharply reduced, usually by a ratio of 3 to 1. When he listened to the sound by means of an auscultation tube applied to the auditory bulla the timbre of a complex tone was noticeably changed by the muscular contraction. The fundamental of a tuning fork was weakened while its overtones came forth more clearly. Similar results were obtained in human cadaver heads when a strong tension was exerted on the tensor by means of a thread that had been attached to it. Lucae also attempted to imitate the contractions of the tensor tympani muscle and to measure the effects. He worked with fresh cadavers and observed the ossicular movements both with mechanical recording and by the use of an auscultation tube. He attached a thread to the tensor muscle and ran the thread over a pulley to various weights. He found that the application of the weights reduced the transmission of sounds applied by bone conduction.

Mach and Kessel (2) repeated this experiment on cadavers and studied the action of both muscles by observing the movements of the ossicles by means of a microscope. They found that a loading of the tensor tympani muscle with a 3 gram weight reduced the ossicular movements by 40 per cent when the stimulus was an aerial tone of 256~ but had hardly any effect when the stimulus was 1024~. This frequency difference was not observed for the stapedius muscle.

Cochlear Potential Studies. Crowe, Hughson, and Witting in 1931 were the first to make use of the electrical potentials of the ear as a method for studying the action of the tensor tympani muscle. Their experiments were made on the anesthetized cat. After a wide exposure of the middle ear they attached a hook to the tendon of this muscle, tied a thread to the hook, and ran the thread directly out of the bulla and around a pulley to a scale pan. They then observed the effects of adding various weights to the scale pan while stimulating the ear with sounds. They found that the weights above 10 grams caused a noticeable reduction of voice sounds and all low tones. Reductions up to 50 db were observed for 1000 and 2000~. Tones above 2000~ were less affected (Hughson and Crowe, 3).

This study of the tensor tympani muscle was continued by

Wever and Bray (3), also with the cat as the experimental animal, and with improvements of technique. The method of applying the tension to the tensor tendon as used by Crowe, Hughson, and Witting does not exactly reproduce the action of the muscle. The normal direction of pull of this muscle is dorsomedial and slightly forward, whereas the direction taken by a thread that is led directly out of the bulla is ventral and thus nearly at right angles to the proper direction. Because the muscle remained attached to the roof of the bulla the effective direction was somewhere between the normal and this ventral direction, but just what it was is difficult to estimate. To obtain the proper direction of pull, Wever and Bray used a small glass tube that was given a rounded angle of about 90° near its tip. The thread, after being tied to the tendon, was passed through this tube and the tube was then lowered into the bulla with its end just above the belly of the muscle, in such a position as to give a normal direction of pull. The thread was then passed over a pulley to a scale pan. The cochlear potentials were recorded with an electrode on the round window membrane while stimulating with pure tones of known intensities. Figure 68 shows results obtained for a number of

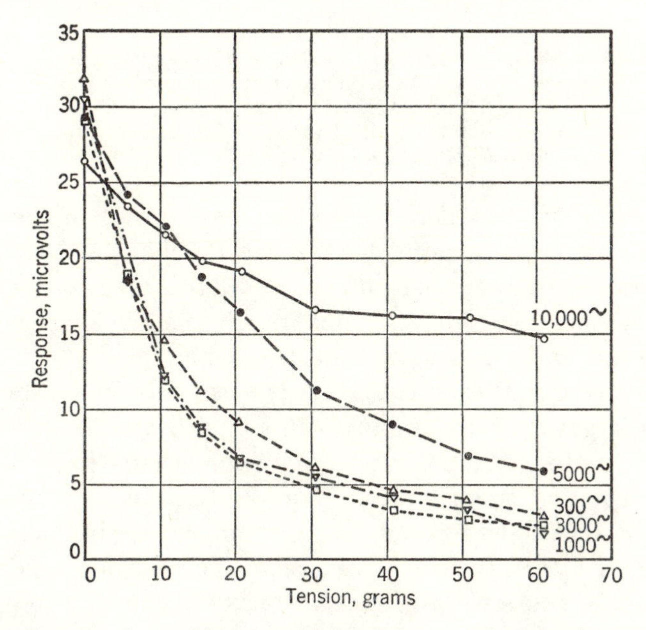

Fig. 68. Effects on the cochlear potentials of the cat of artificial tension on the tensor tympani tendon. From Wever and Bray (3) [*Annals of Otology, Rhinology and Laryngology*].

stimulating tones with weights up to 61 grams. Here it is evident that for all tones the application of tension causes a reduction of responses. The rate of reduction is particularly great for the smaller tensions and then grows less as the tension is increased. The effects are similar for low and intermediate tones, but are significantly smaller for the high tones. Some further relations to frequency are shown in Fig. 69. Here it becomes clear that the

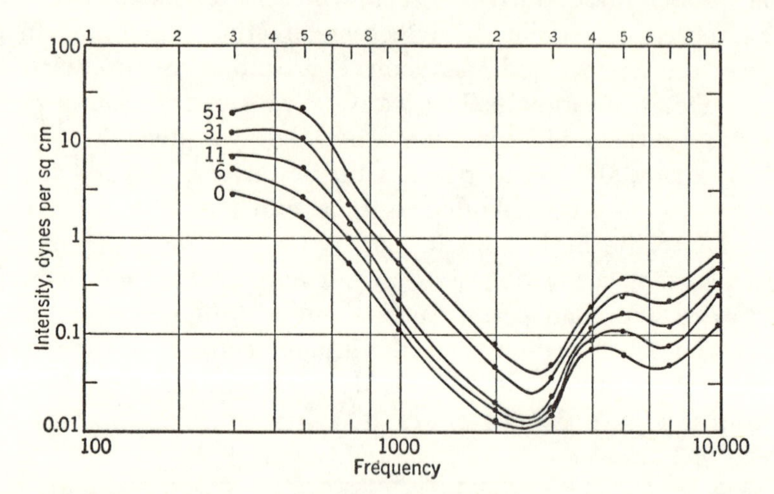

Fig. 69. Effects of tension on the tensor tendon of the cat, shown as a function of frequency. The curves represent the sound intensity necessary to produce a standard cochlear response of 10 microvolts in the presence of different amounts of tension, indicated in grams at the ends of the curves. From Wever and Bray (3) [*Annals of Otology, Rhinology, and Laryngology*].

impairment of transmission, shown here as a raising of the curve, is accompanied by a general shift of the function to the right. It appears that tension applied to the conductive mechanism of the ear not only adds friction and thus impairs its efficiency but also adds stiffness and raises its natural frequency.

A similar study (Wever and Bray, *6*) was made on the stapedius muscle and gave results as shown in Figs. 70 and 71. Here it is evident that the reduction of transmission occurs for all conditions, except that small tensions (up to 4 grams) for tones between 2000 and 3000~ produce a slight improvement. In general, the changes of transmission are greater for the low and intermediate tones and less for tones above 2000~. It is of interest that the reductions of transmission for a given amount of tension are much greater for the stapedius muscle than for the tensor tympani muscle; at 1000~, for example, a tension of 11 grams on the tensor tendon gives a reduction of 8 db whereas this same tension on the stapedius tendon gives a reduction of 30 db.

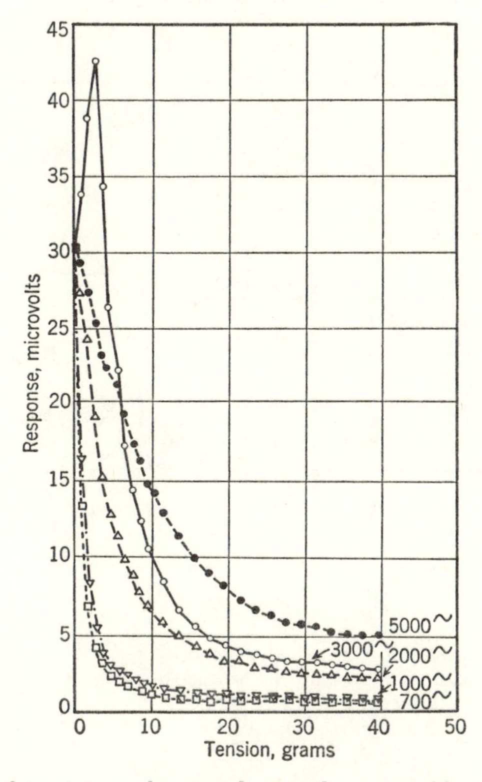

Fig. 70. Effects of tension on the stapedius tendon on cochlear potentials in the cat. From Wever and Bray (6) [*Journal of Experimental Psychology*].

There is no doubt that the larger tensions used in these studies far exceed what the muscles themselves are capable of exerting, though tests showed that the tensions up to 50 grams never produced any injury to the ear. Lorente de Nó measured the strength of the tensor muscle in the rabbit and found it to be only 1.2 grams. One of our recent experiments on the cat gives data on this point. The animal was decerebrated and strychninized and the tensor tendon was cut. The contractions of the stapedius muscle were elicited by pinching the pinna, and these contractions were probably maximal. Yet the reductions of cochlear potentials amounted to only 12 db during stimulation with 200~. As was ascertained by reference to our artificial tension studies, this reduction at 200~ is obtained with a tension of about 1 gram. Other results indicate that the tensor tympani muscle of the cat can exert a maximum tension of about 3 grams.

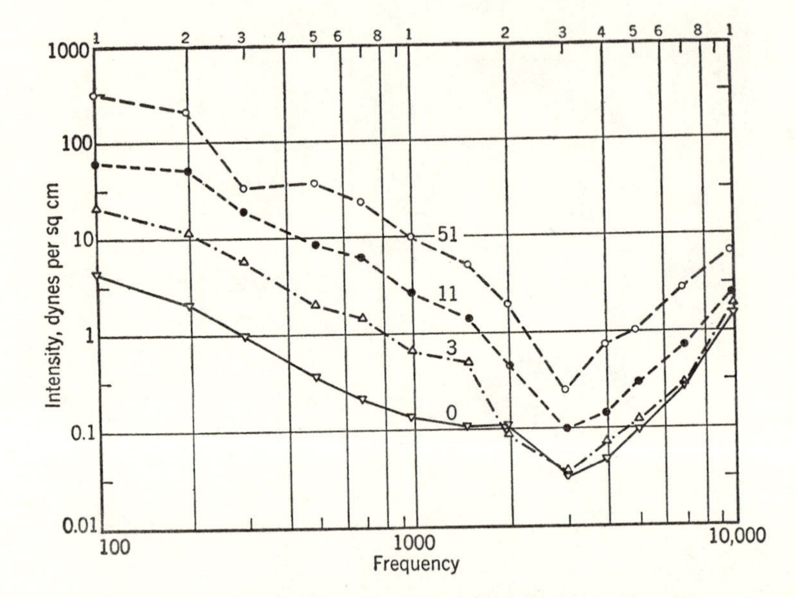

Fig. 71. Effects of tension on the stapedius tendon of the cat, shown as a function of frequency. The curves represent the sound intensity required to produce a standard response of 10 microvolts in the presence of different amounts of tension, shown in grams above the curves. From Wever and Bray (6) [*Journal of Experimental Psychology*].

As already mentioned, Wiggers observed reductions in the cochlear potentials of guinea pigs as a result of spontaneous contractions of both muscles. He stimulated with tones between 60 and 2500~ and obtained the greatest reductions at the lowest frequencies and progressively smaller reductions for tones up to 1200~. Between 1200 and 1700~ he obtained slight improvements, and from here to 2500~ there were no changes. Still higher tones were not investigated. The maximum effects amounted to 40 db at 100~.

In general, the observations on voluntary contractions of the tympanic muscles in man agree with the results just described for the lower animals in indicating a relatively greater effect upon the low tones. Schapringer in 1870 reported his experience that on a strong effort of contraction all the low tones were reduced in intensity and the tones below 70~ were lost altogether.

CONCLUSIONS REGARDING TYMPANIC MUSCLE FUNCTION

The experimental evidence is in firm support of the intensity-control theory. The two tympanic muscles in their contractions cause modifications of the efficiency of sound transmission. These modifications usually are in the direction of reduced transmission, so that we can say that in general the form of this theory known as the protection theory is the correct one. The exception to this protective function appears for stapedial contractions in both the cat and the guinea pig, which cause slight improvements of conduction for tones over a narrow range in the medium high region.

There are limitations to the protection that the muscles can supply. Sudden sounds like explosions cannot be protected against because the energy is carried in a sharp wave front and is fully effective before the reflex contractions can occur. For long-enduring sounds also a part of the protection is lost as the muscles suffer from adaptation or fatigue. Within these limits, however, the protection given by the muscles is of great service to the ear. A protection of 10 db or so is of importance if a sound is near the limit of the ear's tolerance. In this connection Kato showed that the ears of cats and rabbits are more likely to be injured by overstimulation when the reflexes are prevented by narcosis.

The amount of protection evidently varies somewhat with the species, and it varies greatly with frequency. It is fortunate that the greater protection is afforded for the low tones, because the stimulation deafness experiments show that when tones are of equal pressure their destructiveness increases as the frequency becomes lower (Smith and Wever).

The accommodation theory is not supported. The muscular action is a simple reflex and the degree of contraction is not under voluntary control. Nor is there any automatic control that brings about an optimum condition of sound reception in general or any systematic selection of one tone over another. To be sure, the control of intensity shows variations with frequency, but these variations are only incidental to the transmission process.

The labyrinthine pressure theory also must be set aside, from the following considerations. It is true that the muscular contractions cause changes in the pressure of the labyrinthine fluid, but experiments show that such changes do not modify the efficiency of transmission. Békésy (*12*) used a modified auscultation method to measure the transmission of tones of 300 and 1000~ from the external auditory meatus to the round window of a cadaver specimen as a function of the intralabyrinthine pressure. He found that ordinary increases of pressure were without effect, and reductions in transmission of only 10 to 20 per cent resulted from a pressure increase up to 320 grams per sq cm (313,700 dynes per sq cm) above normal. Lempert, Wever, Lawrence, and Meltzer, in experiments described more fully later on (page 363), studied the effects of increased labyrinthine pressure on sound conduction in the living monkey. Cochlear potentials in response to tones of 512 and 2048~ were observed by means of an electrode on the round window membrane when the perilymphatic fluid pressure was raised by various amounts up to 50 mm of mercury (66,600 dynes per sq cm), and never was there any sign of an effect upon the transmission. This result is to be expected on theoretical grounds, for the change of density of the cochlear fluid produced by a pressure of 50 mm of mercury amounts to only 0.03 of 1 per cent.

There seems little question that the tympanic muscles contribute to the strength and rigidity of the ossicular mechanism. Anyone who has severed both the tensor and stapedius tendons is impressed by the extreme fragility then found in the mechanism. Manipulations that formerly could easily be withstood will now cause serious damage, as best shown in an impairment of cochlear potentials. The fixation theory therefore can be accepted along with the protection theory in the explanation of tympanic muscle function.

11

The Effects of Air Pressure on the Ear

A COMMON EXPERIENCE is a feeling of pressure in the middle ear as the atmospheric pressure is changed. Such a change occurs most often from variations of altitude, as in mountain climbing, aerial flight, or riding in elevators. As we gain altitude the outside air becomes less dense and the air within the middle ear cavity tends to expand, pushing outward on the drum membrane. As we lose altitude this process is reversed; the denser air outside then pushes inward on the drum membrane. In either event we perceive the effects as pressure sensations in the ear and also as impairments of hearing.

Usually these sensory effects are slight and transitory because during swallowing the Eustachian tube opens and permits an equalization of the pressure in the middle ear cavity. At times, however, a blocking of the Eustachian tube prevents this equalization. A progressive change of altitude then can build up a large pressure difference between the air of the middle ear and the external air. This difference causes a deformation of the drum membrane that is uncomfortable or even acutely painful and can be so severe as to rupture the membrane. Even without an atmospheric pressure change it is possible for a difference in pressure to arise by a slow absorption of the air in the middle ear cavity.

Impairments of hearing of a more or less temporary nature are often found as secondary effects of congestion in the nasopharynx, as from a common cold, and more permanent impairments come from chronic catarrhal infections and adenoid growths in this region. The problem of altered air pressure in the middle ear cavity therefore has claimed much clinical interest. In the clinical situation it is often difficult to determine to what extent the auditory symptoms are due to pressure changes and to what extent they are due to an extension of the infection and congestion to the ear itself. This problem requires an understanding of the effects of pressure in the normal ear.

Experimentally there are two ways of introducing pressure into the ear, and in each of these the pressure difference may be either positive or negative. One of these ways is to introduce the changes through the external auditory meatus and the other way is to introduce them into the middle ear cavity. A positive pressure in the external auditory meatus is similar to a negative pressure in the middle ear cavity, but is not exactly the same. Either presses the drum membrane inward, but the two differ in their effects upon the inner ear. Positive pressure in the meatus is communicated through the drum membrane and gives positive pressure on the oval and round windows, though reduced somewhat by the resistance of the drum membrane. Negative pressure in the middle ear cavity is exerted alike on the drum membrane and on these two windows. Correspondingly, negative pressure in the meatus and positive pressure in the middle ear cavity have the same effects on the drum membrane but contrary effects on the cochlear windows.

PRESSURE IN THE EXTERNAL AUDITORY MEATUS

The first experimental study of pressure effects was made by Mach and Kessel (*1*) in 1872. Using themselves as subjects, they introduced positive pressures up to 14 cm of water (10.3 mm of mercury) into the external auditory meatus and observed in general a weakening of the loudness of tones. Low tones were affected more than high tones. Many have repeated these experiments, with results that for the most part are in good agreement.

Békésy (*2*) found that both positive and negative pressures of 10 cm of water reduced the loudness of low tones, up to 1000 or 1800~. For a small frequency region around 1800~ there was an increase of loudness. Still higher tones showed slight or irregular effects. He reported that a strong tone was reduced relatively more than a weak tone, and explained this variation as due to an exaggeration of the middle ear's nonlinearity as the pressure difference increased. A further study by Békésy (*3*) included measurements by bone conduction. When the meatus was effectively closed the results were much the same as by air conduction, which was interpreted to mean that the bone-conduction stimulus was really producing aerial sounds in the meatus through compression of its walls. When the meatus was open the effects of air pressure on bone-conducted sounds were considerably less than those on air-conducted sounds.

Dishoeck studied the effects of pressure on normal and defective ears and determined in each the pressure value that gave optimum sensitivity. Usually this value was zero, but some normal ears were best at a slight negative pressure, usually in the range up to −4 cm of water. For the ears of persons suffering from nasopharyngeal congestions the best hearing was nearly always at negative pressures, mostly around −10 but sometimes as high as −40 cm of water. Zöllner suggested that persons whose hearing is improved by application of negative pressures probably have an inequality of pressure in the middle ear to begin with and the application of a negative pressure restores them to a normal condition.

Rasmussen (*1*) extended these observations, using tones from 64 to 11,600~, and obtained similar results. For the low tones his normal subjects usually heard best with a pressure between zero and −6 cm of water, but the total range of optimums varied between +72 and −24 cm of water. For the high tones also the optimum pressures were usually in the region of zero, but a curious variation was found for strong positive pressures. For 8200 and 11,600~ the application of positive pressures reduced the loudness at first, and then as the pressures exceeded about 60 cm of water this trend was reversed and the loudness was increased. With a pressure of +102 cm of water most subjects heard the tone as stronger than at zero pressure. Results for one subject are shown in Fig. 72. It is important to mention that in this experiment Rasmussen used an air-conduction receiver with a perforated diaphragm to prevent effects of the pressure on the receiver itself. This is a necessary precaution, apparently not always taken in earlier studies (see Rudmose *et al.*).

Rasmussen continued his studies with stimulation by bone conduction, applied with a vibrator on the forehead. For the low tones, up to 2050~, the application of both positive and negative pressures usually caused a decrease in the loudness of bone-conducted tones, just as was found for air conduction. For a tone of 4100~ the results were irregular, with many ears showing increases of loudness for both positive and negative pressures. For higher frequencies the loudness was largely unaffected by the application of pressures. In 32 out of 33 otosclerotic subjects the application of either positive or negative pressure had no effect upon the hearing. Rasmussen's results thus support the sugges-

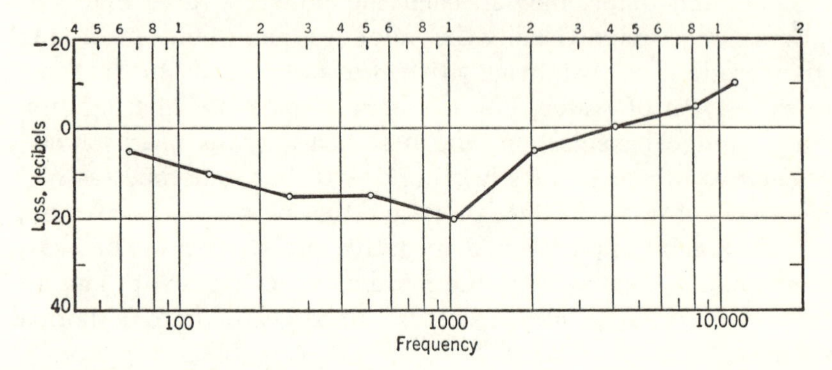

Fig. 72. Effects on sensitivity of a positive air pressure of 102 cm of water in the external auditory meatus of a normal human ear. Above 4000~ the sensitivity is improved. Data from Rasmussen (*1*).

tions of Gellé and others that the effects of pressure on hearing may give a clue to the degree of mobility of the stapes.

PRESSURE IN THE MIDDLE EAR CAVITY

Pressure may be introduced into the middle ear cavity through the Eustachian tube, and in human subjects this is the most feasible method. There are two procedures. One, employed by Valsalva and known by his name, consists of closing the mouth and nostrils and making an effort of exhalation to force air into the Eustachian tube in order to produce positive pressure or a corresponding effort of inhalation in order to produce negative pressure. The other procedure consists of the forcible injection of air through a tube in the mouth or nostrils or applied to the orifice of the Eustachian tube. With either procedure the subject must refrain from swallowing while tests of hearing are carried out.

Experiments on Human Subjects. Bezold and Siebenmann used the Valsalva procedure, and observed that for high tones a "negative Valsalva" caused a loss of hearing whereas a "positive Valsalva" often gave a slight improvement. Fowler introduced pressure through a cannula inserted into the Eustachian tube and observed a reduction of hearing by both air and bone conduction. The reduction was especially marked for low tones.

Loch (2) carried out the most extensive investigation with the Valsalva procedure and with audiometer tests before and after each pressure change. Also he attempted to measure in an approximate way the amount of the pressure change by placing one manometer in the external meatus of the ear being studied and another manometer in the nostril on the opposite side. By this means he was able to ascertain that certain pressures were greater than others, though the determination could not be made in absolute units.

His observations showed that small positive pressures gave slight improvements in acuity for high tones, those above 2048~, without affecting the acuity for low tones. A moderate increase in this pressure reduced the acuity for low tones, and did so about equally over the range from 32 to 1024~. Strong pressures reduced the acuity to these tones by as much as 20 db and caused a reduction in the gain that had been shown for the high tones. These effects are illustrated in Fig. 73.

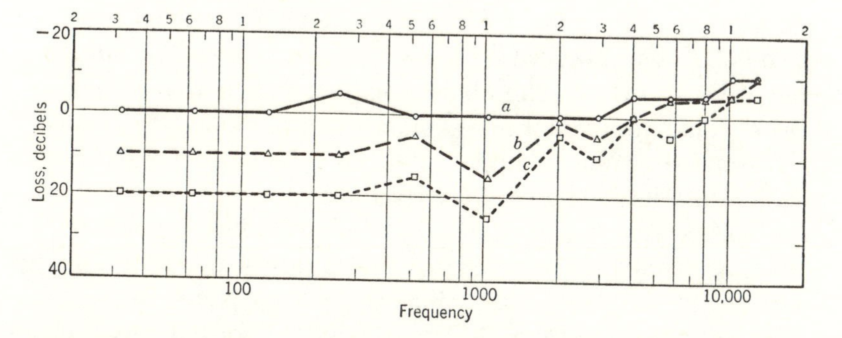

Fig. 73. Effects of positive pressures in the middle ear cavity, produced in human subjects by the Valsalva method. Curve *a* is for slight pressure, *b* for medium pressure, and *c* for strong pressure. Data from Loch (2).

Loch found that negative pressures produced effects similar to those just described for the low tones, but gave different effects for the high tones. A small negative pressure had no effect on tones from 2048 to 5793~ and reduced acuity for higher tones, in contrast to the improvement in this region caused by positive pressures. Stronger negative pressures extended the range of low-tone impairment to 2896~, gave an "island" of no effect around 4096 to 5793~, and then gave impairment for the highest tones. These effects are shown in Fig. 74.

In these experiments the subjects were asked to release the pressure by swallowing. After "positive Valsalva" the acuity usually returned at once to the normal level except for the highest tones. Additional acts of swallowing returned these tones also to normal.

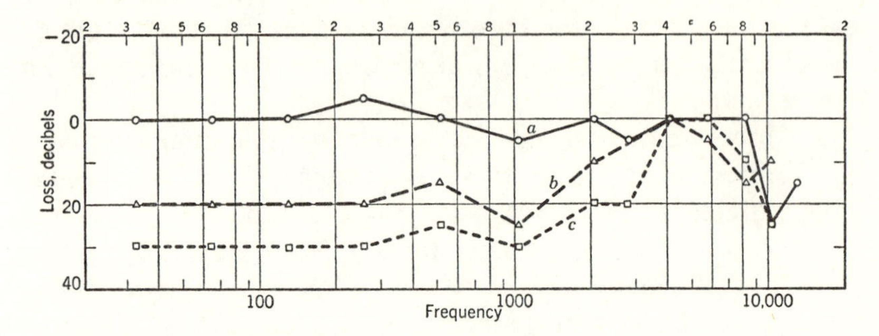

Fig. 74. Effects of negative pressures in the middle ear cavity of man. Curves *a, b,* and *c* represent increasing degrees of pressure from slight to strong. Data from Loch (*2*).

It was found to be more difficult to remove a negative pressure, and sometimes several swallowings failed to remove the effects of a "negative Valsalva." The obvious explanation is that a negative pressure tends to maintain the collapsed condition of the walls of the Eustachian tube whereas a positive pressure inflates these walls and assists in the escape of air.

In another experiment Loch (*1*) used an ingenious method for producing a negative pressure in the middle ear cavity. He inserted a small rubber balloon into the nasopharynx on one side and placed it so that when it was inflated it blocked the passage of air through the nose on this side. There is then a slow absorption of the entrapped air. Tests of hearing were made before the insertion of the balloon and at intervals thereafter over a period of 90 minutes. Slight improvements were observed in the beginning, during the first 10 minutes or so, involving mainly the highest tones as in the negative pressure experiments just mentioned. Then, showing up after 20 minutes and growing marked after 45 minutes, there was a loss of high-tone acuity. This loss continued and was extended to the low tones after 60 minutes. On removal of the balloon and ventilation of the middle ear by swallowing, the acuity returned fully to normal. Some sample results are given in Fig. 75.

Animal Experiments. For more precise results on this problem we turn to experiments on animals, in which the auditory bulla is opened and the pressure is introduced directly into the middle ear cavity. Kobrak (2), working with rabbits, introduced pressure

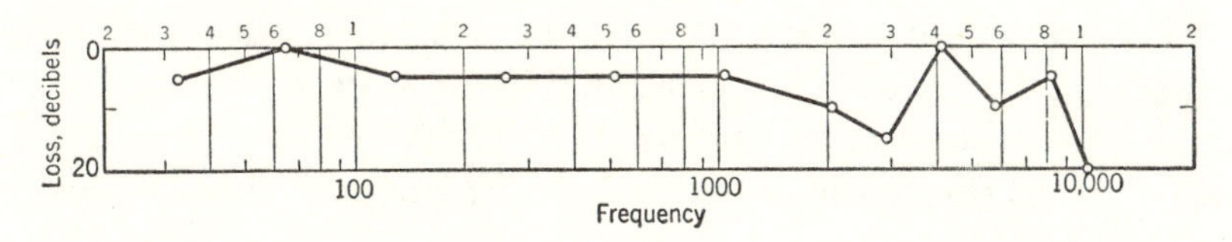

Fig. 75. Effects of blocking the Eustachian tube in man for a period of 75 minutes. Compare with curve *a* of the preceding figure. Data from Loch (*1*).

into the bulla on one side while observing the reflex contractions of the tensor tympani muscle on the opposite side in response to tones. He found that moderate positive or negative pressures had no influence on this reflex, but that strong pressures made it necessary to increase the intensity of the tones in order to elicit the contractions. A pressure of 12 cm of water produced an average impairment of 5.2 db over the range of 256 to 4096~.

Thompson, Howe, and Hughson were the first to use the electrical potentials of the ear in the study of pressure effects. Working on cats, they introduced pressures through a fine tube inserted into the Eustachian tube and recorded the pressures with a manometer connected to a tube sealed in the bulla. They found impairments of electrical potentials for positive or negative pressures exceeding 5 mm of mercury. Pressures up to 32 mm of mercury caused impairments as great as 20 db for low tones and smaller impairments for high tones.

Further experiments on cats by the cochlear potential method were carried out by Wever, Bray, and Lawrence (7). The auditory bulla was exposed and two holes were drilled in it, one about 2 mm in diameter and the other about 5 mm in diameter. The larger hole was used as a viewing window for placing a foil electrode through the smaller hole onto the round window membrane. The wire to this electrode ran through a small rubber stopper that was fitted to the hole. After this placement the larger hole was closed with another rubber stopper through which was passed the small end of a tapered glass tube whose other end was connected by rubber tubing to the pressure apparatus. This apparatus included a rubber bulb for producing pressure changes and two pressure gauges, one for large and another for small pressures. The Eustachian tube was kept closed by pressing on its membranous portion with a spatulate prod. The animals were maintained under diallyl-barbituric acid anesthesia supplemented with curare, and artificial respiration was used throughout the experiment. The effects of positive pressures up to 50 mm of mercury were observed for tones from 100 to 10,000~.

Figure 76 shows the potentials as a function of pressure during

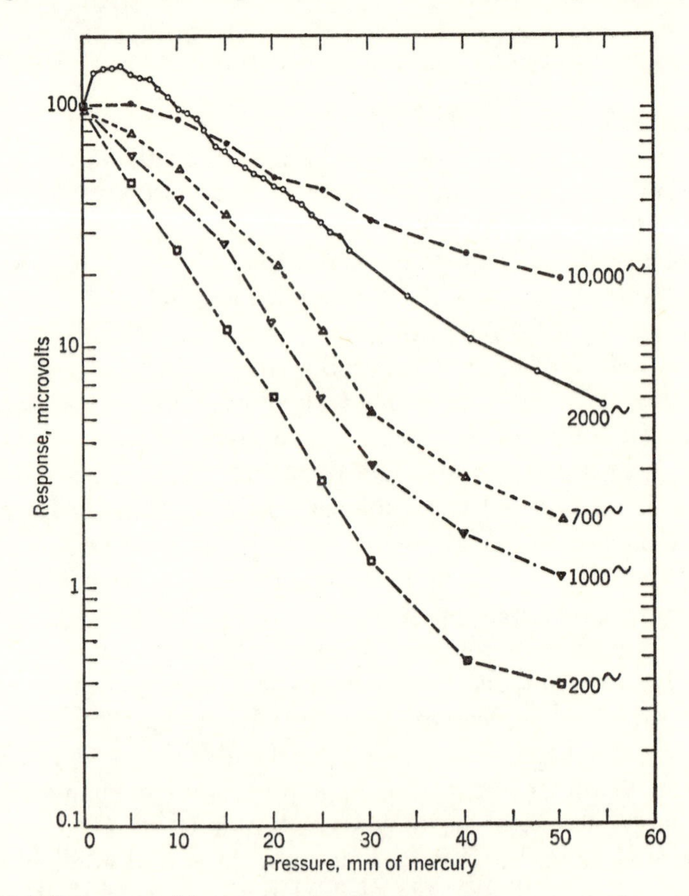

Fig. 76. Effects on cochlear potentials of positive air pressure in the middle ear cavity of the cat. From Wever, Bray, and Lawrence (7) [*Journal of Experimental Psychology*].

stimulation with various tones. Each tone was adjusted initially under zero pressure to an intensity giving a response of 100 microvolts and was kept at that intensity during the introduction of pressure changes. It is evident that in general the responses are diminished on the application of pressure. However, for certain tones there is a preliminary increase of response. This occurs for tones above 2000~ and pressures below 10 mm of mercury. For stronger pressures these tones also show a decline.

The frequency relations are shown more clearly in Fig. 77,

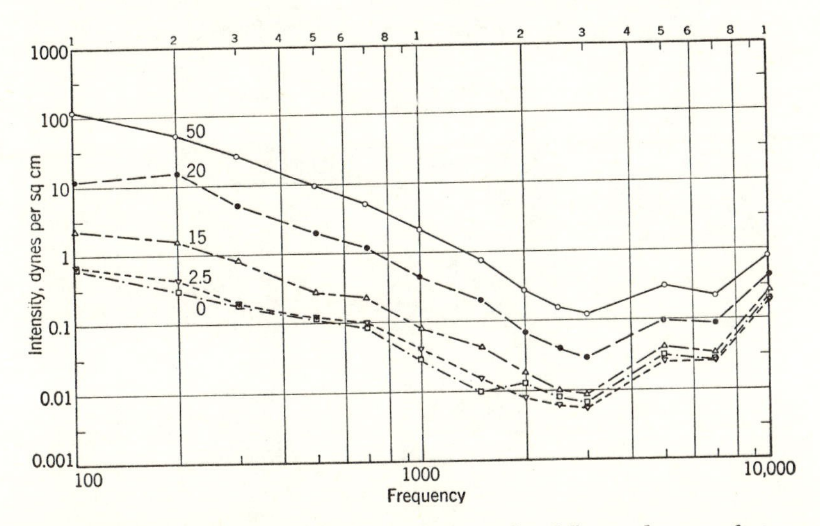

Fig. 77. Equal-response curves for the cat's ear under different degrees of air pressure in the middle ear. For each curve the pressure was constant, and is indicated in millimeters of mercury by the number on the curve. The ordinate shows the sound intensity required under each condition to produce a standard response of 10 microvolts. From Wever, Bray, and Lawrence (7) [*Journal of Experimental Psychology*].

where another method of plotting is used. Here any one curve represents a constant condition of pressure, and at each of several frequencies the tone is presented at the intensity necessary to produce a constant response of 10 microvolts. It is evident that in general the low tones are more seriously affected than the high tones. A pressure of 50 mm of mercury reduces the sensitivity to a 100~ tone by 49 db, whereas it reduces that to a 10,000~ tone by only 13 db.

The effects of pressure were studied also during stimulation by bone conduction. Figure 78 gives results for four tones, with the corresponding air-conduction curves for comparison. The bone-conduction functions are obviously different. They show large impairments at the lower values of pressure and then rise, often rather sharply, and level off at the higher values of pressure.

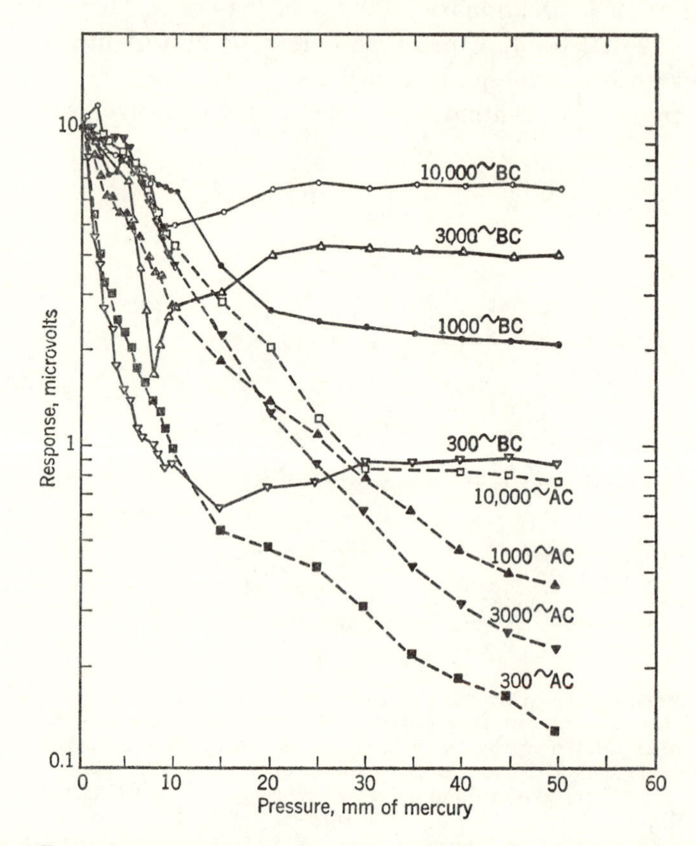

Fig. 78. Effects of positive air pressure in the middle ear cavity on cochlear potentials in the cat in response to tones presented by air conduction (AC) and by bone conduction (BC). From Wever, Bray, and Lawrence (7) [*Journal of Experimental Psychology*].

Similar studies on the effects of negative pressures were carried out by Wever, Lawrence, and Smith (2). Figure 79 shows the pressure curves obtained for a number of tones presented by air conduction at an intensity that initially produced a response of 100 microvolts. Figure 80 gives equal-response curves for the normal ear and for an ear exposed to a negative pressure of 30 mm

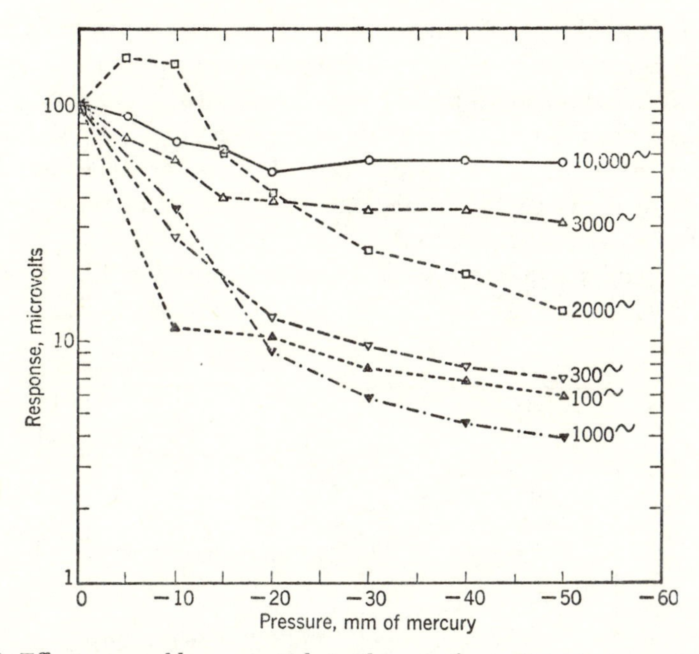

Fig. 79. Effects on cochlear potentials in the cat of negative air pressures in the middle ear cavity. From Wever, Lawrence, and Smith (2).

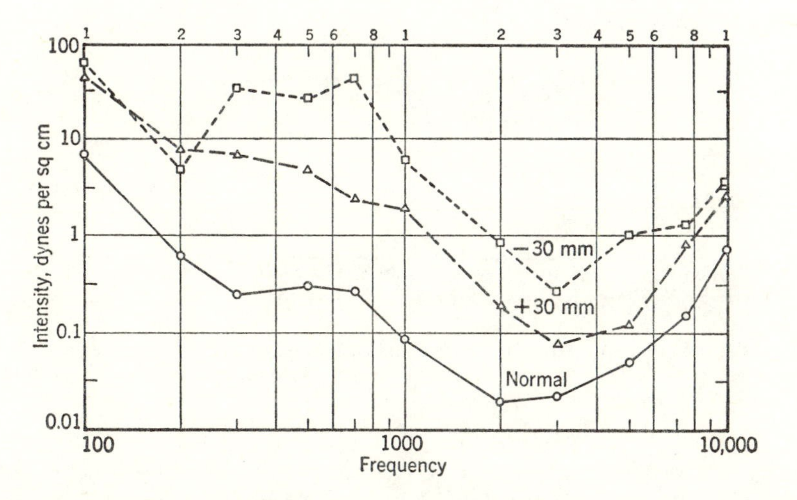

Fig. 80. A comparison of the effects of positive and negative pressures. Each curve shows, for normal conditions or at pressures of +30 mm or −30 mm of mercury as indicated, the sound intensity required to produce a standard response of 10 microvolts. From Wever, Lawrence, and Smith (2) [*Annals of Otology, Rhinology, and Laryngology*].

of mercury. The graph includes for comparison the curve obtained on this same ear for 30 mm of positive pressure. In this instance the effects for negative pressure are greater than those for positive pressure, though in other ears the reverse is found, or there is little difference between the two types of pressure. The relation to frequency is much the same, with the low tones more seriously affected than the high tones.

The results obtained with negative pressures for bone-conducted tones, as shown in Fig. 81, likewise resemble those for

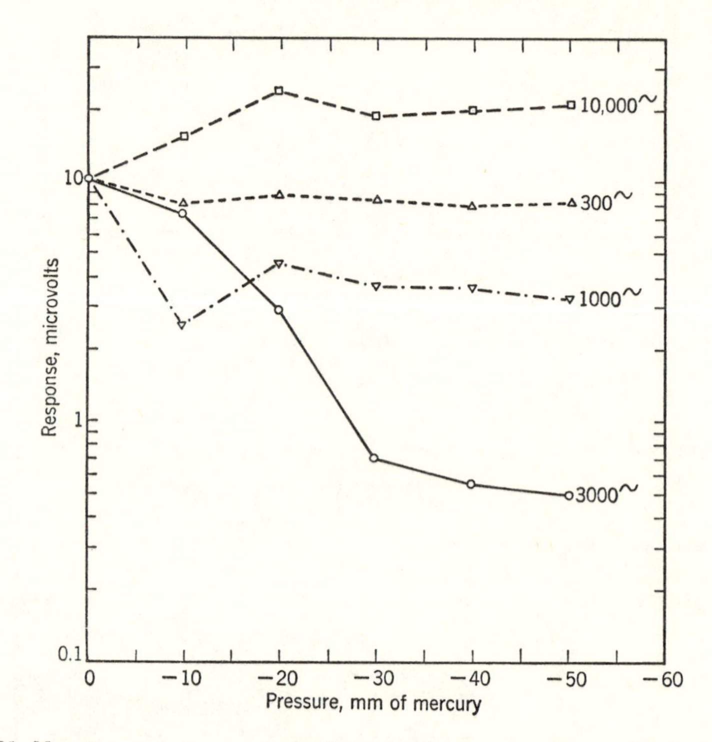

Fig. 81. Negative pressure curves for bone-conduction stimulation observed in the cochlear potentials of the cat. From Wever, Lawrence, and Smith (2) [*Annals of Otology, Rhinology, and Laryngology*].

positive pressures, though with differences of detail. The sharp decline and rebound, which is a prominent feature of Fig. 78, is largely absent, and is seen here in a moderate degree only in the curve for 1000~.

LOCUS OF THE PRESSURE EFFECTS

We still have to consider the locus of the pressure effects. Usually these effects have been attributed to the altered tension of the drum membrane, and there is no doubt that this is the principal condition. The only question is whether there is any effect upon the inner ear.

Effects upon the inner ear might arise in two ways, either directly by action on the oval and round windows or indirectly by movements of the drum membrane that are transmitted through the ossicular chain. It has sometimes been suggested that air pressure in the middle ear might act directly on the stapes and by its displacement cause a change in the intralabyrinthine pressure. Dishoeck obtained evidence against this view by tests on persons with perforations of the drum membrane. Such persons failed to experience any impairments of hearing on the introduction of air pressure into the external meatus. Rasmussen at first entertained the two possibilities that the effects might be due to intralabyrinthine pressure or to increased tension of the round window membrane, but later withdrew these suggestions insofar as they relate to air-conducted sounds. He continued to favor the idea that the effects as seen in bone conduction are due to intralabyrinthine pressure, acting, he supposed, directly upon the organ of Corti. He supported this idea by pointing out that in the otosclerotic ear the application of pressure fails to alter the loudness of bone-conducted sounds, whereas in this ear after a fenestration operation, when mobility of the cochlear fluid is regained, a reduction of loudness occurs.

Results obtained by Wever, Bray, and Lawrence in the study already referred to are pertinent to this question of inner ear effects. In one of the experiments an operation was carried out to exclude the action of the peripheral apparatus. The ossicular chain was severed at the incudostapedial joint and a small portion of the long process of the incus was broken off to prevent a reestablishment of contact between incus and stapes. This operation was carried out by working through the 5 mm hole drilled in the bulla, so that thereafter the bulla could be closed and pressures introduced as before. The results obtained with two amounts of positive pressure, 20 and 50 mm of mercury, are shown in Figs. 82 and 83. In each of these figures the base line represents the normal condition of zero pressure and the curves show in decibels the change of intensity necessary to restore the response to its original level of 1 microvolt after the introduction of pressure. The two uppermost curves in each figure show the effects of pressure on air

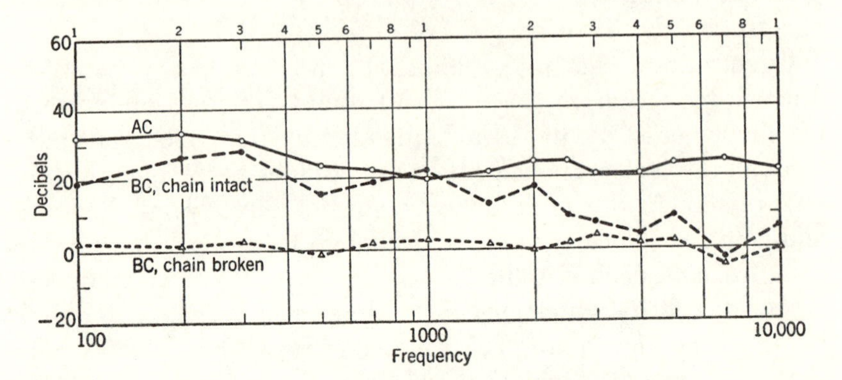

Fig. 82. The effects of a positive air pressure of 20 mm of mercury in the cat's middle ear cavity on cochlear potentials for air conduction (AC) and bone conduction (BC) with an intact middle ear and for bone conduction after breaking the ossicular chain. Positive decibels represent a loss of sensitivity. From Wever, Bray, and Lawrence (7) [*Journal of Experimental Psychology*].

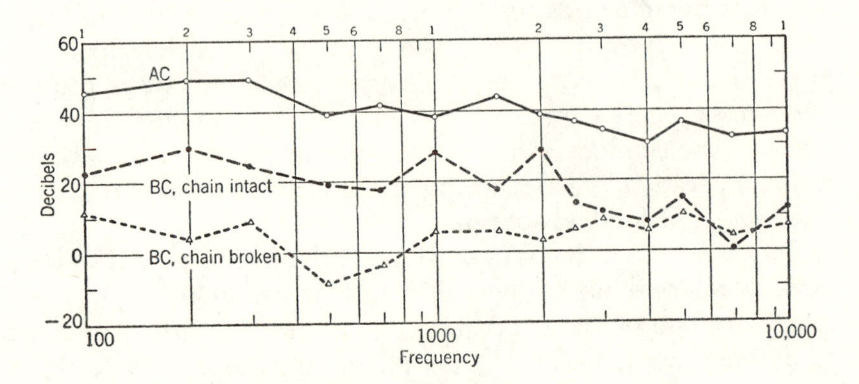

Fig. 83. Results as in the preceding figure for an air pressure of 50 mm of mercury. From the same source.

conduction and bone conduction with the ossicular chain intact and the lowermost curve shows these effects on bone conduction after the chain was broken. The stronger pressure of 50 mm of mercury shows significant changes from the base line under this last condition, though even here the effects are small in comparison with those obtained by air conduction when the middle ear mechanism is intact. It follows that the principal effects of pressure are to be attributed to the action on the drum membrane, but that minor effects are produced directly on the inner ear.

From the form of the changes in sensitivity for air-conducted tones as caused by both positive and negative pressures we can conclude that these pressures increase both the stiffness and the damping of the drum membrane. The increased damping accounts for the general impairment of sensitivity, whereas the increased stiffness causes this impairment to be more severe for the low tones. The changes seen in bone conduction are more complicated but partly reflect these same conditions. The differences between the two bone-conduction curves in Figs. 82 and 83 show that a restraint communicated through the ossicular chain can greatly affect responses to bone-conducted sounds, at times by as much as 25 db. It is notable that this restraint is more prominently shown for the low tones, up to 2500~, and is of slight importance for higher tones. We shall find later, in Chapter 16, that such an effect is typical of artificial immobilizations of the ossicular chain and of the immobilizations of the stapes that occur in the disease of otosclerosis. It is probable also that the slight effects of pressure that are seen for bone-conducted sounds after the ossicular chain is broken are due to a partial immobilization of the footplate of the stapes in the oval window.

PART V

Sound Conduction in the Cochlea

12

The Pathways of Entry to the Cochlea

We have traced the passage of sounds through the peripheral structures of the ear, and now we come to the final and essential step in the mechanical processes, the action upon the cochlea itself. The more peripheral structures may be injured or lost, or as is true of some of the most primitive ears they may never have been present, and yet hearing in some degree is possible. But without the cochlea (or its equivalent as found in the primitive ears) there is no hearing, for deep in the interior of this structure lie the sensory cells whose activities are indispensable to the normal excitation of the auditory nerve fibers. It is now our task to discover how sounds make their final way to these sensory cells and how they act upon them to produce the excitatory processes. Here we meet with fundamental problems that have long been the subject of study and often also the subject of controversy. As we have seen, there are many questions about other aspects of the hearing process and many theories that have been devised to cope with them, but none of these questions has aroused such deep and continued interest as the ones concerned with the cochlea. Indeed, when "auditory theory" is mentioned without any qualification we may be sure that the happenings in the interior of the cochlea are the ones in question.

We can distinguish four problems, closely related and indeed almost inseparable at times, having to do with the mechanical operations of the cochlea. They are (1) the pathways of entry of sounds to the cochlea, (2) the manner of transmission of sounds within the cochlea, (3) the spatial distribution of cochlear action as functions of the frequency and intensity of sounds, and (4) the nature of the local stimulation process. These problems will form the subject matter of this and the two following chapters.

Sounds may enter the ear by either of two routes, by way of the external auditory meatus or by way of the bones of the skull.

The first route concerns aerial waves, and because such waves represent the energy of vibrating objects at a distance this route is much the more important, informing us of acoustic events in our general environment. The second route requires direct contact with a vibrating object and thus gives us information of a much more limited kind.

AIR CONDUCTION

When the ear is normal the aerial sounds entering the meatus impinge upon the drum membrane and pass through the ossicular chain to enter the cochlea at the oval window. In the earlier literature, beginning with DuVerney, there was much consideration of an alternate route for these sounds. This, the aerotympanic route, was traced through the drum membrane and by way of the air of the tympanic cavity to the round window. As we have seen, DuVerney was of the opinion that the ossicular route was the more important, but he considered that there might be an advantage in receiving sounds by both routes at once. In later discussions the aerotympanic route received further emphasis until at times it was regarded as the chief portal to the cochlea. This view has not been wholly abandoned even now, as witness the rather remarkable hypothesis of MacNaughton-Jones (1940). Alexander is one of the later writers to take a compromise position, holding that the two routes share the responsibility for sound transmission, with the ossicular route serving for low tones and the aerotympanic route serving for high tones.

Early experiments dealing with this question of an aerotympanic route were carried out by Mach and Kessel (3), Buck, Burnett, and Weber-Liel. These experiments consisted mainly of the microscopic observation of the round window membrane under strong illumination, often after strewing starch grains on its surface, while exposing the ear to slow pressure variations or to intense sounds. Movements of the membrane were visibly demonstrated under these conditions. However, it was disputed whether these movements were caused by sounds affecting the membrane directly or only after traversing the ossicular chain and cochlear fluid.

A more recent study was made by Link, who placed small mirrors on the round window membrane in cadaver ears and observed deflections of a beam of light when the ear was exposed to sounds. He was able to obtain effects for all tones up to 3000~, with a maximum around 500 to 900~. Link carried out various tests to ascertain whether the movements of the round window membrane resulted from sounds communicated to it directly from the air or by way of the ossicular chain. He found that breaking the ossicular chain caused the movements to disappear. Withdrawing fluid from the cochlea reduced the movements and then eliminated them when the fluid removal was complete. Partially filling the tympanic cavity with water so as to impede the movements of the drum membrane caused the movements to be reduced. On the other hand, he found that the movements continued unabated when a wall of paraffin and cement was built about the round window to shield it from the sound while still permitting the sound to have access to the drum membrane. He concluded that sounds do not affect the round window membrane directly.

Results that have already been presented in our consideration of the ear as a mechanical transformer prove the essential correctness of Link's observations and also show the reason why the round window route of entry for sounds is unimportant in the normal ear. The reason is simply that the middle ear mechanism provides a suitable matching of impedance for aerial waves and thereby facilitates a transfer of energy to the inner ear fluid, whereas the round window membrane, stiffened by the fluid behind it, presents a high impedance to sound waves and reflects a greater part of their energy. The difference is great enough to make the round window route negligible by comparison. This is true for all tones throughout the audible range, as shown in Fig. 84. On the average the round window route is inferior to the ossicular route by about 30 db, and even at 100~ where the difference is least it is still inferior by 14 db. From these data we can easily calculate the maximum effect of having the sound reach the cochlear fluid by the round window route as well as by the ossicular route. If for a given sound the pressure conveyed to the cochlear fluid by the ossicular route is, let us say, 1 dyne per sq cm, and we suppose as a maximum figure that the sound reaches the round window with the same intensity that it has at the drum membrane, then the pressure conveyed to the cochlear fluid by the round window route will be on the average 30 db less, or 0.03 dynes per sq cm. If these two pressures are in such a phase relation as to be fully additive in their effects upon the hair cells the total pressure will then be 1.03 dynes per sq cm, whereas if the phase relation is the contrary of this so as to make the effects

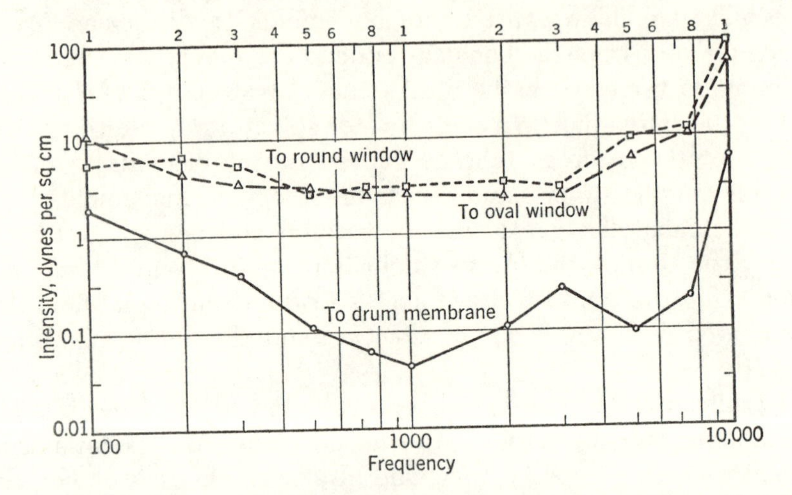

Fig. 84. Comparison of the ossicular and aerotympanic routes to the cochlea. The curves show the sound intensity necessary to produce a standard response of 10 microvolts in the cat when the sound was delivered to the drum membrane in the ordinary way and when it was delivered directly to the oval window or to the round window after removing the peripheral structures and sealing a tube over the window. From Wever, Lawrence, and Smith (*1*) [*Annals of Otology, Rhinology, and Laryngology*].

subtractive the total will be 0.97 dynes per sq cm. In either case the combined pressure will vary from 1 dyne per sq cm by only 0.3 db, which is imperceptible. If we make the calculation for 100~ where the difference between the two routes is the least the total pressure will be 1.2 dynes per sq cm at the maximum and 0.8 dynes per sq cm at the minimum, and the variation from the pressure produced by the ossicular route alone is +1.6 or −1.9 db, which still is negligible for practical purposes.

When the middle ear is impaired so that its transformer action ceases this overwhelming advantage held by the oval window route is lost. The two windows now operate on more nearly the same level of efficiency, and we can no longer ignore the round window as a route of entry. The relative efficiency of the two windows, when the sounds reaching them are of equal intensity, is seen in a comparison of the two uppermost curves of Fig. 84.

How the oval and round window routes interact in the defective ear and what forms the sensitivity function takes will depend upon the particular conditions, such as the degree of access of the sound to these windows and the remaining mobility of the stapes. The types of auditory impairments appearing in this situation will be discussed later, in Chapters 15 and 16. Here we are concerned only with the basic principles of the interaction.

Interaction of Oval and Round Window Pathways. That sounds entering the cochlea by oval and round windows combine their effects in a particular manner has already been shown in our study of phase distortion produced by the stapes. There we used a method involving the cancellation of the effects of a tone along one pathway by the same tone introduced along the other pathway at the proper intensity and phase relation. We may now formulate the general principle governing this combination of pathways: the stimuli applied in this manner combine their effects like vectors, just as any forces do that have direction as well as magnitude.

Figure 85 shows the nature of this vectorial addition under the simplest conditions: when the two stimuli have the same frequency and are adjusted in intensity so as to give the same response when presented separately. The stimulating frequency was 1000~ and the intensities were adjusted to give 10 microvolts of response for each route. On stimulating by both routes at once the resulting response varied according to the phase relation, as shown in the figure. Here the maximum response is 20 microvolts, at a phase relation that is arbitrarily called zero phase. Other phase relations gave smaller responses as shown, with a minimum at 180°. This minimum was 0.5 microvolts, representing a slight imbalance of the separate responses plus a small amount of background noise.

When the responses obtained by the separate pathways differ greatly in magnitude the combined effects still vary with the phase relation and give maximum and minimum values, but the range of variation is less than that shown here. The combination still represents the vector sum of the separate responses.

This generalization fails, however, when the stimulus intensity

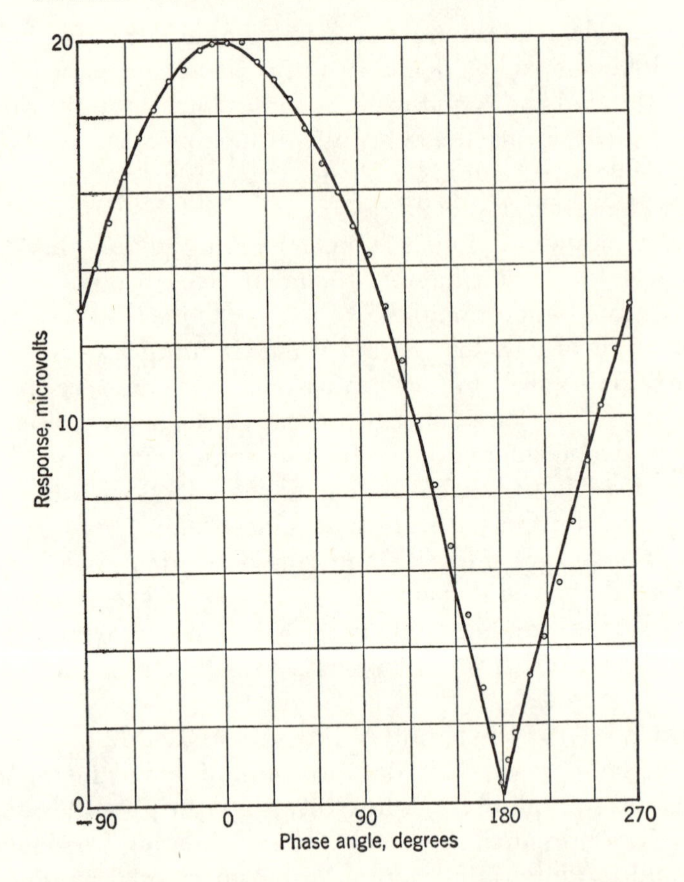

Fig. 85. Vectorial summation of stimuli introduced simultaneously by way of the oval and round windows, shown in the cochlear responses of a cat's ear. The phase relation of the two pathways was varied over a range of 360° as shown on the abscissa. Here the points represent the observations and the curve represents the theoretical summation. From Wever and Lawrence (6) [*Journal of the Acoustical Society of America*].

is raised to the point where the ear is overloaded. Then, as we know, the response is nonlinear and an increase of intensity does not lead to a proportionate increase in response but may even lead to a decrease. Overloading appears in the combined response for two pathways just as it does for one pathway. Figure 86 gives an example. As for the preceding figure the stimulating frequency was 1000~, but now the intensity was adjusted to give a response of 100 microvolts by each route separately. The two routes in combination then gave the responses plotted, which depart widely from the vector sum at high levels and never exceed 110 microvolts.

We have already concluded from other evidence (Chapter 9) that overloading does not occur in the fluid pathways of the

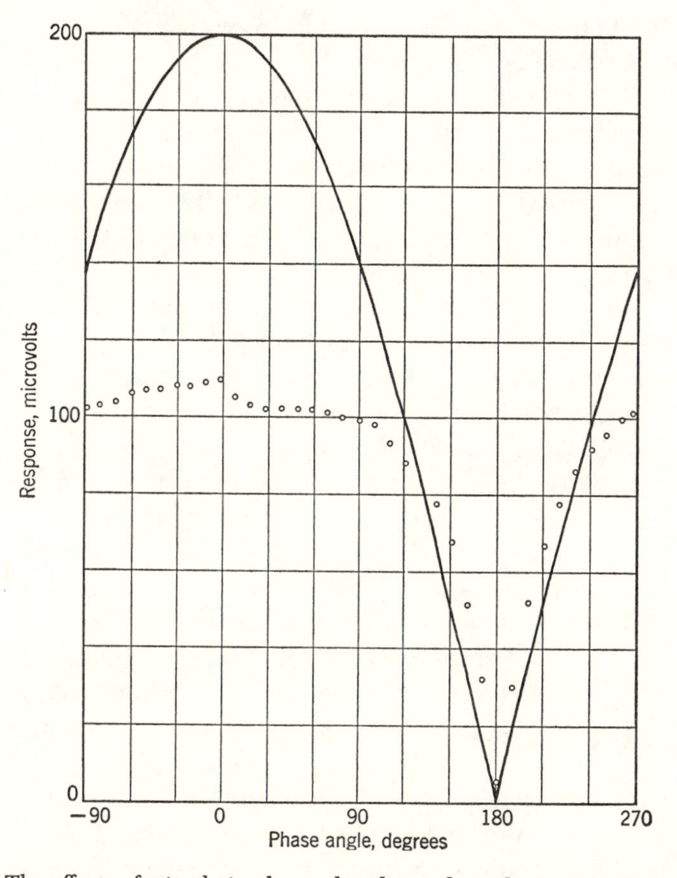

Fig. 86. The effects of stimulation by oval and round windows as in the preceding figure except for the presence of overloading in the cochlear response. As before the points represent the observations and the curve represents the vector sum of the separate responses. From Wever and Lawrence (6) [*Journal of the Acoustical Society of America*].

cochlea but takes place in the sensory structures and probably in the hair cells. The results just presented therefore indicate that sounds entering by the oval and round windows are reaching the same points of overloading in the cochlea and are acting upon the same distorting hair cells. The following experiments extend this suggestion and prove that the overlapping of stimulation is true for all points in the cochlea and for all hair cells.

Let us first examine our problem a little more closely. Specifically, the problem is whether the pattern of stimulation on the basilar membrane is the same when the sound enters by way of the round window as when it enters by way of the oval window. Consider first the consequences of the negative case. If these patterns are different then we should have to explain the results just given as follows. A sound entering by one route excites a certain group of sensory cells and elicits a certain magnitude of cochlear potentials, whereas a sound entering by the other route excites another group of sensory cells and elicits a magnitude of potentials that by a suitable adjustment of the sound intensity may of course be made the same. When both routes of stimulation are used the two potentials are radiated from their cells of origin through the surrounding conductive tissues and are combined in this radiating field. An electrode introduced anywhere in this field picks up the combined effect. Because each separate potential represents the phase as well as the magnitude of its stimulus the combined effect will be the vector sum and a function like that shown in Fig. 85 will result from a variation of the phase relation. However, if the two groups of cells are situated at different places in the cochlea as has just been supposed, this function will be found for some given conditions of stimulation at only one position in the radiating field. If the electrode is moved to some other position nearer to one group of cells and farther from the other group the former balance of magnitudes will be upset, for now the potentials from the nearer group will be enhanced and those from the farther group will be diminished. It is easy to show, on stimulation by a single route, that moving the electrode to new positions about the cochlea causes large variations in the recorded potentials. Therefore, on the assumption that we have made, we can restore the balance of our two routes only by a new adjustment of the stimulus intensities. On the other hand, if the two routes of stimulation involve the same sensory cells the balance will remain for all points in the field. Moving the electrode to a new position may change the magnitude of potentials recorded if this new position is farther away in general from the active cells or if tissues of greater resistance are interposed, but this reduction will be true for both pathways alike and the original balance will remain. Hence from these considerations we can make a crucial test of our two possibilities.

For this test we used the cat as the experimental animal, with one sound tube connected with the external auditory meatus and another sound tube sealed over the round window. The apparatus arrangement was that described previously and shown in Fig. 44. For recording the potentials a needle electrode was used instead of the usual foil electrode. At first the needle was placed at the basal end of the cochlea immediately adjacent to the sound tube leading to the round window. As before, while stimulating with some chosen tone, say 1000~, the sound intensities were separately adjusted in the two stimulating channels to produce some desired level of response. Then with stimulation by both routes the phase was adjusted for a minimum of response. The electrode was then moved to some new position on the surface of the cochlear capsule, say, toward the middle of the cochlea or to its apex. Invariably it was found that no new intensity or phase adjustments were necessary for a minimum at this new position. The original settings continued to produce a minimum for any electrode position whatever. This procedure was repeated for other stimulating tones and the condition just described was found to hold at every frequency throughout the auditory range. Tones of different frequency of course required different conditions for balance because of the sensitivity and phase variations of the two routes, but once these conditions were obtained for one electrode position they held for all others from which the potentials were picked up.

The necessary conclusion is that a tone produces the same pattern of action in the cochlea regardless of whether it is introduced by way of the oval window or by way of the round window. The two forms of stimulation involve the same sensory cells and when acting with equal effectiveness they involve each one of these cells to the same extent.

BONE CONDUCTION

We speak of hearing by bone conduction when vibrations reach the cochlea by way of the bones of the skull. Much attention has been given to the problem of the particular routes of travel of these vibrations from their point of application on the surface of the head. Initially this is a problem of anatomical connections. The pathway considered to be the primary one is altogether osseous from the first skull bone to the cochlear capsule. The other pathways, generally designated as osseotympanic, involve a transfer of the vibrations from the bones of the skull to parts of the middle ear mechanism. A recent outline of these pathways was given by Guild (*4*).

There is a danger of thinking about these pathways in too simple a manner. It is not enough that the sound vibrations merely reach the cochlea. They must do so in such a way as to set up relative motions between the sensory cells and surrounding structures. Only by this means are the sensory cells stimulated. This problem will be considered in detail a little later on, where proof of this principle of stimulation will be given. It was the failure to perceive this principle of relative motion that led to confusion in nearly all of the early discussions of bone conduction.

One example will suffice to show the sort of confusion that was prevalent before this principle was recognized. Lucae in 1864 attached a lever to the ossicular chain of a cadaver specimen and on applying the stem of a tuning fork to the skull he observed vibratory movements of the lever. He thereupon concluded that the ossicular chain is involved in bone conduction. His conclusion happens to be correct for the usual conditions, but it does not follow from his observations. Indeed, these observations—showing that both ossicles and skull are vibrating—lead more logically to the opposite conclusion. We can go still further and say that if the head and all its contents are executing the same movements then there can be no stimulation at all.

A number of writers, including Mach and Bezold, came very near to an appreciation of the importance of relative motion in bone conduction, but the first to achieve a clear formulation of the principle was Bárány in his penetrating treatment of 1938. With this principle in mind let us return to our problem of the pathways of bone conduction.

The sounds transmitted along the osseous pathway to the cochlear capsule can act upon this capsule and its contents in at least two ways, which we call translatory and compressional modes of vibration.

The Translatory Mode of Bone Conduction. In translatory bone conduction the cochlear capsule is shaken bodily back and forth. All of its particles move in the same direction at any instant. However, this motion is not followed exactly by the cochlear contents. Because of inertia these contents tend to follow the capsular movements only with a time lag. They are able to depart from the capsular movements because the capsular walls are incomplete at the oval and round windows and the fluid can pass in and out of these windows to a slight extent. This manner of stimulation is represented in Fig. 87.

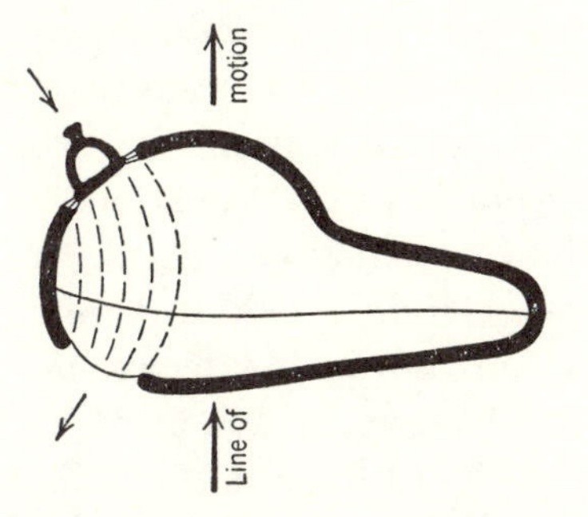

Fig. 87. The translatory mode of bone conduction. When the head is moved along the line indicated by the heavy arrows the cochlear fluid is displaced toward the round window and the basilar membrane is depressed. From Wever (*6*) [*Annals of Otology, Rhinology, and Laryngology*].

The inertia that opposes the vibratory force exerted by the capsule is of two sorts, the inertia of the cochlear contents themselves and the inertia of the ossicular chain. These two inertia forces operate in slightly different directions.

(1) When the motion is in line with the oval and round windows the fluid moves alternately toward one window or the other, now bulging out the membrane of the round window and then pressing out the footplate of the stapes. The basilar membrane lies in the path of this fluid displacement and is displaced in the process.

(2) The inertia of the ossicular chain as a whole restrains the movement of the stapes so that as the cochlear capsule is moved the stapes lags behind. The result is much the same as in ordinary aerial stimulation when the capsule remains stationary and the stapes vibrates; in either case the cochlear contents are displaced relative to the capsular walls. The displacement takes a path

through the basilar membrane on its way to the round window and in doing so it stimulates the sensory cells. This form of stimulation requires a translatory oscillation along the axis of free movement of the ossicular chain, which axis as suggested earlier is probably perpendicular or nearly perpendicular to the footplate of the stapes. An oscillatory movement in any other direction is effective only so far as it has a component in this direction.

It is important to note that the part taken by the ossicular chain in the translatory form of bone conduction is not a simple one. The ossicular system is not merely a mass, but its movements involve elasticity and friction as well. Such a complex system possesses distinctive resonance characteristics that enter into its actions. The system is more free to move at some frequencies than at others and its movements assume various phase relations to the driving forces as the frequency is altered.

The Compressional Mode of Bone Conduction. In compressional bone conduction the cochlear capsule undergoes periodic variations in volume. Herzog, to whom this explanation of bone conduction is due, seems to have implied that the whole capsule expands and contracts in synchronism with the rarefactions and condensations of the sound waves. Actually, this mode of vibration can be realized in a solid only if it is contained in a denser solid whose flexions compress it equally from all sides. That this condition is satisfied for the cochlear capsule is at least doubtful. It is more likely that the cochlear capsule along with the whole petrous bone takes part in the flexural movements of the side of the head (Békésy, *18*). These movements act upon the cochlear capsule mainly from two sides and thereby alter its volume. Because the cochlear contents for all practical purposes are incompressible the changes of volume cause fluid displacements in and out of the cochlear windows. A compression forces the fluid out and a rarefaction draws it back.

It is essential for stimulation that these fluid movements be asymmetrical with respect to the two windows and thus with respect to the basilar membrane. There are two reasons for such asymmetry. The quantity of fluid lying on the vestibular side of the membrane exceeds that on the tympanic side, in man by a ratio of 5 to 3. Also, the oval window is partially restrained by the presence of the stapedial footplate and its attachment to other parts of the ossicular system. Therefore a large fraction of the displaced fluid takes a path through the basilar membrane toward the round window, and it moves the basilar membrane up and down in the process. This form of action is shown in Fig. 88.

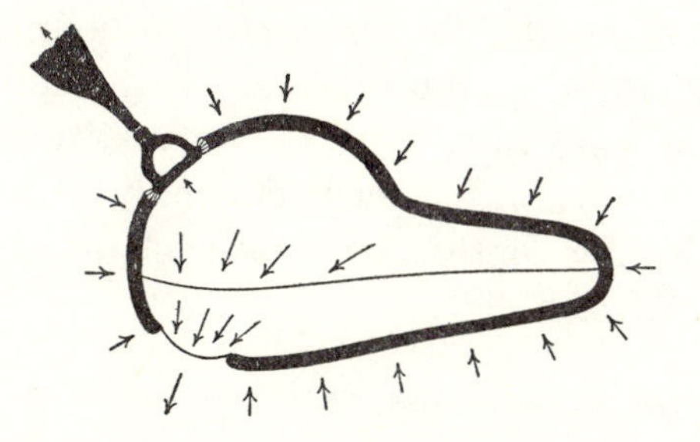

Fig. 88. The compressional mode of bone conduction, as conceived by Herzog. When the cochlear capsule is compressed from all sides its fluid contents are pressed out, mainly taking the path of least resistance through the round window. From Wever (6) [*Annals of Otology, Rhinology, and Laryngology*].

Secondary Pathways of Bone Conduction. Of the various possible secondary pathways the ones with strongest claims for special consideration are the following:

(1) The vibrations of the skull may radiate sound into the surrounding air, and some of this sound may find its way into the external meatus. Alternatively, the vibrations may pass to the walls of the meatus and here produce aerial waves. In either case the sound thereafter acts on the drum membrane like any other aerial stimulus.

(2) The vibrations may pass to the walls of the tympanic cavity and set up waves in its contained air. These waves have often been thought of as entering the round window directly, but in the normal ear, for reasons already made clear, they will be far more effective in acting upon the tympanic membrane.

(3) The movements communicated to the walls of the external meatus and tympanic cavity may move the tympanic membrane through its annulus or move the ossicles, especially the incus, through their suspensions. In this course there is no aerial stage. This method is essentially the same as the translatory form of bone-conduction stimulation except for possible differences in the phase of the movements of the annulus or ossicles and the movements of the cochlear capsule.

(4) Another form of inertia stimulation, proposed by Békésy (3), is based on the idea that as the skull moves, the lower jaw remains relatively stationary and effectively produces an alternating compression of the external auditory meatus.

Unquestionably many pathways operate in the transmission of bone-conducted sound. The problem is to determine the relative importance of these pathways and their mutual interactions. When multiple pathways exist they can summate in either a favorable or an unfavorable sense depending on their phase relations. We shall now consider the experimental evidence pertaining to these pathways.

EVIDENCE ON THE MODES OF BONE CONDUCTION

Bárány obtained evidence on the production of aerial waves during bone-conduction stimulation. He measured these waves by means of an acoustic probe placed close to the opening of the external meatus, and found their effective amplitude (measured in terms of their stimulating effect) to be about half that of the bone-conducted sound under his conditions, for a frequency of 435~. These aerial waves were produced by the bone-conduction receiver itself, by vibrations set up in the clamp that held the head, and by vibrations of the head. It was not possible to determine exactly the contributions of these different sources, but precautions were taken to minimize the first two, which left the vibrations produced by the head as the most important. Under Bárány's conditions the aerial sound had a phase relation such as to reduce the effective magnitude of the bone-conducted sound by about 40 per cent.

Convincing evidence for the existence of translatory bone conduction was obtained by Bárány. He stimulated with a bone-conduction receiver at various points on the surface of the head and measured the relative effectiveness and also the phase of the resulting sounds. He did this by balancing the bone-conduction sound against an air-conduction sound in the ear being tested, while the opposite ear was excluded by a constant masking noise. His observations were made mostly with a 435~ tone and with the head held firmly by having the person bite upon a cast of the lower jaw. The results varied in detail in different ears, but were fairly consistent for a given ear. Some sample results are given in Fig. 89. Here the head is represented as seen from above, the stimulated ear is on the right, the masked and hence inactive ear is on the left, the forehead is above, and the occiput is below. Each arrow has its origin at a point of stimulation on the head, has a length that represents the relative effectiveness—the loudness—of the tone, and has a direction according to its phase angle

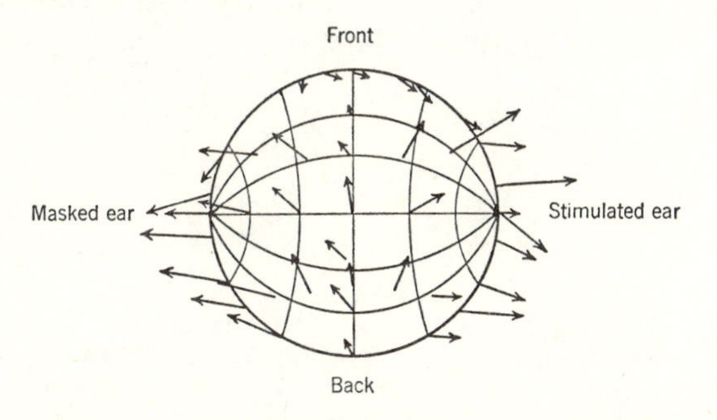

Fig. 89. Effect of the point of application of a bone-conduction stimulus to the surface of the head. The head is seen from above, with the forehead at the top of the drawing. The left ear was masked and the effective loudness of the sound was measured in the right ear. The origins of the arrows represent the points of stimulation, their lengths represent the relative loudness, and their directions represent the phase relations. From Bárány [*Acta Oto-Laryngologica*].

with respect to an air-conduction tone in the same ear. The phase angle is measured as the counterclockwise angle from a line drawn to the right from the origin of each arrow.

It is first to be noted that the longest arrows are on the two sides of the head and the shortest arrows are at the front and back. Thus the most effective bone-conduction stimulation is roughly in line with the axis of the ossicular chain. When the direction of the applied vibrations is at right angles to this axis, as when the bone-conduction receiver is applied to the forehead or the occiput, the effects are at a minimum. The contrary direction of the arrows on the two sides simply means that when the stimulation is contrary in direction the phase is altered by 180°. The stimulation at points on top of the head gave results that varied systematically as shown, with arrows directed to the right or within about 45° of the right for application to the right of the midline and arrows inclined more or less to the left for application at the midline or left of the midline. The pattern is difficult to analyze in further detail.

Results obtained by Békésy (*17*) on the form of vibration of the head are consistent with the ones just mentioned for low frequencies of stimulation. He used a bone-conduction receiver on the forehead and picked up the resulting head vibrations both at the forehead and at the occiput. He found that for all low tones the head behaved as a unitary mass: the forehead and occiput were always found to be moving in the same direction at any instant. This type of motion is illustrated in *A* of Fig. 90. However,

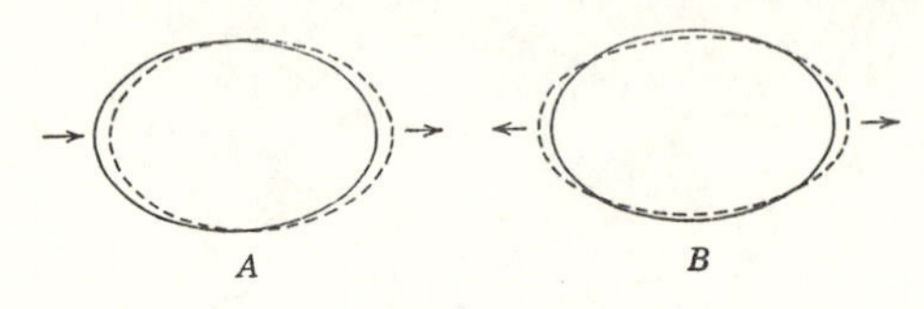

Fig. 90. The form of motion of the head in bone conduction according to Békésy (*17*). For low tones the motion is as represented at *A*, with the head moving as a whole, and for high tones it is as represented at *B*, with opposite surfaces moving in contrary directions.

when the driving frequency was raised above 1000~ the vibratory movement of front and back regions of the head began to get out of phase with one another and at 1500~ they became of opposite phase. This means that as the forehead moves forward the occiput moves backward; the sides of the head then necessarily must be moving inward. This form of movement is shown in *B* of Fig. 90. In the same report Békésy also stated that the translatory motion continued to 1500~ and turned to compressional motion at 1800~. No doubt there are considerable variations among different heads on account of their individual characteristics of mass and rigidity. Still higher frequencies gave forms of compressional vibration that varied in detail because of resonances at certain points. Békésy's results thus show that the translatory form of bone conduction should prevail at low frequencies and the compressional form should prevail at high frequencies. Both forms should be present at intermediate frequencies, those around 1000 to 1500~. This conclusion requires the assumption that the petrous bone takes part in the regular movements of the skull as these are represented in Fig. 90. Actually, because of the peculiar position of this bone in the skull, its movements may not follow any simple pattern. Indeed, it is easy to see from Bárány's observations, shown in Fig. 89, that the vibratory pattern is very complicated.

Smith (*1*) made a study of bone conduction in the cat as affected by a mechanical fixation of the stapes. He measured the change in magnitude of the cochlear potentials produced by a standard sound when by means of a thread the stapes was brought into firm contact with a steel needle that was anchored in the middle ear cavity. The degree of fixation varied in different trials according to the tension on the thread, but some average results are given in Fig. 91. For comparison the effects upon air-con-

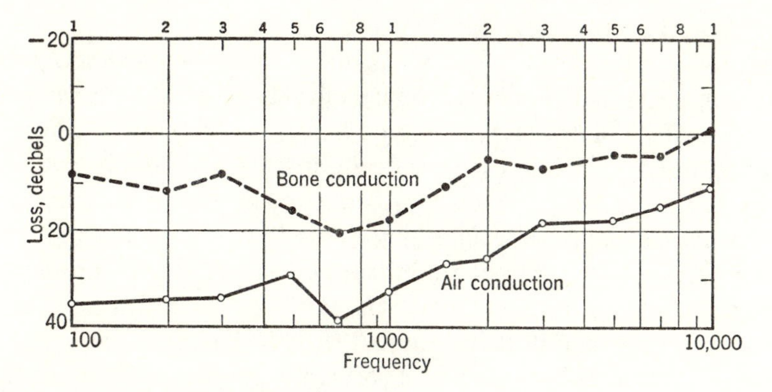

Fig. 91. Changes of cochlear responses for air and bone conduction caused by a mechanical fixation of the stapes. From observations in the cat by Smith (*1*) [*Journal of Experimental Psychology*].

ducted tones are also shown. It is apparent that fixation of the stapes reduces bone conduction for all tones except those at the extreme high-frequency end of the range, and has the greatest effect in the region of 500 to 1000~.

This evidence points especially to the importance of the translatory mode of bone conduction. It is this mode that requires the participation of the stapes and would be impaired when the stapes is fixed. The compressional mode, on the other hand, should be affected little if at all. It depends upon a differential action at the two windows, which in its simplest form is a yielding at one window and an absence of yielding at the other. Theoretically another form of differential action in compressional bone conduction is possible in which both windows yield but with different phases relative to the driving force. However, the following considerations lead us to a discounting of this possibility.

The action just suggested should show certain distinctive variations with frequency, mainly on account of the mechanical characteristics of the middle ear mechanism. The displacement of a simple tuned system changes from phase agreement to phase opposition relative to the driving frequency as this frequency rises through the resonance point. At the same time the phase of the velocity of motion changes from leading to lagging, passing through phase agreement at the resonance point. The form of variation with frequency, under this assumption of an out-of-phase yielding at both windows, will therefore depend on whether stimulation is caused by the displacement or the velocity of basilar membrane movement. If it is caused by the displacement, then the compressional mode of bone conduction should normally be poor for tones below the resonance point, for the stapes should be moving outward when the round window is moving outward, and little or no discharge of fluid will occur across the basilar membrane. At the same time, this conduction should be favorable for tones above the resonance point, for the stapes should be moving inward when the round window is moving outward, and a maximum discharge should occur across the basilar membrane. A fixation of the stapes then would produce a progressive impairment of the responses to the higher tones, which decidedly is not the case. If stimulation is caused by the velocity of basilar membrane movement, then the compressional mode of bone conduction should be the poorest at the resonance point of the middle ear system, and a fixation of the stapes ought to improve the responses in this region. The chief resonance point of the cat's middle ear is around 1000~, and this is just the region where stapes fixation has its most deleterious effects for bone conduction. This hypothesis of an out-of-phase yielding at the two windows thus is not in agreement with the observations.

From all these considerations it appears that the compressional form of bone conduction is unimportant or is not much affected by a fixation of the stapes.

SENSORY STIMULATION BY AIR AND BONE CONDUCTION

A question of fundamental importance is whether the pattern of stimulation of the sensory structures is the same for bone conduction as it is for air conduction. Ordinary observations provide a preliminary answer. A tuning fork is perceived as having the same pitch whether we hold it opposite the external auditory meatus and receive its sounds through the air or whether we press the stem of the fork against the head and receive the sounds by bone conduction. In general, a pure tone has the same quality when heard by either means of conduction, and our ability to discriminate its pitch from others is the same. We can likewise receive complex sounds such as music and speech by both methods, and these sounds too have the same general character in each case. Yet usually a complex sound is not perceived as absolutely identical by air and bone conduction. There are variations in timbre. These variations are due to distortion in the ear of the types mentioned in Chapter 8, which is suffered in different degree by the two forms of conduction. Frequency and phase distortion are the most important conditions here. They operate in different ways for these two types of conduction. Thus the middle ear mechanism imposes its peculiarities of sensitivity and phase upon the aerial sounds, as shown in Figs. 41 and 47, and the various tissues forming the bone-conduction pathway, chiefly the skull, impose their peculiarities upon the bone-conducted sounds.

Actually, it is difficult to obtain complex stimuli for air and bone conduction that are truly identical, for distortion is introduced by the apparatus used to convert aerial into mechanical vibrations. If a high level of stimulation is required the bone-conduction sounds are likely to suffer not only from frequency and phase distortion but from nonlinear distortion as well because of the heavy energy requirements of this type of stimulation.

It is fortunate that auditorially we can tolerate a considerable amount of distortion without too serious a result. A pattern of complex sound remains recognizable in spite of changes in relative intensities and phases of the components. Even the introduction of additional harmonics from nonlinear distortion is not serious unless carried rather far, because the harmonics themselves maintain the interrelations originally present in the fundamental components of the sound. Hence we are able to comprehend speech and music when represented by bone conduction practically as easily as by the more familiar aerial means. In dealing with speech we are doubtless aided by the fact that we hear our own voices about equally by both methods (Békésy, *18*).

Further evidence for the similarity of the cochlear processes in air and bone conduction is provided by experiments carried out by Békésy, Lowy, and ourselves. It has no doubt been observed many times that an air-conducted tone can be made to beat with a bone-conducted tone when both tones, differing a little in frequency, are presented to the same ear. Békésy (3) carried out what amounts to a refinement of this demonstration. He presented a bone-conducted tone of 400~ to the middle of the forehead and the same tone by air conduction through receivers opposite both ears and then varied both intensity and phase relations until the tone disappeared. From these results he concluded not only that the two forms of stimulation involve the same final auditory pathways but that both produce the same patterns of displacement in the cochlea.

Evidence that the cochlear patterns are the same comes from measurements of the cochlear potentials produced by these two forms of stimulation. The cochlear responses show the same functional relations to the intensity of the stimulus. This functional identity was first pointed out by Wever and Bray (2) and used as a basis for the conclusion that air and bone conduction involve the same sensory receptor cells in the inner ear.

Lowy in experiments on cats and guinea pigs was able to obtain a mutual cancellation of cochlear potentials produced by air and bone conduction. He led an aerial tone into the meatus and stimulated at the same time with a vibrator in contact with a bone of the skull, and he found that with suitable adjustments of intensity and phase the responses could be made to disappear. He showed further that once the conditions were established for this cancellation the recording electrode could be moved to a new position on the cochlear capsule without altering the cancellation. These observations were made for various tones over a range of 250 to 3000~.

We have confirmed these observations, using tones from 100 to 15,000~ and various stimulating intensities up to the entrance of overloading. In general it may be said that the two stimuli combine their effects vectorially, provided that their total intensity is below the level of overloading. It is therefore clear that any given tone involves the same sensory cells in the same pattern of stimulation regardless of whether it is applied by air or bone conduction.

THE FUNCTION OF THE ROUND WINDOW

Eduard Weber more than a century ago presented the correct view of the function of the round window. He opposed the older notion that this window is a path of entrance of sounds to the cochlea, and made the proper argument that sounds traveling by this route would be seriously attenuated in their successive transfer through media of different densities. Instead, he said, the round window is a relief opening to the labyrinth that permits the contained fluid to move under the influence of the stapes. The thrusts of the stapes are transmitted to the fluid and cause it to move bodily, and the membrane of the round window is displaced correspondingly. In this process the sensory structures are stimulated.

Unfortunately, the experiments carried out subsequently in efforts to test this theory gave varied and confusing results. Many early attempts, some of which have already been referred to, were made to ascertain whether the round window really moved in response to sounds communicated along the ossicular chain as Weber's view required. The careful investigation made by Link, already described, led to the clear conclusion that in the normal ear a sound has little direct effect upon the round window membrane. As Weber had asserted, this membrane moves in response to the movements of the cochlear fluid.

EFFECTS OF BLOCKING THE ROUND WINDOW

Further studies of round window function were made by observation of the effects of blocking this window. In rare instances such blocking has been observed in a histological examination of the ears of persons considered to have had normal hearing. Moos in 1871 and Habermann in 1901 reported cases in which the round window niche was filled with bone. Oppikofer described one in which the window was occluded by a mass of fatty tissue. Hallpike and Scott described one in which otosclerotic bone covered about 70 per cent of the membrane and dense fibrous tissue covered the remainder. Yet on the basis of simple tests—watch, whisper, and fork tests—most of these ears were regarded as functionally normal.

Link described another case in which deafness was complete in both ears, and examination after death revealed a bony closure of both round windows. However, ankylosis of the ossicles was present as well, and no doubt was responsible in large part for the functional losses.

Link reported also the results of tests on six patients with perforations of the drum membrane large enough to permit access to the round window niche. He inserted a plug of oil-soaked cotton in the niche and found no alterations of acuity. However, when he inserted a bent probe into the niche and pressed lightly on the cotton plug he found that four of the six persons tested in this way obtained a slight but unmistakable improvement of hearing for tones between 256 and 1024~. The other two experienced no change of acuity.

A number of experiments on round window blocking have been carried out on animals. A series of experiments dealing with this problem by the cochlear response method, with cats as the experimental animals, was initiated by Hughson and Crowe. In 1931 they reported that pressure exerted on the round window membrane by means of a plug of moist cotton caused a marked increase in cochlear potentials. However, a simple blocking with cement had no appreciable effect. Later they reported that ears in which periosteal grafts were implanted in the round window niche showed clear improvements of responses when tested after varying periods from 2 days to 7 weeks. Hughson then used fascial tissue grafts in the round window as a means of improving the hearing of partially deafened persons. His analysis of the results obtained in 25 operations led him to conclude that significant improvements had been obtained. Wishart, however, in a discussion of these data, regarded the observed changes as within the usual variations of repeated audiograms. Two facts, that the improvements developed gradually after the operation and that such improvements were almost as common in the unoperated as in the operated ear, suggest the presence of a practice effect as the chief factor in the variations of hearing.

Culler, Finch, and Girden repeated the blocking experiments on dogs, with slight changes of technique. They filled the round window niche with a plug of soft gum wrapped in gauze and attached a thread so that the plug could be quickly withdrawn. Their measurements of cochlear responses with the plug in place were followed at once by measurements after the plug was removed. Invariably they obtained poorer responses when the plug was in the window, to an amount of about 10 db. Moreover, when they used a conditioned response method to test the hearing of the animal with and without the plug they obtained the same difference of 10 db in favor of the normal condition.

On the basis of further investigations these authors suggested that Hughson and Crowe's results in their experiments on the cat may have represented an electrical artifact, a modification of the surface conductivity of the cochlea from the application of the moist cotton. The manipulations, they thought, may have provided a more suitable path for the leakage of currents from the interior of the cochlea to the electrode, which in Hughson and Crowe's experiments was remotely located in the cranial cavity in the vicinity of the internal auditory meatus. When, as in the experiments of Culler, Finch, and Girden, the electrode was placed in contact with the cochlear capsule, the results varied according to the particular procedures; sometimes the potentials were increased and sometimes they were decreased by the application of saline solution in the round window region. These authors concluded that the actual hearing of the animal was always diminished by the damping or immobilization of the round window membrane.

Milstein carried out experiments with cats in which the round window was plugged with various tissues, such as bone, fat, muscle, and fascia. He reported that all of these tissues except the fascia led to the appearance of labyrinthitis or destruction of the round window membrane. His histological examinations after the fascia implantations revealed a normal membrane, except in two instances. Therefore he resorted to injuries about the membrane in an effort to obtain a tissue growth that would block the window. Actually, infection set in and produced this desired result in four animals, as later histological study proved. These four animals were exposed to an intense tone of 8000~ for 12 hours daily over a period of 8 to 12 weeks. A histological study then revealed serious injury to the organ of Corti in the unoperated, normal ear, but no such changes were found in the ear in which the round window was blocked. Milstein interpreted these observations as showing an impairment of sound conduction as a result of the thickening and possible fixation of the round window membrane. There is also a strong possibility that ossicular conduction was impaired by the infective processes.

Further Experiments on Round Window Blocking. We have used a number of procedures to study the effects of blocking and immobilizing the round window membrane (Wever and Lawrence, 2).

Series 1. The simplest method of immobilization is the manual application of a probe to the membrane. We used a rod whose end was half the diameter of the membrane and applied it to the center of the membrane with sufficient pressure to produce a noticeable indentation. When this was done during the observation of cochlear potentials in response to a sound not the slightest effect was seen. Increasing the pressure within the range that the membrane is capable of withstanding without rupture caused in some instances a very slight reduction of the responses.

In carrying out this experiment it was found necessary to use a probe of nonconducting material in order to prevent short-circuiting the potentials. Our probe was made of a synthetic plastic material. It was also found necessary to keep the shank of the probe small to avoid a disturbance of the sound field.

Series 2. In a second method a slender plastic tube was placed over the round window. This tube was 100 mm long and 2 mm in outside diameter, and its bore was equal in size to the round window niche. It was tipped with rubber to make an air-tight seal against the bone and its other end was connected with a rubber tube about 4 meters long.

It was found that the application of this tube had little or no effect on the responses to low tones but caused diminutions of responses to high tones. These changes were perceptible around 750~ and increased progressively with frequency to about 10 db at 10,000~, as shown in Fig. 92. The farther end of the rubber tube could be either open or closed, and the pattern remained the same. When the tube was closed at the end and an air pressure up to 25 mm of mercury was introduced the responses were affected only slightly.

What is happening is that the cochlea is being loaded with an air column whose inertia is added to that of the cochlear fluid. This mass loading produces a progressive impairment of responses to high tones. Raising the air pressure by 25 mm of mercury only

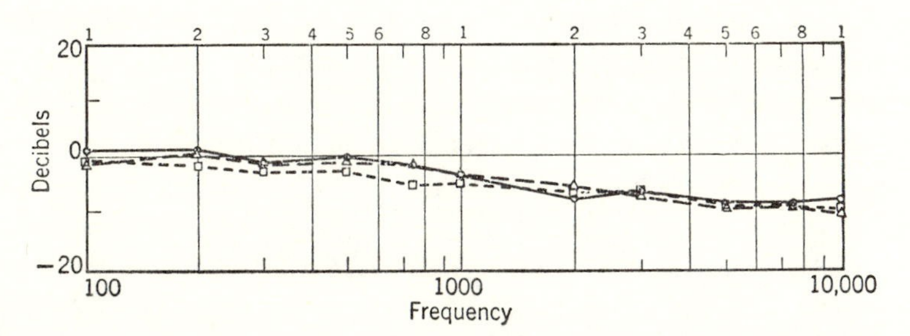

Fig. 92. Effects of occlusion of the round window with a tube. Negative decibels represent a loss of sensitivity. Results are given for three cat ears. From Wever and Lawrence (2) [*Annals of Otology, Rhinology, and Laryngology*].

increases the air density by 3.5 per cent and thus has little effect.

Series 3. A third method of immobilizing the round window consisted of packing the round window niche with bone wax, usually with a few cotton fibers intermingled to make it easier later to remove the plug as a unit. The plug was pressed firmly into the niche and held by a prod attached to a mechanical manipulator. The results of this procedure are shown in Fig. 93.

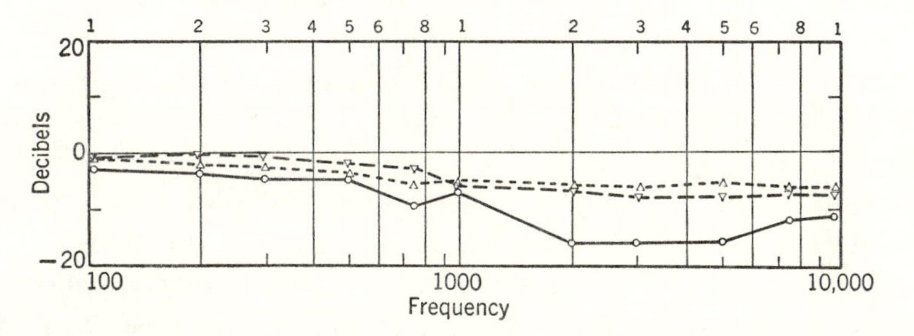

Fig. 93. Effects of plugging the round window niche with wax when the middle ear apparatus was intact. Results are given for three cat ears. From Wever and Lawrence (2) [*Annals of Otology, Rhinology, and Laryngology*].

As before, the low-tone responses are affected but little and the high-tone responses are reduced in progressive fashion. Again it appears that plugging the round window niche adds mass to the responsive system.

Series 4. In the three series just mentioned the ear was intact except for the necessary opening of the bulla. In a fourth series the lateral wall of the tympanic cavity was dissected away and the drum membrane and outer ossicles were removed. Sometimes the stapes was left intact in the oval window and at other times its crura were removed and only the footplate remained. The sound tube was sealed over the oval window and stimuli delivered directly to this site. A blocking of the round window then gave the results shown in Fig. 94.

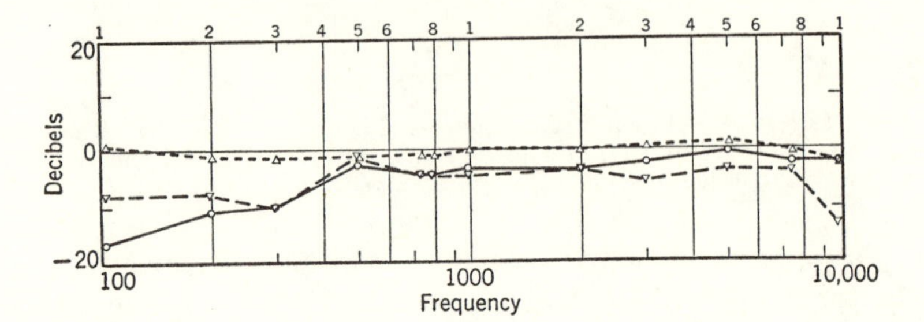

Fig. 94. Effects of plugging the round window niche with wax after removal of the peripheral structures of the middle ear. Results are given for three cat ears. From Wever and Lawrence (2) [*Annals of Otology, Rhinology, and Laryngology*].

The pattern of changes now differs from that obtained by the preceding methods. A slight loss of 2 or 3 db appears throughout the frequency range, and in two out of the three ears there is a greater loss for low tones.

Series 5. We have come to the conclusion that the slight effects shown in all the blocking experiments heretofore are due to the difficulty in producing a true blocking of the round window. Ordinary plugs inserted into the round window niche simply vibrate bodily with the membrane and do no more than add to the mass of the vibrating system. Even a heavy layer of substance over and around the window is of slight effectiveness if, as can easily happen, it entraps a minute amount of air next to the membrane, for then the membrane can move in and out of this air bubble. The amount of air required to permit this movement can be estimated from a consideration of the volume displacement of the cochlear fluid for some representative sound. For a 1000~ tone of moderately strong intensity (1 dyne per sq cm) this volume displacement is at most about 3 ten-millionths of a cubic millimeter—the volume of a sphere of about the diameter of a red blood corpuscle. An air bubble need be only a million times this size—about a third of a cubic millimeter—to allow almost complete freedom of the fluid movements.

In further experiments (Wever, *6*) we have improved our technique of blocking. In this effort we have used firm wax and have packed it into the niche after thorough drying of the surfaces. We have used a large quantity of wax because the effectiveness of the blocking increases progressively with the mass of material. Figure 95 shows two degrees of this heavy blocking.

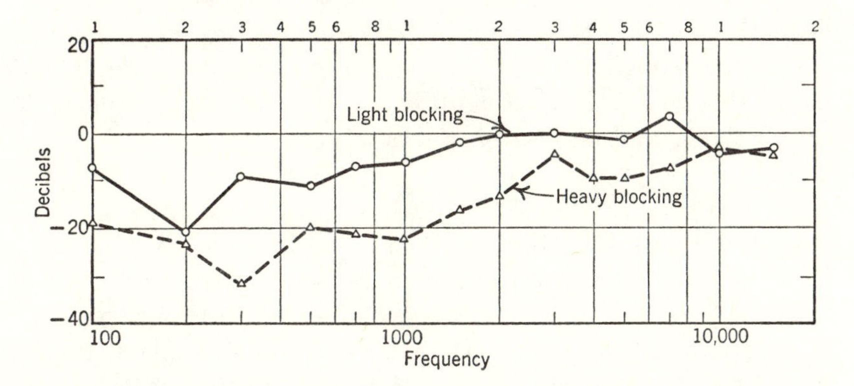

Fig. 95. Effective blocking of the round window of the cat after removal of the middle ear structures. Two degrees of blocking were used. From Wever (*6*) [*Annals of Otology, Rhinology, and Laryngology*].

As in Series 4 the ossicular mechanism was removed and the sound was delivered directly through a tube sealed over the oval window.

These results again show that the principal effects are on the responses to low tones. For the heavier blocking the impairment reaches 20 db for all tones below 1000~.

It is now clear that the pattern of changes resulting from a blocking of the round window is different according to the presence or absence of the middle ear mechanism. The effects of the blocking evidently depend upon the properties of the whole mechanical system.

From the observed changes we infer, as already suggested, that the effect of the blocking when the middle ear is present is mainly one of adding mass to the vibratory system. When the middle ear is absent the blocking, whether light or heavy, seems to have the effect of adding stiffness. These seemingly contradictory assumptions can be reconciled if we say that the procedure really adds both mass and stiffness but the relative importance of these two properties is altered by the presence of the middle ear mechanism. If we suppose that the middle ear mechanism contributes greatly to the stiffness of the cochlear system and relatively little to its mass, then the presence of the block will significantly increase the mass and add only slightly to the stiffness; hence the high-tone responses are impaired. When the middle ear mechanism is absent the stiffness supplied by the block looms large, and the low-tone responses are impaired. This explanation is consistent with the observation already made (Chapter 5) that removal of the middle ear most seriously impairs the sensitivity to middle and high tones.

OVAL WINDOW BLOCKING

Further information on this general question is gained by blocking the oval window and stimulating by way of the round window. These experiments, like the foregoing, were performed on cats. The peripheral portion of the middle ear was removed, leaving the stapes in the oval window. Wax was then packed into the exposed tympanic cavity, with care to embed the stapes thoroughly without tearing it out of its emplacement in the oval window. Because the stapes already fills the oval window and it may be firmly embedded, this blocking can be made more complete than the round window blocking. The sound was introduced by a tube sealed over the round window, and the cochlear potentials were recorded as usual with a foil electrode on the round window membrane.

The results are shown in Fig. 96. All responses are impaired, especially for the middle frequencies where the amount of impairment exceeds 30 db. Evidently the blocking has added greatly to both the mass and the stiffness of the vibrating system.

These observations of the effects of blocking either oval or round window while stimulating by way of the other window prove the correctness of Weber's theory of cochlear action. On stimulation by the normal route, the round window serves as a place of yielding in response to pressures exerted at the oval window. As the stapes vibrates the round window membrane undergoes oscillations of equal volume displacement.

The question has often been raised whether other points of yielding exist in addition to the round window. It has even been suggested that the cochlear fluid itself is capable of being compressed sufficiently to allow the stapedial movements. Some have suggested that blood vessels and lymphatic vessels could act as avenues of alternating fluid flow. The cochlear aqueduct and endolymphatic duct also have been looked to as possible avenues.

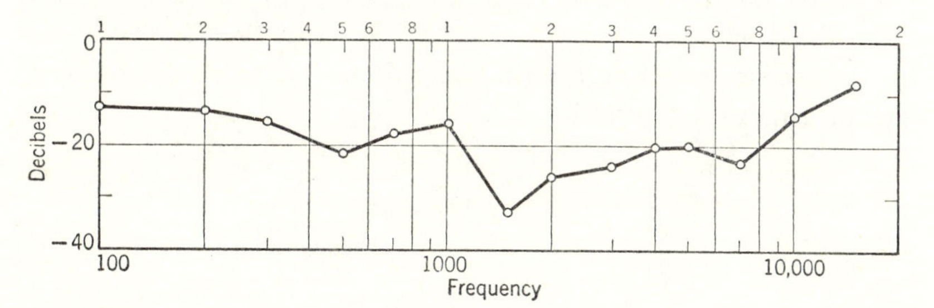

Fig. 96. Effects of blocking the oval window of the cat after removal of the middle ear structures. From Wever (*6*) [*Annals of Otology, Rhinology, and Laryngology*].

Pohlman accepted these suggestions and went still farther in supposing that the cochlear capsule itself could expand and contract under the influence of stapedial pressures.

We can at once abandon the suggestion that the fluid can be compressed sufficiently to permit the necessary volume changes. Water is compressed by only 4.9×10^{-13} of its volume by a pressure of 1 dyne per sq cm. Hence the cochlear contents, which (in man) measure approximately 0.1 cubic cm, would yield under this pressure by only 4.9×10^{-14} cubic cm. Yet our calculations, already mentioned, show that a 1000~ tone at 1 dyne per sq cm produces a fluid displacement of about 3×10^{-10} cubic cm, and thus nearly a thousand times the displacement that there is room for through compression of the fluid. The idea of an expansion and contraction of the bony walls is equally indefensible, for the yielding of bone is even less. For all practical purposes the fluid and the cochlear capsule may be considered as fixed in volume.

The suggestion regarding the blood vessels and other ducts as alternative avenues of pressure discharge does not appear very reasonable in view of the small dimensions of these vessels and the fact that they are already filled with fluid and lead only to fluid-filled and tissue-filled regions. We have finally the evidence of Figs. 95 and 96 that when one window is blocked the sensitivity of the responsive system may be reduced as much as 20 or 30 db. A more complete blocking, if we could achieve it, would no doubt cause still more striking reductions. Indeed, we have an indication of this fact in otosclerosis where, as we shall see later, the impairment of hearing caused by the bony closure of one window may attain 100 db, of which perhaps 55 db is attributable to the immobilization effect proper. This evidence proves beyond question that these proposed alternative avenues of pressure discharge have no functional significance.

13

The Passage of Sounds Through the Cochlea

We come now to consider the finer details of the propagation of sound within the cochlea, which leads us further into the field of auditory theory. Our primary concern is with the actions of aerial sounds, and our theoretical conceptions have been developed almost wholly for such sounds, but it is clear from what we have learned in the foregoing chapter that sounds that reach the cochlea by bone conduction must operate within it in essentially the same manner.

Initially, at the very basal end of the cochlea, the movements of the stapes can be communicated to the interior of the cochlea only by way of the cochlear fluid, but beyond this initial stage there are three possibilities, represented in different theoretical formulations. The propagation farther along the cochlea can continue to be carried out exclusively by the fluid, or it can be taken over by the basilar membrane, or it can occur as an interaction between fluid and membrane. The first of these possibilities is usually assumed or implied by the resonance theories, whereas the second and third are represented in different forms of the traveling wave and standing wave theories.

THE RESONANCE THEORIES

The simplest view of the action of sound is that the movements of the stapes are communicated immediately to the whole fluid contents of the cochlea. This is a mass-action theory; the fluid is thought of as vibrating bodily between the stapes and the round window. This description probably represents what Eduard Weber had in mind in the presentation of his view in 1841, when the details of the action were of no particular concern. Later, with the further development of the resonance theory by Helmholtz and others, it became necessary to alter and refine this conception of mass movement.

The Helmholtz theory conceived of every discriminable tone as having a specific locus of action on the basilar membrane. Necessarily, then, the fluid movements must be distinctive for every tone, following a particular path from stapes to round window through the tone's own region of the basilar membrane. This idea was further defined by Lux, Budde, and Wilkinson, who regarded the double columns of fluid—from the stapes to the vibrating region of the membrane and from this region to the round window—as assisting in the differentiation of the basilar membrane. These columns, they pointed out, vary systematically in length according to the frequency of the stimulating tone and produce a mass loading of the membrane. For a high tone, whose tuned segment of the basilar membrane is in the basal region, the double fluid column is short and the pressure discharge is fairly direct from stapes to round window. For a low tone, whose tuned segment is more apically situated, the column is long and more circuitous. According to this view, then, the wave of displacement initiated at the stapes does not spread widely throughout the scala vestibuli and cochlear duct but takes a particular path of travel according to the tonal frequency and the local impedance of the basilar membrane. Despite its restriction, however, the column is thought of as moving as a single mass.

In this situation, we can properly speak of a mass movement only as a convenient designation of a progressive transfer of motion from molecule to molecule in which the time of transfer is negligibly short and the phase of the motion varies little between initial and end points of the path under consideration. In other words, the speed of wave propagation is great relative to the distance concerned.

Though mass movement in this sense is usually indicated or implied in formulations of the resonance theories as the simplest representation of the fluid action, it is not at all essential to such theories. Any speed of conduction will serve, within limits of the time required for tonal perception. The same is true of a simple frequency theory like Rutherford's in which there is no spatial differentiation in the cochlea and all parts of the basilar membrane are stimulated by every sound. This matter is different, however, for two other classes of theories, the traveling wave and standing wave theories. In these theories the speed of conduction in the cochlea assumes a critical importance, as we shall see.

These two types of wave theories have arisen in opposition to the resonance theory mainly from a consideration of basic limitations imposed by the principle of resonance itself. As Wien pointed out in 1905, the manner of operation of a resonator bears a very intimate relation to the degree of damping to which the resonator is subjected. If the resonator is lightly damped it will readily respond only to its tuned frequency and will be little affected by other frequencies. Such a resonator therefore will achieve a high degree of frequency discrimination. Our ability of distinguishing simultaneously existing tones seems to demand such a low degree of damping in the cochlear resonators. This same undamped resonator, however, will be slow to reach its maximum amplitude on exposure to a driving frequency and will likewise be slow in dying down to zero after this driving frequency is removed. If we were equipped with such resonators we should be unable to distinguish rapidly successive changes of pitch as they occur in speech and music. To distinguish such changes the resonators must be highly damped. Yet here we come back to our first problem that was solved by light damping. We thus are in a dilemma, with simultaneous and successive discriminations of pitch demanding widely different degrees of damping. The requirements are so severe that neither form of discrimination is satisfied by a compromise value for the damping characteristic.

The dilemma can be resolved by another sort of compromise as discussed at length elsewhere and treated briefly a little later here, in which more moderate demands are made on the principle of place representation in the cochlea. Heavy damping then can be admitted. This limited use of the place principle is possible if the frequency principle is adopted as a supplementary basis for pitch discrimination.

Many, however, have sought a different solution, one that appeals to them at the outset as being simpler. Adhering to the place principle as the sole basis of tonal differentiation, they have tried to find another mode of cochlear operation that would avoid the difficulties imposed by resonance. This quest led first to the formulation of standing wave or sound pattern theories and later to the development of traveling wave theories. Because a standing wave in the cochlea can only be established by the reflection of a traveling wave we shall reverse the historical order of these developments and consider the traveling wave theories first.

THE TRAVELING WAVE THEORIES

The traveling wave theories receive the name from their conception of a wave of displacement that progresses in a systematic way along the basilar membrane and produces a local stimulation in its path. As already indicated, such theories vary accordingly as the progression of the wave is ascribed to the membrane alone, with the fluid assuming only a passive role, or is ascribed to a continuing interaction between fluid and membrane. Most of the theories are of the latter type.

Twelve forms of these theories have appeared so far. Their authors and dates of appearance are shown in the listing below.

TRAVELING WAVE THEORIES

Hurst, 1894
Bonnier, 1895
Ter Kuile, 1900
Watt, 1914
Békésy, 1928
Ranke, 1931
Reboul, 1937
Zwislocki, 1946
DeRosa, 1947
Peterson and Bogert, 1950
Huggins, 1950
Fletcher, 1951

An aid to our understanding of this type of theory is the analogy of a loosely held rope, one end of which is given a quick flip. A wave begins at this end and travels along the rope. If the rope is slack enough the wave moves slowly, declines in amplitude as it travels, and finally dies away to zero. The particular conditions in and about the rope can be controlled to give wide variations in the speed of the wave and its changes of amplitude.

According to all the theories mentioned the movement of the stapes produces a fluid displacement in the basal region of the cochlea and this displacement is communicated to the basal end of the basilar membrane. The displacement wave thus started is then propagated up the cochlea. At this point the theories begin to diverge, especially with regard to the particular form of the wave and the means of its propagation. In these respects the earlier theories vary more than the recent ones. The theories vary also as to what feature of the wave action is taken as determining the excitation of the cochlear nerve fibers.

Only brief indications will be given here of the forms of the first seven theories, for all of these have been treated in detail elsewhere (Wever, 5). The remaining theories, which are relatively recent, will be discussed more fully. Even these, however, will not be covered in complete detail for reasons that will be clear when later on we present experimental results that apply to these theories.

HURST'S THEORY, 1894

In Hurst's theory, an inward movement of the stapes sets up a wave of displacement in the basal portion of the basilar membrane. This wave is a narrow bulge that moves along the membrane toward the apex. When the wave reaches the apex it runs around the terminal wall to Reissner's membrane and returns along this membrane to the basal end of the cochlea where it disappears. If the stapes makes a second inward movement before the first wave has run its course a second wave will be sent up the basilar membrane and somewhere will encounter the returning first wave. At this meeting place of upgoing and downgoing waves a stimulation is said to occur.

The location of the meeting place along the cochlea will depend upon the speed of wave transmission and the time interval between successive stapedial movements. If the time interval is long the first wave will have covered most of its circuit and the meeting place will be near the basal end of the cochlea. If the interval is short the meeting will occur in the apex soon after the first wave has made its turn. Thus the low tones are localized in the basal end of the cochlea and the high tones progressively toward the apex. It should be noted that this form of localization is contrary to most other conceptions since Cotugno (1760).

The basic assumptions as indicated here provide a limit to the speed of wave propagation. If we take the total cochlear pathway as 6.3 cm and the lowest perceptible pitch as 15~, then the speed must not exceed $15 \times 6.3 \times 10^{-2} = 0.94$ meters per second. If the speed is greater than this the first wave will vanish before the second has appeared on the membrane and no meeting can occur for this low frequency.

BONNIER'S THEORY, 1895

According to Bonnier, the stapedial vibrations produce to-and-fro movements of the basilar membrane and these movements are freely communicated as successive waves over the spiral course of the membrane. The wave is thus spread out in graphic form over the membrane, which facilitates our perception and analysis of it. Stimulation occurs all along the membrane without particular spatial differentiation. This theory is peculiar in that it is a traveling wave theory but not a place theory; the theory resorts to the frequency principle for the appreciation of pitch.

For the graphic representation of even one complete wave at 15~ as indicated in this theory the speed of propagation must not exceed $15 \times 3.15 \times 10^{-2} = 0.47$ meters per second.

TER KUILE'S THEORY, 1900

Ter Kuile's theory states that as the stapes moves inward and displaces a quantity of fluid a bulge is formed on the basilar membrane that is at once extended apicalward. The extension of the bulge is brought about both by conduction in the membrane and by additional fluid displacement by the stapes as it continues its inward thrust. When the stapes reaches its most forward position the depressed portion of the membrane extends from the basal end of the membrane to a position determined mainly by the time of this stapedial movement, which is half a wave period. As the stapes moves outward the fluid is sucked backward and the bulge is erased. This erasing action begins at the basal end and extends apicalward. Meanwhile the original bulge has continued its movement along the membrane. At the moment when the stapes reaches its most outward position the original bulge will have progressed a distance of one wave length up the cochlea and the erasing action will just have caught up with it, whereupon the bulge vanishes. The length of membrane covered by the bulge, measured from the basal end to the vanishing point, determines the pitch. This length increases with the wave period, hence high tones are represented by short stretches of stimulation and low tones by long stretches.

It is clear that the speed of propagation must be low enough to permit a full wave length to exist on the basilar membrane for the lowest discriminable tones. Hence the speed of propagation of the original bulge must not exceed $15 \times 3.15 \times 10^{-2} =$ 0.47 meters per second as in the preceding theory. A curious feature is that the erasing action has half a period to cover the same distance that the primary bulge traverses in a full period, and hence its wave must travel at double the speed of the other. This difference in speed is unexplained.

WATT'S THEORY, 1914

In Watt's theory the manner of formation of the original bulge is the same as in ter Kuile's. However, the basal portion of the bulge is not erased as in that theory. A balance between the elastic forces of the membrane and the forces exerted through the fluid as the stapes moves backward causes the basal portion of the bulge to diminish in amplitude, and to diminish more the nearer to the basal end, but a partial amplitude remains. Also, as the bulge moves on from the point reached at the most inward position of the stapes it steadily falls away in amplitude. The result is that at the end of a complete period of the in-and-out movement of the stapes a bulge is present on the membrane over a distance determined by the length of the period. Also, this bulge has a maximum amplitude at its midpoint. The bulge does not vanish at the end of a period, though subjected to a continuing decrement. Enough of it persists to give an accumulation of amplitude over several periods. The persisting maximum is the determiner of pitch. Its locus varies with the period, extending farther toward the apex for the low tones.

This theory requires the same speed of propagation for the primary bulge as was calculated for ter Kuile's theory.

BÉKÉSY'S THEORY, 1928

Békésy's hypothesis and the observations on which it was based mark the beginning of the modern era of traveling wave theories. He worked first on mechanical models of the cochlea and reported the observation of traveling waves. He then studied cadaver specimens in which suitable operative exposures had been made and found that the cochlear structures were set in motion by a sound in a manner consistent with the traveling wave theory.

In the models the wave produced in the artificial basilar membrane was seen as traveling from the basal end upward, varying continually in amplitude, and usually expiring before it reached the apical end. As the wave proceeded the amplitude of its undulations changed progressively, rising to a maximum value somewhere along the membrane, after which this amplitude rather rapidly diminished toward zero.

Two successive views of the wave produced by an outward movement of the stapes are shown in curves *a* and *b* of Fig. 97.

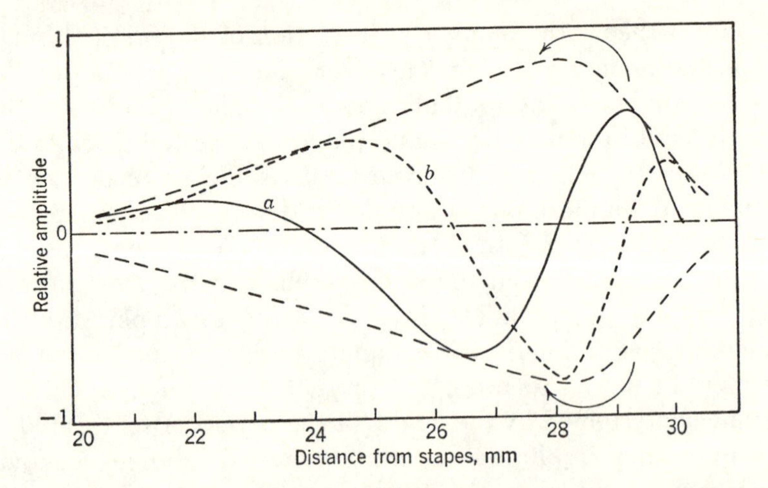

Fig. 97. Békésy's theory of basilar membrane movement. Curve *a* represents the form of the membrane at one instant and curve *b* the form an instant later. In its full progression the maximum of the wave outlines the area enclosed by the long-dashed lines. The curved arrows represent eddy currents set up in the fluid. Adapted from Békésy (*1*,*7*).

Curve *a* represents the wave at one instant and curve *b* represents the same wave a little later. The complete course of this progressive wave outlines an area enclosed by the long-dashed curves. An inward movement of the stapes will produce a wave likewise progressing within the envelope shown by these dashed lines, but at corresponding instants this wave would take forms that are mirror images of those shown.

This same picture is conceived for the actual basilar membrane, though the conditions of exposure and observation in the cadaver specimens prevented its being seen as a whole. Only one region of the cochlea could be exposed and examined at any one time. It was found easier to make this study in the apical region. Here Békésy made several measurements of the amplitudes of movement produced by tones of various frequencies. The results are shown in Fig. 98. The solid curves represent the regions over

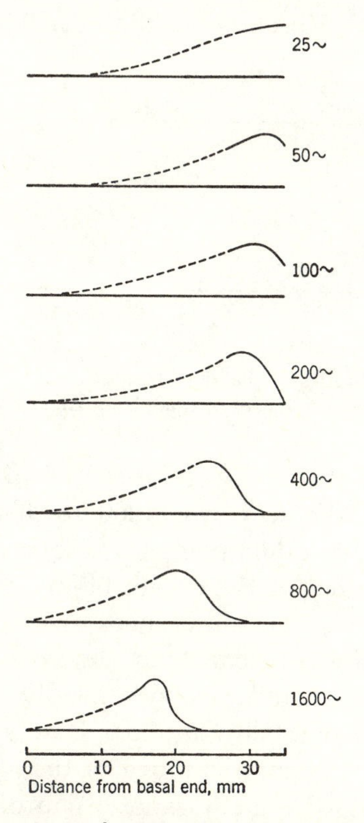

Fig. 98. Békésy's measurements of amplitude of motion in the cochlea, for several frequencies. The solid lines of the curves represent the measurements and the broken lines an extrapolation based on further observations. From Békésy (*14*) [*Acta Oto-Laryngologica*].

which the measurements were made and the broken lines show extensions of these curves made according to observations still to be described.

Observations in middle and basal regions were exceedingly difficult and were made only by removing the more apical portion of the cochlea. This dissection was continued until only the basal end was left. Observations made on mechanical models indicated that this procedure produced only minor modifications in the behavior of the structure that remained.

An alternative procedure was to make measurements of amplitude at a single point along the basilar membrane while stimulating with a continuous series of frequencies of constant amplitude. The results of this procedure, for the most basal point exposed, are shown in Fig. 99. Here we see that a maximum response was

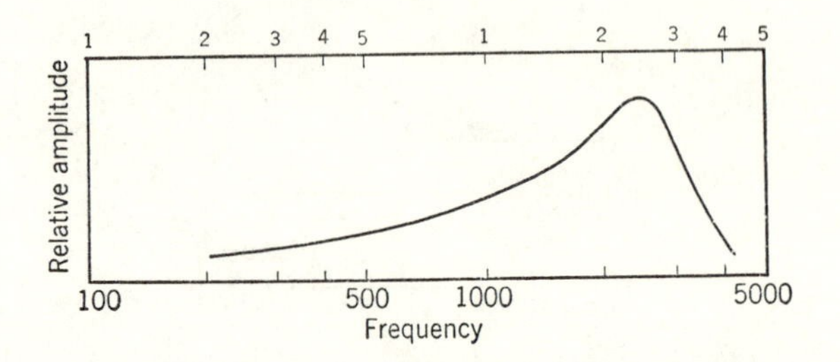

Fig. 99. Maximum amplitudes of movement of the basilar membrane as measured in the basal region of the cochlea for various frequencies. After Békésy (*13*).

produced by a tone of 2500~, yet noticeable effects were produced for frequencies all the way from 200 to 4000~. Similar results were obtained at five other points. These results gave the basis for the dashed portions of the curves of Fig. 98.

Measurements of phase differences were made similarly for single regions of the cochlea. After the cochlea was opened its contents were illuminated stroboscopically and the structures were made to appear in slow motion. It was first noted that all parts of the sensory apparatus move in the same phase; the tectorial membrane, basilar membrane, organ of Corti, and usually Reissner's membrane as well were seen to vibrate as a unit.

The results on phase that were obtained by observations through an opening at the extreme apical end of the cochlea are shown in Fig. 100. The solid curve represents for various frequencies the phase difference between the movements of the stapes and the movements of the exposed region of the sensory structure. Also shown is the relative amplitude observed for this region over the same frequency range. We see that this apical portion of the structure vibrates in phase with the stapes at low frequencies and then, from about 80~ on, it progressively lags in

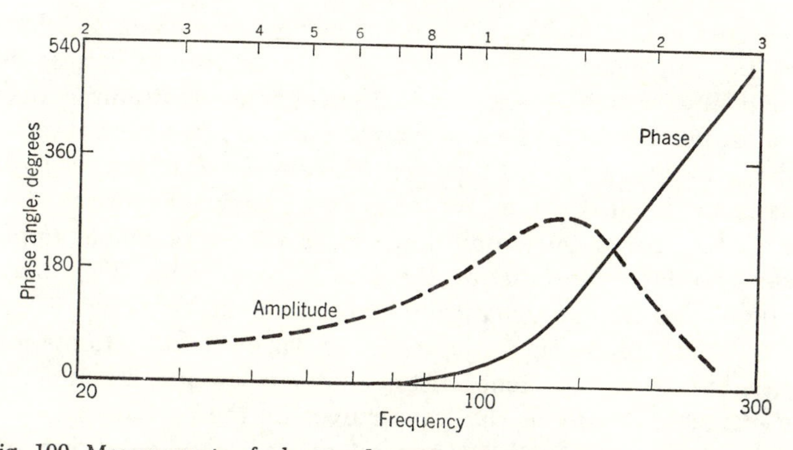

Fig. 100. Measurements of phase and maximum amplitude of basilar membrane movement at the extreme apical end of the cochlea. Data from Békésy (*16*).

phase as the frequency is raised. The phase difference finally reaches 540° for a 300~ tone.

From results like those described, showing patterns of displacements and phase relations for single regions of the basilar membrane when set in motion at various stimulating frequencies, it is possible to infer the forms of the patterns for one stimulus frequency acting upon various regions of the cochlea. This prediction is indicated for four frequencies in Fig. 101.

These patterns are considered as proving the presence of travel-

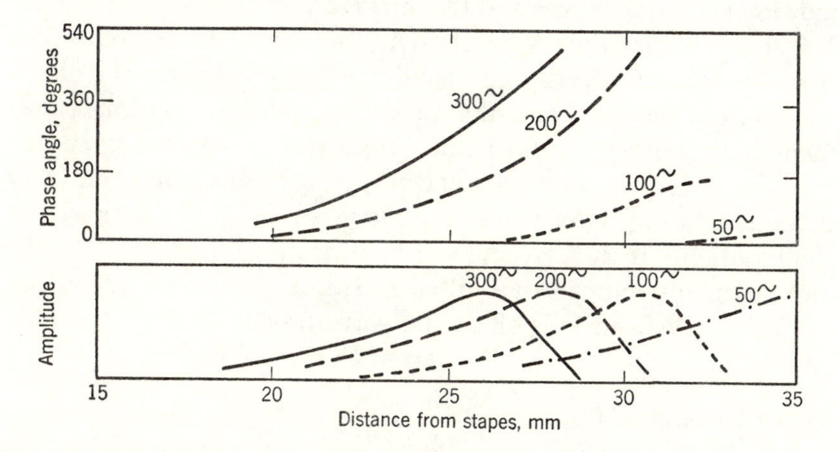

Fig. 101. Phase angle and maximum amplitude of movement as the wave progresses up the cochlea, plotted for four frequencies according to Békésy (*16*).

ing waves in the cochlea. The point along the cochlea at which the traveling wave reaches its maximum amplitude varies systematically with frequency, with the high tones attaining their greatest maximums in the basal region and the low tones attaining theirs at positions toward the apex. The envelopes of the response curves are broad for the low tones and somewhat more peaked for the high tones, but for all they extend with appreciable amplitude over large portions of the basilar membrane. The phase relations also vary systematically, at least for the low tones. There is hardly any phase lag in the traveling wave for the extreme low tones, but as the frequency rises a notable lag appears and this lag increases rapidly as the wave moves up the cochlea.

Békésy also conceived that just apicalward of the point of maximum amplitude the undulations of the membrane cause an eddy movement on either side of the membrane. This eddy movement is represented by the curved arrows in Fig. 97. He suggested that the eddy movement, which is much more restricted in scope than the wave motion as a whole, provides the actual stimulating force upon the sensory cells.

Both in his mechanical model and in cadaver specimens Békésy measured the time required for waves to travel from the stapes to six selected points in the apical region of the cochlea. His method consisted of placing a small mirror on the "cochlear partition," stimulating with the sound of an electric spark, and recording the movements of the stapes and the mirror on an oscillograph. Figure 102 shows his results when translated into speeds of conduction. Two curves are given, one (solid line) representing the mean speeds from the basal end of the cochlea to the points of measurement and another (broken line) representing the mean speeds over the distances between adjacent points of measurement. Only by extrapolating these curves can we get an idea of the speeds at the basal end of the cochlea; but extrapolating from the last two measurements indicates an initial speed of about 790 meters per second. This speed falls off as the wave proceeds up the cochlea and becomes 2 meters per second at the apical end.

RANKE'S THEORY, 1931

Ranke treated the cochlea as a tube with yielding walls and attempted to solve the problem of the effects of sound by use of the mathematical principles worked out by Frank (3) for the transmission of pulse waves through the blood vessels. Certain modifications of Frank's development were necessary because the cochlea must be regarded as a tube bent back upon itself with the basilar membrane forming a common elastic wall.

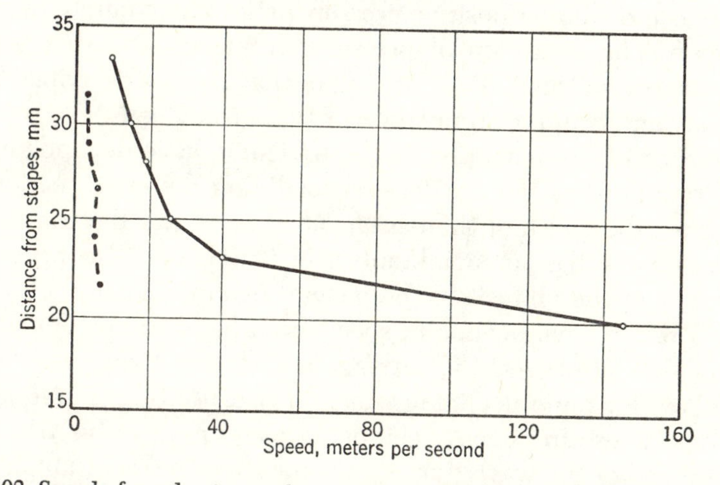

Fig. 102. Speed of conduction in the cochlea, according to Békésy (*13*). The solid curve represents the mean speed over the path from the basal end to the points indicated. The broken curve represents the mean speed over the distances between adjacent points.

According to this theory, the motion of the stapes sets up pressure waves that travel along the two scalae from base to apex and produce waves of displacement in the basilar membrane. These displacement waves consist of a number of undulations that vary in form as they travel. The displacement is small near the basal end of the cochlea and rises progressively as the wave attains a maximum amplitude at some place along the cochlea. The course of the wave up to this place defines what Ranke called the initial zone of the action. Thereafter the amplitude undergoes relatively rapid changes, passing through a fairly prominent minimum and a second maximum smaller than the first, and then showing other minor variations as the amplitude declines rapidly toward zero. These rapid variations occur in what is called the transition zone.

These patterns of membrane movement are similar in form for

all tones but vary as to locations along the basilar membrane according to the stimulating frequency. The patterns for the low tones extend far toward the apex and those for the high tones go only short distances.

It is further suggested that a negative pressure arises on both sides of the basilar membrane just at the beginning of the transition zone and a larger positive pressure arises immediately beyond this zone. The transition of pressures from negative to positive may give rise to the eddies in the fluid that Békésy described and may play a part in the stimulation of the sensory cells.

The speed of propagation of the traveling wave depends upon local conditions in the cochlea, especially on the cross section of the scalae, the width of the basilar membrane, and the transverse tension on the basilar membrane, and it depends also upon the frequency of the vibration. The general conditions are such as to cause a progressive decline in speed from basal to apical ends of the cochlea. However, the propagation of any given wave does not follow a simple declining function because of the frequency relations. These frequency relations cause a progressive decrease in the speed of the wave during the course of its rise of amplitude and a very rapid increase in speed after the wave has passed its maximum. At the basal end of the cochlea, for a tone of 10,000~, the speed is about 160 meters per second. This speed falls to 42 meters per second as the wave travels the distance of about 4 mm to the maximum point for this tone, and then the speed increases rapidly to about 700 meters per second. For lower tones the speed in this same basal region is smaller, varying as the square root of the frequency until the very low frequencies are reached, for which the speed becomes nearly constant. The speed falls further as the low-frequency wave proceeds up the cochlea until at the apex, for a frequency of 20~, it is only 5 cm per second.

REBOUL'S THEORY, 1937

Another form of tube-resonance theory was developed by Reboul. Much as was done in the two preceding theories, he regarded the pressure changes produced by sounds as moving up the cochlea and giving rise to a wave of displacement of the basilar membrane. The form and locus of the displacement depend upon the frequency of the stimulating wave and its speed of propagation. The speed of propagation is determined by properties both of the basilar membrane and of the surrounding fluid; these include the thickness and elasticity of the membrane, the density and compressibility of the fluid, and the diameter of the scalae. These properties vary along the cochlea in such a way as to cause an increase in propagation speed as the wave moves from base to apex. This speed is estimated as 38 meters per second at the basal end and as 415 meters per second at the apical end. It should be noted that this variation is contrary in direction to that assumed in other theories. Its effect is to make the wave increase in wave length as it travels up the cochlea.

According to Reboul's development of this theory, the low tones, up to about 750~, move the basilar membrane almost as a whole and fail to produce any maximum on the membrane. Higher tones, up to about 4000~, produce a single maximum, whereas still higher tones produce two or three maximums. For these uppermost tones, however, the most basal maximum is always much more prominent than the others. The locus of the point of greatest displacement moves farther toward the basal end of the cochlea as the frequency becomes higher, and this locus determines the pitch. For tones below 750~, the pitch depends upon the frequency of nerve stimulation. Reboul therefore incorporates the principle of the volley theory.

ZWISLOCKI'S THEORY, 1946

Zwislocki's theory follows the ones just discussed in general outline, for like these it pictures a wave that travels up the basilar membrane, rising in amplitude in its course until it reaches a maximum and then falling rapidly to zero. The location of the maximum varies systematically with frequency in the usual way, with the maximums for high tones at the basal end of the cochlea and those for low tones located toward the apex. However, Zwislocki's theory differs from the others as to the cochlear variables that are emphasized. Here the most significant conditions are the elasticity of the basilar membrane and the damping. Of less importance are the dimensions of the canals and the density and viscosity of the perilymph. Fluid friction and the mass of the basilar membrane and of other parts of the cochlear duct are considered to play so small a role that they can be ignored altogether.

The damping is so great that most waves—all except those for the lowest tones—fall to zero before they reach the apical end of the cochlea. Therefore reflections fail to appear, and there are no standing waves.

The speed of propagation of the waves depends upon the elasticity of the basilar membrane and the cross-sectional area of the scalae, and these vary in such a way as to cause the speed to decline rapidly as the wave proceeds up the cochlea. The speed is calculated as 45 meters per second at the basal end and is considered to fall off continuously to about 4 meters per second at the apical end.

DEROSA'S THEORY, 1947

In DeRosa's theory the wave motion in the cochlea is worked out on the basis of somewhat simpler assumptions than those of other recent theories. Constancy throughout the cochlea is assumed for the cross-sectional area of the scalae and for both the width and elasticity of the basilar membrane. The mathematical treatment gives results that are interpreted as showing the existence of two traveling waves in the cochlea. As a result of the stapedial movements, one pressure wave appears at the basal end of the cochlea and travels in the fluid of the scala vestibuli. This wave is responsible for initiating a second pressure wave that also begins at the basal end of the cochlea. This wave, however, travels in the basilar membrane, and at a speed that is considerably slower than that of the first wave.

These two waves produce nodes of compression along the basilar membrane, but because they are traveling at different speeds their nodes are differently spaced. Stimulation occurs at the points where the nodes coincide.

A determination of the points of coincidence for the various tones depends of course upon the particular speeds of wave propagation. As a preliminary estimate DeRosa adopted a constant figure of 30 meters per second for the speed of the primary wave. Further consideration, however, showed that for tones below 1000~ this speed ought to be reduced by friction and viscosity until at 60~ it was estimated as only 1.5 meters per second. For 60~ this speed gives a wave length of 2.5 cm and thus permits a little more than one complete wave to appear on the basilar membrane. It will of course be necessary for the speed to decline further to about 0.4 meters per second in order that 15~ shall similarly be represented on the membrane.

The membrane wave is generally estimated as traveling with about half the speed of the other wave. The illustration given by DeRosa shows that in his calculations for tones over the range of 500 to 8000~ he used speeds varying from 12 to 47 meters per second for the primary wave and half these values for the secondary wave.

At the higher frequencies the waves show multiple coincidences and hence produce multiple stimulations. However, the damping is regarded as severe, amounting to at least 9 db per wave length as the wave travels toward the apex. Therefore the most basal stimulation will be predominant and will determine the pitch.

PETERSON AND BOGERT'S THEORY, 1950

This theory is a further application of hydrodynamic principles to the problem of cochlear function. It bears a close similarity to the formulations of Ranke and Zwislocki, but makes fewer simplifying assumptions and thus concerns itself with mechanical properties more closely resembling those of the actual cochlea. Even so, it is found necessary to idealize the cochlear structure somewhat to facilitate the mathematical treatment. The two scalae are taken as uniformly tapered from basal to apical ends and as equal to one another at every point. The basilar membrane is regarded as increasing in width at a uniform rate and as falling off regularly in stiffness. The mass of the vibrating system is regarded as the combined mass of the basilar membrane and of the endolymph contained in the cochlear duct, and this mass is taken as increasing uniformly along the cochlea. Resistance is neglected in the mathematical treatment, but later work by Bogert with an electrical model indicated that its introduction would not seriously affect the results.

According to this theory, the stapedial movements set up a pressure wave that travels through the scala vestibuli at the usual speed of sound through water, given as 1430 meters per second. This wave undergoes changes in amplitude as it travels, with the form of the changes varying according to the frequency. For the very low tones the amplitude declines progressively. For medium low tones the amplitude declines for a way and then rises as the apex is approached. For high tones the changes are more complicated and the wave presents a secondary maximum as shown in Fig. 103.

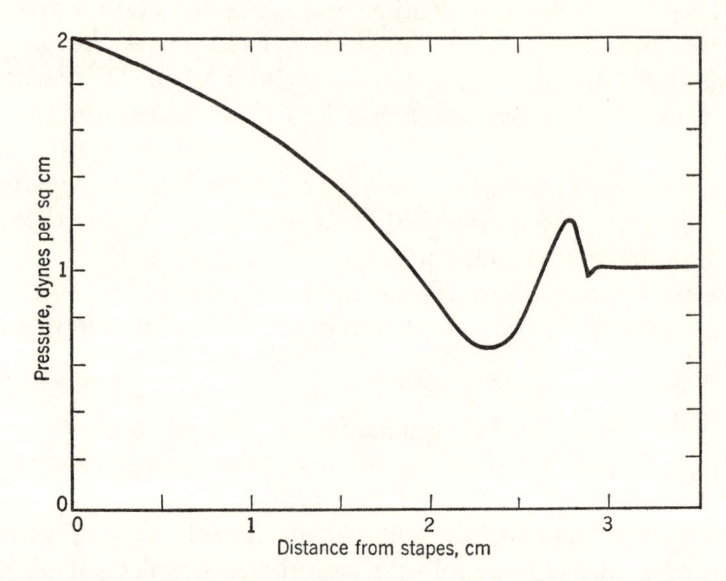

Fig. 103. The form of the pressure wave in the scala vestibuli for a tone of 3160~ according to the Peterson and Bogert theory.

At the same time another pressure wave is traveling up the scala tympani. For most tones this wave is a mirror image of the one in the scala vestibuli. For the very high tones, however, an asymmetry enters in the basal region and beyond this region the two waves are both rising.

It is the pressure difference between these two waves that causes a displacement of the basilar membrane. A displacement wave therefore moves along the basilar membrane. This wave moves at a comparatively slow rate, which varies along the cochlea according to the width of the basilar membrane and the cross section of the scalae. The speeds, calculated for low frequencies (below 100~) vary from 350 meters per second at the basal end of the cochlea to 7 meters per second at the apical end. For higher frequencies these speeds will be considerably slower because of the effects of mass, but exact indications are lacking.

The displacement wave is described as rising in amplitude as it travels up the basilar membrane, attaining a maximum at some place, and then falling rapidly to zero. The maximum is broad for the low tones and relatively sharp for the high tones. The low tones have their maximums in the apical region and the high tones have theirs in the basal region. A feature of interest in this localization is a close crowding of all tones below 1000~ in the last 3 mm of the cochlea.

HUGGINS' THEORY, 1950

This theory has been reported only briefly as yet, but is mentioned here as representing a special development of the traveling wave hypothesis in an effort to explain the fine pitch discrimination displayed by the ear. The wave of displacement of the basilar membrane is derived in much the same manner as in the Peterson and Bogert theory, but this displacement is not regarded as directly responsible for stimulating the outer hair cells. Rather, it is suggested that the basilar membrane displacement is communicated without change of form to the tectorial membrane through the connections afforded by the cilia of the hair cells. The tectorial membrane is considered as having rigidity in the longitudinal direction, and hence it is accorded the properties of a rigid beam. Such a beam when displaced from its resting position exerts counterforces along its length that are proportional point by point to the fourth derivative of the displacement. It is these forces that are regarded as stimulating the outer hair cells. The inner hair cells are considered to be excited by the displacement itself.

After determining by calculation the expected form of the displacement produced in the basilar membrane by a 1000~ tone, Huggins then obtained the fourth derivative of this displacement and found that its form was much more sharply peaked. A local excitation of outer hair cells is thereby obtained.

The theory continues in a consideration of the time relations of stimulation of outer and inner hair cells. These relations are said to vary for any given tone as the displacement and "beam" (i.e., the fourth derivative) waves proceed along the cochlea. For a 1000~ tone, the "beam" wave precedes the displacement wave from the basal end of the cochlea to a point just beyond the maximum displacement where the two waves are in phase. Beyond this point the displacement wave is leading. It is further assumed that when the displacement wave is leading, and therefore when the nerve fibers supplying the inner hair cells are firing ahead of the fibers supplying the outer hair cells, the impulses mediated by these outer hair cell fibers are inhibited. Hence the "beam" wave becomes ineffective beyond the resonance point. Further assumptions regarding the details of this action are made in order to explain such observations as those of Galambos and Davis on interactions found in the cochlear nuclei.

It will be obvious that the assumptions made by this theory are of a highly critical nature. Any slight departure from the true beam characteristics as postulated for the tectorial membrane would change radically the form of the fourth derivative function. Similarly, any variations from the interrelations assumed for the nerve fibers supplying outer and inner hair cells would have a serious result. It is always precarious to build a physiological theory on assumptions that do not allow of great individual variations, for in physiological processes variation is the rule, and the ear in particular is distinguished by its ability to continue to function despite many kinds of accidental change.

FLETCHER'S THEORY, 1951

Fletcher's theory employs the same fundamental equation of motion as that used by Zwislocki and by Peterson and Bogert, but because of different interpretations of constants and of limiting conditions the calculations lead to somewhat different results. All three impedance factors—mass, stiffness, and friction—are taken into account in the development. Moreover, the theory makes use of modern data on the dimensions of the cochlear structures, and with only moderate simplifications. A point of particular interest is that only the fluid in the immediate vicinity of an element of the basilar membrane is considered as moving in phase with the element and thus of weighting it. The effective mass therefore is only moderately greater than (2.33 times) the mass of the element itself. Accordingly, the vibration is considered as independent of the size and shape of the scalae.

The pressure wave produced in the cochlear fluid by the movements of the stapes travels with the speed of sound in water and reaches all parts of the basilar membrane in a short time. This pressure wave, however, undergoes a progressive reduction in magnitude. The reduction is particularly rapid for the high tones, so that for these tones the pressure becomes negligible beyond the basal region of the cochlea.

This pressure wave gives rise to a traveling wave of displacement of the basilar membrane, which begins at the basal end and moves progressively toward the apex. As this wave moves apically it rises in amplitude to a maximum and then falls somewhat more rapidly. The amplitude patterns have nearly the same form for all frequencies but are located in the basal region for high tones and toward the apex for low tones.

The speed of propagation of the displacement wave diminishes as the wave proceeds up the cochlea. According to Fletcher's computations of conduction time, this speed is about 320 meters per second a third of the way up the cochlea and about 15 meters per second at the apical end. These values are in fairly good agreement with Békésy's measurements.

We have now surveyed the various forms of traveling wave theories. Before entering into a general discussion and evaluation of these theories, we interpose a brief consideration of the older standing wave theories.

THE STANDING WAVE OR SOUND-PATTERN THEORIES

If a traveling wave strikes an obstruction that causes it to be reflected back along its original path without loss of amplitude the incident and reflected waves combine to form a particular pattern of disturbance called a standing wave. Such a pattern is easily seen in a rope that is anchored at one end and whose other end is held under moderate tension and moved up and down at a suitable rate. The rope breaks into segments, with certain portions showing up-and-down movements and other portions remaining stationary. There is no longer any evident progression of motion along the rope. If the reflection is only partial, or if for any reason the reflected wave is smaller than the incident wave, we shall obtain a complicated pattern consisting of a standing wave and a forward-moving (traveling) wave of reduced amplitude. If the incident and reflected waves are not constant but fall off in amplitude continuously along their paths their resultant will give patterns in which the features of traveling and standing waves are intermingled in a rather complicated way.

Let us now consider the possibility of standing wave patterns in the cochlea. If a traveling wave passing along the cochlea reaches the apex without having suffered very serious attenuation in its path it may be reflected backward to the basal end. Indeed, if the attenuation is slight a given wave may be reflected back and forth between the two ends of the cochlea several times before being extinguished. A standing wave then appears.

EWALD'S THEORY, 1898

The most famous standing wave theory and the only one that has been developed in much detail is Ewald's. He called it a pressure-pattern theory, and in formulating it he had in mind as an analogy the patterns that Chladni had exhibited in metal plates by strewing sand or powder over them and then setting them in vibration.

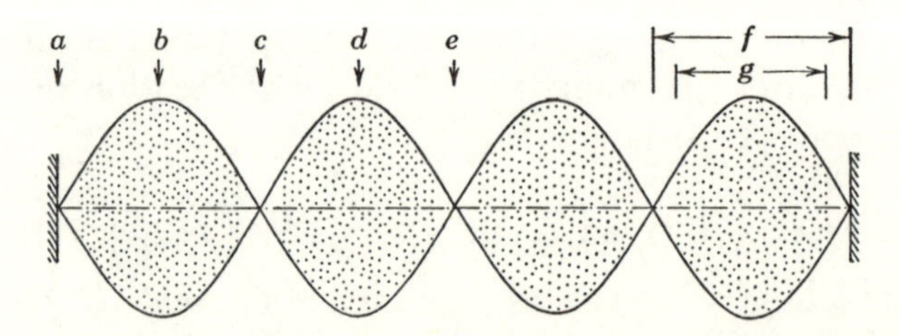

Fig. 104. Ewald's pressure-pattern theory. The pattern shown here is that assumed by the basilar membrane when stimulated by a tone of 40~. Points *a, c, e* are nodes; *b, d* are loops. A loop covers half a wave length *f*, and is effective over only a part of its length *g*. From Wever, *Theory of Hearing*, 1949, John Wiley and Sons.

Ewald was not explicit about the particular conditions of attenuation and reflection that he assumed for the cochlea, but from his description of the patterns themselves it is easy to see that attenuation must be negligible and reflection almost perfect. Therefore repeated reflections occur. It follows further that the standing waves must always bear an integral relation to the length of the basilar membrane. Indeed, Ewald assumed that the lowest perceptible tone produced just two loops, each a half wave length long. Higher tones must produce three, four, five, or larger integral numbers of loops. Figure 104 represents the action of a 40~ tone, with a pattern of four loops.

In this pattern of cochlear action Ewald indicated two features that could serve for the perception and discrimination of tones. One is the spatial separation of the loops, which decreases stepwise as the frequency rises. The other is the width of the band in the region of each loop where the amplitude of movement is sufficient to cause nerve excitation. This width likewise decreases as the frequency rises.

The speed of transmission of motion along the basilar membrane determines the number of loops formed at any frequency. Contrariwise, we can derive the speed from Ewald's specification of the number of loops for certain tones. If the length of the membrane is 3.2 cm and 20~ is represented by two loops or one wave length, as Ewald said, the speed of transmission is $20 \times 3.2 \times 10^{-2} = 0.64$ meters per second.

The principal attacks upon Ewald's theory have centered on its assumption of a generalized pattern involving the stimulation of portions of the entire basilar membrane by every tone. The Helmholtz type of theory, in which every tone has its place, has made a stronger appeal. Some therefore have sought to modify Ewald's theory in the direction of a more localized action. Lehmann suggested that damping in the cochlea may restrict the area within which the standing waves are strong enough to be effective. Koch and Gildemeister similarly assumed a considerable amount of damping which grows progressively greater toward the apex so that the apical loops are suppressed and only the basal ones can be effective. Such damping, however, will prevent wave reflection at the apical end of the cochlea and thereby will prevent the formation of standing waves. These modifications of Ewald's theory therefore turn out to be forms of traveling wave theories and not standing wave theories.

Some of the traveling wave theories admit the presence of standing waves as a minor complication of the action. Thus in Ranke's theory a standing wave is described as superimposed upon the traveling wave in the initial zone of the cochlea. It is produced by a partial reflection of the traveling wave at the beginning of the transition zone. In Reboul's theory the traveling wave reaches the end of the cochlea after only moderate attenuation. A large portion of its energy therefore is reflected backward and forms a standing wave pattern. These standing wave patterns, however, are considered to be of only secondary importance, and are accorded no special role in the determination of nerve stimulation.

AN EVALUATION OF THE ABOVE THEORIES

We have purposely refrained from making any detailed criticisms of the individual theories presented above. The reader will already have seen that many of the assumptions made by these theories are arbitrary and difficult to justify. More particular criticisms of most of them will be found in the extended treatment already referred to (Wever, 5). In this treatment will be found also a discussion of other theories not described here, especially the simple frequency theories of Rutherford, Ayers, and Hardesty and the frequency-analytic theories of Meyer and Wrightson. Here we shall be concerned further with the general principles underlying the wave theories. We shall pay special attention to the traveling wave theories because they have enjoyed much popularity and extensive development in recent years.

The first four theories mentioned—those of Hurst, Bonnier, ter Kuile, and Watt—are chiefly of historical interest and need not be given further consideration here. The others, beginning with Békésy's theory, have a number of features in common, no doubt because to a large extent they have been inspired and guided by Békésy's work.

Let us begin with the basic objective of this type of theory. We recall that the theory was developed because of the limitations that were found to be inherent in a resonance theory when that theory is forced to account for fine pitch discrimination and also for the differentiation of rapidly successive tones. It was seen that in the presence of damping a resonator cannot provide the necessary pitch discrimination, yet without damping it cannot respond selectively to successive sounds. The hope was then raised that some principle other than resonance might be found whose application would avoid this difficulty.

A consideration of hydrodynamic principles, when the cochlea was regarded as a tube with yielding walls, led to a formulation of the traveling wave theories, as we have seen. It is unfortunate for the basic objective of this development that such theories have failed to achieve any fine differentiation of cochlear action. As these theories have been worked out, the fundamental response to any tone is always broad, extending over a considerable portion of the cochlea. Especially is this true for the low tones, the very ones for which our pitch discrimination is most keen.

Békésy, facing this discouraging fact, tried to find a way to escape its limitations. For this purpose he proposed the eddy hypothesis, as we have seen. An eddy caused by the basilar membrane movement, not the movement itself, is said to provide the excitatory force to the hair cells. These cells are supposed to remain insensible to the vigorous motions of the basilar membrane and yet to respond readily to the eddy currents that this membrane is said to set up in the adjacent fluid. The eddy effects are thought of as cumulative over several cycles of the wave.

Against this sort of hypothesis we can set forth the evidence of the cochlear potentials, if it is agreed that these potentials constitute the excitatory agency for the nerve fibers or at any rate are a direct function of this agency. These potentials follow faithfully the periodicity of the stimulus waves at all frequencies. There is no indication of a slow building up of action over several cycles, but every cycle, and indeed every momentary state within a cycle, reveals itself in the electrical activity. It is even more significant that the nerve responses themselves reproduce the periodicity of the stimulus at the low frequencies and bear a simple fractional relation to this periodicity at the high frequencies. No relations such as these could be expected on the hypothesis of eddy stimulation. We shall return to this problem in the next chapter.

BASIC ASSUMPTIONS OF TRAVELING WAVE THEORIES

Let us now examine the traveling wave theories further in respect to their basic assumptions. These assumptions are three in number, having to do with the form of the traveling wave, its variation in form with frequency, and its speed of progression along the basilar membrane.

(1) The wave proceeds up the cochlea, changing in amplitude as it goes. More particularly, it rises in amplitude rather gradually

for a time, attains a maximum, and then rapidly falls away, usually vanishing completely before the apex is reached.

(2) Tones of different frequency produce waves that rise in amplitude at different rates and attain their maximums at different places in the cochlea before falling to zero. High tones rise rapidly, attain their maximums in the basal region, and expire there. Low tones progress farther up the cochlea. All tones thus have their effects distributed over the cochlea in a manner that varies systematically with frequency. The perception and discrimination of pitch depend upon this distribution of action.

(3) The speed of transmission of the wave along the cochlea is relatively slow. This slow speed is essential for a distribution of the wave pattern over the basilar membrane. The wave length of the disturbance in any region is always equal to the speed of propagation divided by the frequency. In order that several waves shall coexist in the cochlea, as most traveling wave theories require, it is essential that the quotient, speed of propagation over frequency, shall be small relative to the lineal extent of the basilar membrane. The speed of propagation is not necessarily constant along the course of propagation of a wave. Indeed, most of the theories assume a systematic variation in this speed. The result of such a variation is a progressive alteration in wave length as the wave proceeds up the cochlea.

It is clear that as an irreducible requirement the speed of propagation must be slow enough to allow a little over one-fourth of a wave length to appear on the basilar membrane for the lowest discriminable tone. Any faster speed would prevent a maximum being developed on the membrane by this tone. Most of the theories in fact assume that at least one complete wave length is present on the membrane, which requires that the speed of propagation be even slower.

Table 2 presents some data of interest in this connection. This table shows the maximum and minimum speeds that are assumed or implied by the different theories, and indicates also the mean speed for a wave traversing the whole membrane. This last quantity determines the conduction time along the membrane and thereby determines the lowest tone that is discriminable in terms of the spatial pattern, on the assumption that one-fourth of a wave length will suffice for this purpose.

TABLE 2. SPEEDS OF WAVE TRANSMISSION ALONG THE COCHLEA ACCORDING TO VARIOUS THEORIES

Theory	Maximum speed (meters per second)	Minimum speed (meters per second)	Mean speed for 31.5 mm (meters per second)	Conduction time for 31.5 mm (milliseconds)	Lowest discriminable pitch (cycles)
Békésy	790(?)	2	12	2.6	96
Ranke	700	0.05	21*	1.5	167
Reboul	415	38	95	0.33	750
Zwislocki	45	4	22	1.43	175
DeRosa	30	0.4(?)	1.9*	16.6	15
Peterson & Bogert	350*	7	31.5	1.0	250
Fletcher	500(?)	15	14.3	2.2	114

* For low frequencies

We see from these results that Békésy's theory cannot account in terms of place of action on the basilar membrane for the perception and discrimination of tones below 96~ and that other recent theories, except DeRosa's, are in even greater difficulty. DeRosa's theory is extended to the lowest tones by the arbitrary assumption of extremely slow conduction speeds.

AN EXPERIMENTAL TEST OF TRAVELING WAVE THEORIES

The particular assumptions underlying the traveling wave theories make it possible to develop an experimental test of crucial significance. For the purpose of such a test we have carried out a series of observations on cats by use of the electrical potential method.

We stimulated these ears in two ways. One way was the ordi-

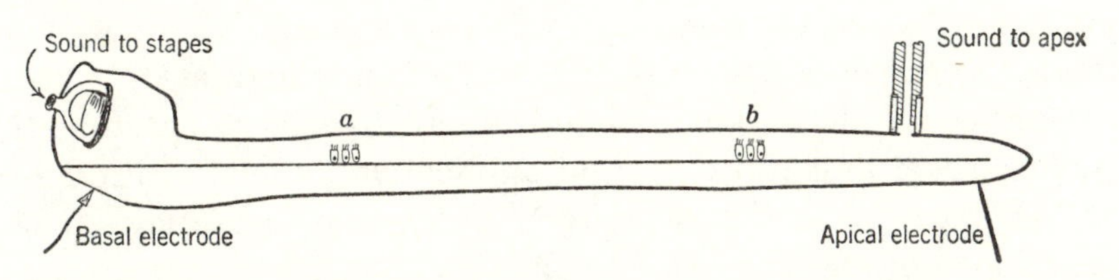

Fig. 105. Sketch showing the method of stimulating the cochlea at both its ends and recording its responses from two electrode positions. The cochlea is pictured as unrolled. At *a* and *b* are two groups of hair cells, referred to later in the discussion.

nary one, with an aerial sound introduced by a tube into the external auditory meatus and thus reaching the cochlea by way of the ossicular chain and the oval window. The other means of stimulation was by way of a small hole drilled through the cochlear wall in the apical region. Sound was introduced into this hole by one of two methods, either as aerial sound through a small tube fitted over the hole or as mechanical vibrations of a needle inserted into the hole and driven by a crystal vibrator. The hole was 0.020 inches in diameter and the needle was 0.018 inches in diameter, so that the needle vibrated in the hole like a piston in a cylinder. This method when carried out properly gives very effective stimulation, but there is always a possibility of lateral contact between the needle and the wall of the hole thus giving bone conduction of the sounds, and therefore we have employed the alternative procedure of aerial stimulation for all crucial tests. We also found it desirable at times, in order to prevent any leakage of sounds to undesired points, to remove all parts of the middle ear except the stapes and to introduce the sounds at the oval window by means of a tube sealed tightly over this window.

In the two forms of stimulation, that by the usual oval window route and that by way of a hole in the apex of the cochlea, we used the same tonal frequency from a single oscillator. A divider network delivered the oscillator output to two control channels in which the stimuli could be varied independently in intensity. A provision was made also for independent control of the phase relation between the stimuli of the two channels. A phase changer gave phase variations for 15 representative frequencies between 100 and 15,000~ over the full range of 0 to 360°. The two channels led to two loudspeakers or alternatively to one loudspeaker and the crystal vibrator mentioned above. We recorded the electrical potentials by means of an electrode on the round window membrane, except in certain instances to be mentioned later.

Our results show that a sound is effective in stimulating the cochlea regardless of its point of application. A sound introduced at the apex elicits the same sort of cochlear potentials as one introduced by the usual route, and these potentials bear the same functional relations to the stimulus. Simultaneous stimulation at the two ends of the cochlea gives results that can be stated very simply: the observed potential at all times represents a vector sum of the potentials obtained from the two stimuli acting separately.

The following simplified form of procedure was found advantageous in most of the experimental tests. One stimulus was applied by itself and the intensity was adjusted to give some standard amount of response, say 10 microvolts, and then this stimulus was removed and the other turned on and adjusted for the same amount of response. The two stimuli were then applied together and the phase relation between the two was varied by small steps (usually by steps of 5°). It was found that at a certain phase relation the two stimuli combine their effects to produce a response just double that of either stimulus acting alone. As the phase is altered from this relation the response falls in magnitude until a point is reached just 180° away from this relation, at which point the response becomes zero. This type of summation has been found for all tones between 100 and 15,000~ and at all intensities of stimulation below the level of overloading.

Possible Loci of Cochlear Summation. We still have to determine just where this summation takes place. Our theoretical application of the results depends upon this determination. There are three possibilities. The summation might occur in the paths of sound conduction up to the sensory processes, in the sensory processes themselves, or in the paths of electrical conduction from the hair cells to our recording electrode.

As we consider the first possibility further we see three particular regions in which the summation of sound might occur. They include the air outside the cochlea, the bone of the cochlear capsule, and the fluid within the cochlea.

A purely aerial summation requires that the apical stimulation shall produce aerial waves and that these waves shall find their way along the same path as the waves introduced through the external auditory meatus. When the apical stimulus consisted of tones from a loudspeaker any serious leakage was prevented by the use of a tube tightly sealed over the hole in the cochlea. When this stimulus was obtained with the vibrating needle it was found that withdrawing the needle to a point just above the hole (which hardly changes its production of aerial sounds) caused a loss of response to this stimulus amounting to about 60 db. Another test consisted of applying an acoustic probe at various places about the ear to measure the amount of aerial sound when only the vibrator was acting. This sound was always negligible relative to the sound that had to be applied at the external meatus in order to produce the same amount of response as the vibrator. Aerial summation is therefore excluded.

We now consider whether the summation might occur in the bone of the cochlear capsule. A vibration pickup was placed on the bone to record any movement there. During aerial stimulation at ordinary levels by way of the external meatus no bone vibration could be observed. Raising the aerial stimulus to an extraordinarily high level gave a faint but measurable signal from the vibration pickup on the bone. This signal was 45 db below that observed when we placed the vibrating needle on the bony capsule and adjusted it to produce the same cochlear response as the aerial stimulus. When an aerial sound was applied at the apical hole it was not possible at any available level of stimulation to pick up a vibration from the bone. The tests therefore exclude a summation in the bony capsule.

We turn now to the last possibility mentioned, summation in the electrical pathways, which was tested by the following experiment. The usual cancellation was obtained, with one sound applied to the oval window and another to the apical hole, and with the usual round window electrode. Then, without any change in the stimulating conditions, a needle electrode was applied at some point on the surface of the cochlear capsule. For all the low tones, up to about 1000~, it was found unnecessary to make any adjustment in the conditions in order to maintain the cancellation. The magnitude of response was usually altered for each stimulus, but equally, so that the balance between the two remained undisturbed. The previous phase relation also remained unaltered, at least within our ability to determine it.

The needle electrode may be located at any point on the cochlear capsule, and this observation holds true. It is important to mention also that the vectorial summation holds equally well and maintains the same constancy with electrode position when the two stimuli are adjusted in phase to give a maximum rather than a cancellation of responses. When high tones are used in this situation some adjustments of intensity and phase are necessary, as will be described presently.

The results so far permit us to form the preliminary conclusion, subject to a little modification later on, that for the low tones the summation of effects of our two stimuli does not occur in the pathways of electrical conduction. If this summation occurred in these pathways our balance would be upset by moving the electrode to a new position that involves different distances of conduction and different resistances of electrical paths.

It is easy to prove that the conditions of electrical conduction change with electrode position. The amount of change varies greatly with the circumstances: with such things as the particular position of the electrode, the thickness of the bone, the moistness of the surface, and in this experiment the presence of a hole through the cochlear wall. Figure 106 shows the results obtained

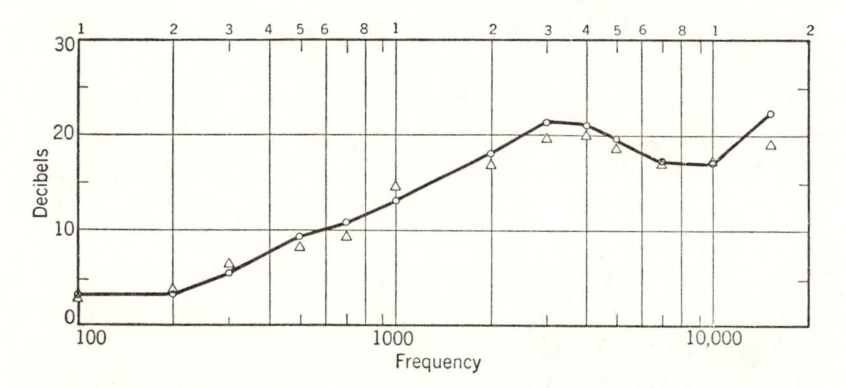

Fig. 106. Differences in the magnitudes of potentials recorded at the two ends of the cochlea of the cat, for both a basal stimulus (circles) and an apical stimulus (triangles). The potentials at the round window always exceed those at the apex, by amounts varying with frequency. From Wever and Lawrence (8) [*Annals of Otology, Rhinology, and Laryngology*].

in an animal of this series when one electrode was on the round window membrane and the other was immediately adjacent to the sound tube that was applied over the apical hole. In this figure are two sets of points, one set (circles) expressing in decibels the ratio of responses picked up by round window and apical electrodes when a constant sound was delivered by way of the oval window and the other set (triangles) expressing this same ratio when the sound was delivered by way of the apical hole. We see in both these sets of data that the responses obtained from the round window electrode always exceed the others, though for the low tones the difference is small. Even for the low tones, however, this difference would be enough to upset the balance required in our cancellation situation if it were not that the same difference appears for both forms of stimulation.

Of the possible sites of summation mentioned in the beginning only two remain, the sound paths in the cochlear fluid and the sensory structures themselves. If the summation occurs in the fluid paths it must be at the terminal ends of these paths, for the initial portions are obviously different. That is to say, the summation must be at the place where the paths reach the sensory structures. Thus for practical purposes these two possibilities reduce to one. We conclude that for the low tones at least the two forms of stimulation, through the oval window and through a hole at the apex of the cochlea, involve the same regions of the basilar membrane and the same hair cells of the organ of Corti.

We have now to examine the results more closely and to consider the high tones. Figure 107 shows the phase changes necessary to restore the cancellation condition when it was first established for an electrode position at one end of the cochlea and the electrode position was then changed to the other end of the cochlea. Points are shown for two different ears. For the low tones, up to 1000~, the readings seem to be randomly distributed about the zero line, with variations of the order of the experimental errors involved in this situation, which are about 5°. Above 1000~ the phase readings rise progressively, indicating that for these tones, and increasingly so the higher the frequency, the two forms of stimulation are not identical. The two forms of stimulation do not differ very much in the magnitude of responses elicited, as Fig. 106 has already shown, but they differ in phase. The phase difference rises rapidly beyond 4000~ and reaches 35° at 10,000~.

We shall now seek to discover more precisely what this difference means. As an aid to our thinking on this problem we begin with a conceptual model in which we consider two groups of hair cells, well separated in the cochlea, which are included in the pattern of some given tone. In this relation, see Fig. 105 above, where the two groups of cells are pictured as *a* and *b*. Sounds introduced at both the oval window and the apex reach these two groups of cells and combine their effects. Let us now assume,

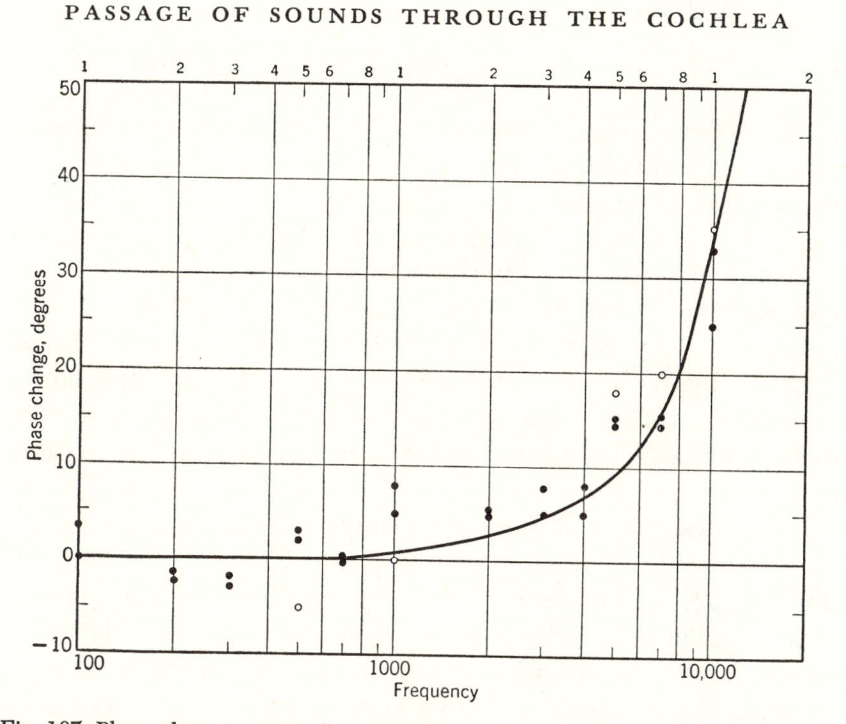

Fig. 107. Phase changes required to restore a null after the recording electrode was moved from the round window to the apex of the cochlea. Results from two cat ears are represented by the two kinds of points. From Wever and Lawrence (8) [*Annals of Otology, Rhinology, and Laryngology*].

for the moment, that the responses recorded from the round window electrode chiefly represent the actions of group *a* and those recorded from the apical electrode chiefly represent the actions of group *b*, and that for each electrode position we can neglect the action in the more remote group of cells.

Let us first record from the round window position and adjust our two stimuli for a cancellation of responses. Essentially what we do is to make the two stimuli equal in amplitude and opposite in phase at point *a*. The sound from the oval window, let us say, tends at a given instant to depress the basilar membrane at this point *a* while the apical sound tends equally to elevate it, and the result is no movement at all and no hair cell stimulation. Our electrode at the round window therefore reports a null.

Now if we turn to the apical electrode we shall not expect a null if the sound waves in traversing the path between *a* and *b* are altered either in amplitude or in phase. Theoretically we should expect both attenuation and phase lag in this passage. Attenuation occurs if any friction is present. A phase lag occurs because the wave takes time to travel this distance. Therefore we should need to readjust the stimuli to reestablish the null. More particularly, if we keep the apical sound constant we should expect to raise the oval window sound by double the amount of the attenuation produced in the path from *a* to *b* and to advance it in phase by double the amount of lag from *a* to *b*; these amounts are doubled because in the first situation the apical sound suffered these changes in passing from *b* to *a* and was adjusted correspondingly, and now the oval window sound must be altered by these amounts plus what it suffers itself in traversing the same path.

Four Basic Considerations. This simplified analysis has brought forward two assumptions and two particular questions for solution. The assumptions are (1) that any tone in its response pattern includes many hair cells spread out over the cochlea and (2) that an electrode best records from the hair cells in its immediate vicinity. The questions concern (3) the degree of attenuation of sound in its passage along the cochlea and (4) the amount of phase lag (or time required) in this passage. These four points are now considered further.

(1) The fact that a sound introduced at one end of the cochlea produces effects that can be recorded with varying effectiveness from electrodes located at various points on the cochlea can mean either that hair cells are stimulated all along the cochlea or that hair cells stimulated at one place radiate their electrical effects through the conducting tissues to all cochlear points. Actually, both these things are true to some extent.

We can produce evidence to show that any sound stimulates every hair cell, though not necessarily in equal degree. All tones spread throughout the cochlea, but the low tones spread with broad and flat response curves and the high tones spread with most of their energy concentrated in one region. The conclusive evidence for a complete spreading of the action of all tones is to be found in the linear form of the cochlear potential curve as a function of intensity. As pointed out elsewhere, this curve could not remain linear over its extensive course of as much as 70 db if a rise of stimulus intensity brought in increasing numbers of active hair cells.

Further evidence on the form of the response curves and the variation of this form with frequency will be presented in a later section.

(2) We see in Fig. 106 that for either source of sound—stimulation by the oval window or through a hole in the apex—the round window electrode always gives larger responses than the apical electrode, but the amount of this difference varies systematically with frequency. This difference varies from 3 db for the low tones to as much as 21 db for the high tones. We can explain this variation with frequency only as due to a different form of distribution of responses for low and high tones and a local bias of the recording electrodes. For the high tones the responses as recorded from the round window electrode are much greater than those recorded from the apical electrode. This is mainly because the principal area of action of these tones is nearer the round window. For the low tones the difference is small because these tones spread broadly over the membrane and affect large numbers of hair cells that are accessible to both electrodes. To strictly local conditions of conductivity at the two electrodes, namely, the difference between the resistance of the round window membrane on the one hand and the resistance of the path through or along the bony capsule on the other hand, we can attribute the difference of 3 db found for the low tones, or even a little more than this if we are willing to say that the difference shown here between high and low tones ought really to be greater than it is.

(3) The close similarity between the two curves of Fig. 106 shows that the form of distribution of cochlear action is the same regardless of the point of introduction of the sounds. This similarity also shows that there is slight attenuation of these sounds as they pass through the cochlea or, more exactly, as they pass over the distance between the areas effectively tapped by the two electrodes. Indeed, the difference between these two curves is a measure of the acoustic attenuation. A more obvious measure is the difference in the responses observed at one electrode for the same oval window and apical stimuli that give equal responses at the other electrode. Such data show the same differences as

already seen in Fig. 106. We note that the acoustic attenuation is positive, as expected. That is to say, the remote stimulus is the less effective to the extent indicated.

The mean difference shown in Fig. 106 over the range 500 to 9000~ (omitting 1000~ for which the results are more variable) is 1.04 db. This value must be halved to obtain the attenuation occurring in a single traverse of the distance between the two electrode areas.

(4) We see in Fig. 107, as already described, the phase shifts necessary to restore a null after changing from an electrode at one end of the cochlea to one at the other end of the cochlea. We see that for the low tones, those below 1000~, the balance between the two stimuli is disturbed but little, or below the limits of experimental variations. For the high tones a readjustment of the phase is necessary. It is instructive to translate these phase changes into time. This we do on a basis of the smoothed curve of Fig. 107 and show the results in Fig. 108. We find that our phase

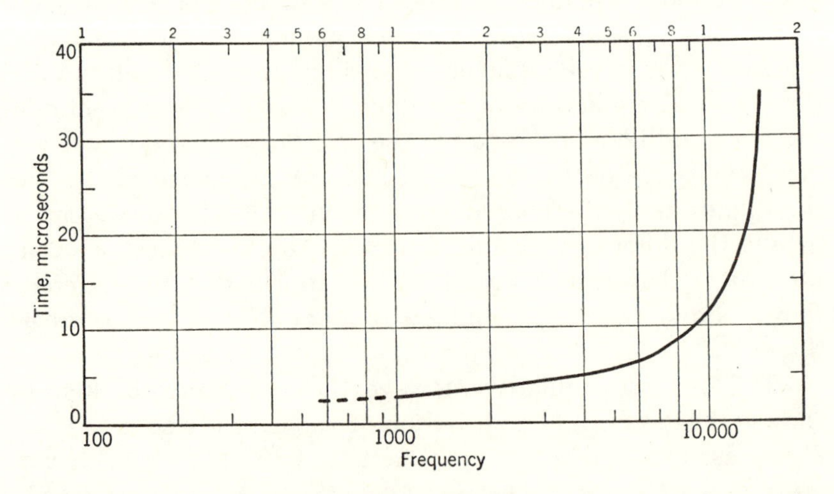

Fig. 108. Time differences corresponding to the phase lags shown in the preceding figure.

lags are not representing a constant time delay. Rather, the delay increases progressively with the frequency. This fact evidently reflects the changing forms of the response patterns and the increasing distances between the areas that are being tapped by the two electrodes.

To make this point clear, let us return to Fig. 105, whose conditions we now treat more rigorously than we did initially. At a given moment the oval window stimulus reaches point *a* with a certain intensity and phase, and it also reaches point *b* after a little delay and a slight diminution of intensity. The apical stimulus similarly reaches point *b* and point *a*, with delay and attenuation at *a*. If our electrode is at the round window we are emphasizing the responses from *a*. If we could neglect the responses at *b* as we did at first, then to obtain a null we should have to adjust the apical stimulus to an intensity giving an effect at *a* equal to that of the basal stimulus at *a* and make its phase contrary to that of the basal stimulus. Actually, the responses at *b* cannot altogether be neglected, but still our adjustments are made primarily with regard to the situation at *a*. Now when we shift to the other recording electrode and place the emphasis upon the action at *b* we shall no longer expect to find a null, and to regain it we expect to make adjustments of intensity and phase.

The fact is that we do not need to make any adjustments for the low tones. This means that for these tones the areas tapped by the two electrodes are too close together or that the differences are too small to detect by our method (i.e., are less than 5°). We do find a difference for the high tones as shown.

One further point needs to be clarified. If, as this analysis indicates, our two electrodes are recording from somewhat different areas of action, and if, as we find for the high tones, it is necessary to make a readjustment of phase when we change to a different electrode, it follows that for each electrode separately a null is obtained for these tones by a balancing of phase relations that differ somewhat among the active cells. More specifically, we are balancing a departure from zero in one direction in the cells nearest our electrode against a contrary departure in the more remote cells. The departures from zero phase relation in the bulk of these cells is probably of the same order of magnitude as the phase differences found for our two electrode positions. The summation for these different cells occurs in the paths of electrical conduction.

Obviously, this discussion has represented the selective action of our electrodes in too simple a manner. We need to think of a continuous gradient of receptivity: an electrode records most effectively from the cells in its immediate vicinity and progressively more poorly for more distant cells.

The Electrical Conductivity of the Cochlea. We have carried out experiments to discover the degree to which electrical potentials are attenuated as they are conducted through the cochlea. In this study we treated the living tissues simply as a conductive material and measured the electrical fields established by externally applied potentials. The fields obtained depend greatly upon the manner and site of application of the potentials, because tissues in general, and especially those making up the ear, have complex electrical properties. Of interest for our present problem are the results obtained by applying voltages to the basal end of the cochlea either at the round window or between oval and round windows.

For one series of measurements, which we shall call the round window series, the voltages were applied by means of concentric electrodes made by inserting a fine insulated wire into a hypodermic needle and grinding the tip off flat. The radial separation of inner and outer electrodes was 0.18 mm. This pair of electrodes when placed in contact with the round window membrane produced an electrical field of the general order of that shown in *A* of Fig. 109. The resulting voltages were recorded with a similar pair

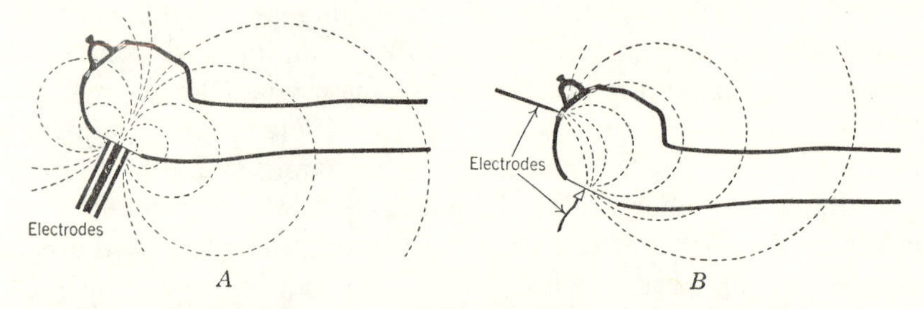

Fig. 109. A conceptualization of the electrical fields set up by our two methods of introducing voltages into the cochlea. At *A* we used a concentric electrode on the round window membrane and at *B* we used two separate electrodes at oval and round windows. The broken lines represent lines of current flow.

of hypodermic needle electrodes applied to the bone of the cochlear capsule either adjacent to the round window or near a small hole drilled in the apex of the cochlea (the same size and location of hole as in the experiments already described).

For the other series of measurements, called the two-window series, one electrode was a platinum foil on the round window membrane and the other electrode was a needle in contact with the stapes. The voltages were then passed between the two windows, and produced a field something like *B* of Fig. 109. These voltages were picked up by means of an "active" needle electrode on the bony surface either near the round window or near the apical hole and an "indifferent" electrode inserted into the masseter muscle, just as in our usual recording of cochlear potentials resulting from acoustic stimulation.

The results of these two series are shown in Fig. 110 as the

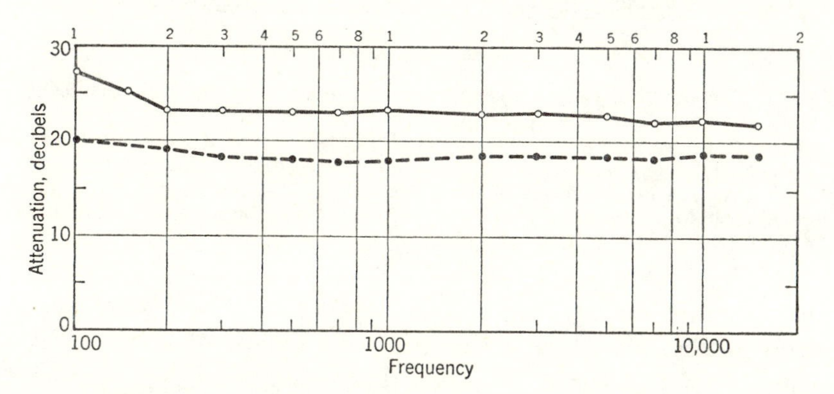

Fig. 110. The electrical attenuation between base and apex of the cochlea, measured for two methods of introducing potentials at the basal end, by a concentric electrode at the round window (solid line) and by a pair of electrodes at the two windows (broken line).

differences in decibels between the effects at basal and apical positions, for various frequencies of input between 100 and 15,000~. The differences thus represent the loss of potentials from one end of the cochlea to the other. There is no obvious relation to frequency, which signifies that the conductive path is almost wholly resistive. For the round window series the mean loss is 23.7 db and for the two-window series the mean loss is 19.0 db. On the basis of these results we estimate that potentials originating within the cochlea at its extreme basal end suffer a loss of about 20 db (a tenfold reduction) in passing along the cochlea to an electrode placed near a hole in the apical end.

Effects of the Potential Gradient. We can obtain a clearer idea of the effect of this gradient along the cochlea if we take a response curve for some tone—either an assumed curve or one representing the best available data—and show how the effective form of the curve ought to change according to the electrode position. Figure 111 represents first, in the uppermost curve, the form of

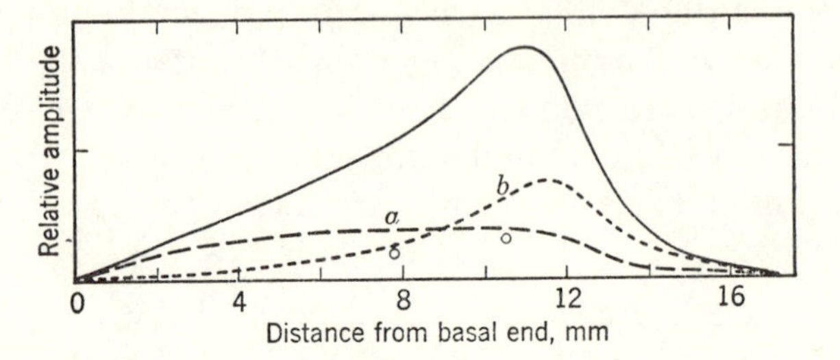

Fig. 111. Hypothetical view of the effects of the potential gradients in the cochlea on recorded potentials. The uppermost curve shows the form of cochlear stimulation as represented by Békésy (*14*) for a tone of 800~. Curve *a* shows the effective form as seen by an electrode at the basal end of the cochlea and curve *b* shows the form as seen by an electrode at the apical end. The centers of area of these derived curves are indicated by small circles, one 7.8 mm from the basal end and the other 10.5 mm from the basal end.

the response function for 800~ according to Békésy's results as given in Fig. 98, rescaled for the cat's length of cochlea. The ordinates of this curve have now been multipled by a gradient that varies continuously from unity at the basal end of the cochlea to one-tenth (−20 db) at the apical end. This weighting gives curve *a* of this figure, which represents the action as seen by our round window electrode. The ordinates of the original curve have also been multiplied by the reverse of this gradient, giving curve *b* of the figure, which represents the action as seen by our apical electrode. If we determine the centers of area of the two curves we find that they lie at points 45 per cent and 61 per cent of the distance from the basal end of the cochlea, and these points are separated by a distance equal to 16 per cent of the length of the cochlea.

If we carry out this procedure in the same manner for 10,000~, using a curve from another source, namely, our estimations from data on stimulation deafness and maximum responses in the cochlea (Wever and Lawrence, *3*), we obtain centers of area of 10 and 52 per cent respectively, with a separation equal to 42 per cent of the length of the cochlea. It is clear that the progressive change in form of the high-tone functions, leading toward an increasing participation of the basal portion of the cochlea, has brought about an increasing separation of the centers of the effective electrode areas. For a middle tone like 800~, whose response is spread out over the middle portion of the cochlea, the two electrodes effectively record from adjacent areas. For a high tone like 10,000~, the round window electrode effectively records from the main response area at the basal end, whereas the apical electrode chiefly represents the extended "tail" of the function. The two areas are then well separated.

Our results as given in Fig. 106 show a differential between electrodes at the two ends of the cochlea that amounts to 18 db or 8-fold for the high tones. This change represents two traverses of the effective distance between electrodes, and thus the change for one traverse is 4-fold or 12 db. We note that this 4-fold change is in agreement with the calculation of the effective separation of electrode areas just made for 10,000~, for a 4-fold change bears the same relation to our estimate of 10-fold for the whole gradient that a separation of 42 per cent bears to the whole length of the basilar membrane.

If provisionally we accept this distance of 42 per cent as representing the effective separation of the areas tapped by our electrodes during stimulation with 10,000~, we shall be able to calculate the speed of sound conduction through the cochlea from the measurements of phase shift. In the average cat's cochlea this percentage represents a distance along the basilar membrane of 8.9 mm. However, because of the spiral form of the cochlea, the mean distance through the fluid pathways is less, as Fig. 112 shows. This distance of 8.9 mm along the basilar membrane reduces to 6.8 mm in the fluid pathways. At 10,000~ the phase shift was measured as 35°, and considered for a single traverse is one-half of this amount or 17.5°, which represents a time lag of 4.86 microseconds. The distance of 6.8 mm covered in 4.86 microseconds represents a speed of 1400 meters per second. This figure of course is not to be taken very seriously, yet it is of the order of magnitude of sound conduction through open water, which at body temperature is 1563 meters per second.

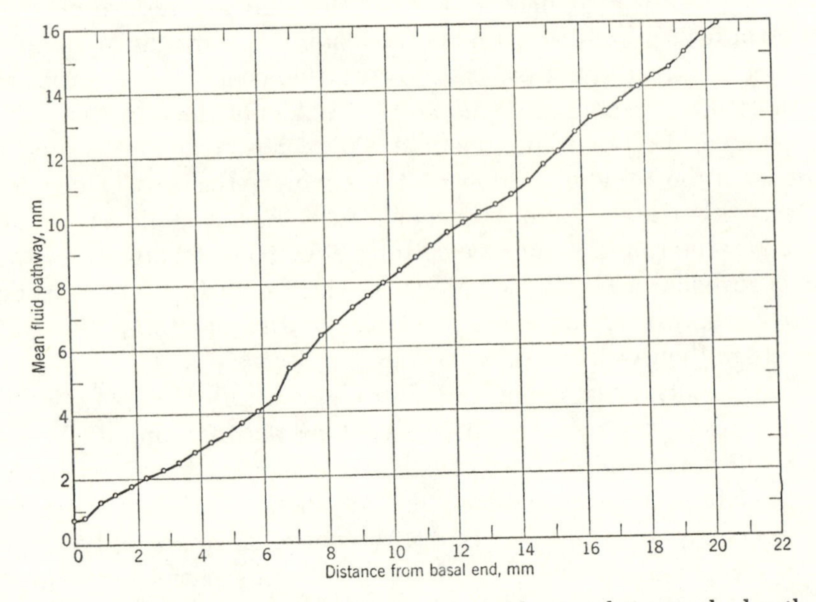

Fig. 112. Length of fluid pathways in a cat's cochlea in relation to the length of the basilar membrane. Both dimensions are measured from the basal end.

FINAL THEORETICAL CONSIDERATIONS

We are ready to return to our evaluation of the traveling wave theories. It will now be apparent that the results of this experiment are quite out of harmony with such theories. According to the basic assumptions of these theories our introduction of two tones at opposite ends of the cochlea should produce a pair of waves traveling in opposite directions. With heavy damping, as these theories usually assume for high tones, the waves should expire without meeting, as shown in *A* of Fig. 113. With lighter damping, as assumed for low tones (or assumed for all tones in certain of the theories), these waves should meet to form a standing wave pattern, as in *B* of this figure. In either event our two electrodes at opposite ends of the cochlea have equally advantageous positions relative to the wave pattern and should record the same response. Here we must bear in mind that we have compensated for any local variations of impedance at the two ends of the cochlea by adjusting the tones separately to produce the same response as measured by the electrode at the adjacent location.

Yet the responses obtained for the two tones presented either singly or together always differed at the two electrodes, and for the high tones they differed by as much as 18 db.

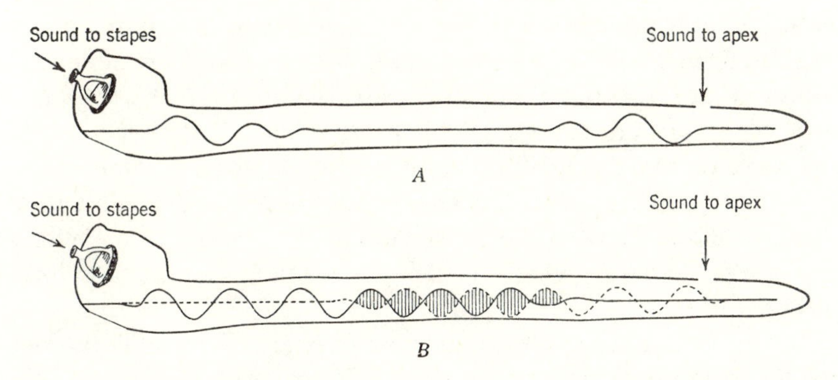

Fig. 113. Cochlear patterns resulting from simultaneous stimulation at both ends of the cochlea, according to a traveling wave theory, under conditions of high damping (*A*) and moderate damping (*B*).

We have to conclude that no such patterns as those of Fig. 113 are formed. Both oval window and apical stimuli pass freely and rapidly through the cochlear fluids and affect the basilar membrane in the same region and in the same fashion. The two stimuli together produce a pattern of action differing only in magnitude from that produced by one of them alone. This pattern varies in form according to frequency as already described. For the low tones the pattern is so broad that the two electrodes at opposite ends of the cochlea record it about equally well. For the high tones the basal electrode has an advantage in being nearer the region of greatest activity. For these tones, therefore, the two electrodes give detectable differences in terms of intensity and phase.

These intensity and phase differences are small, however. For the high tones the measured intensity difference is of the order of 1 db. We must bear in mind that this difference is for a double traverse between the centers of the effective electrode areas. A single traverse, with which we are of course concerned in normal stimulation, would suffer only half this attenuation. We must also bear in mind, however, that a whole electrode area determines the value as measured, and when we say that the center of one area differs from the center of another area by some amount we must allow regional departures from this amount, especially for portions of the area remote from both centers. However, by the very nature of the summation that is envisaged here, a summation in which regions near an electrode are enhanced in their effects, it follows that the fringe regions—regions remote from the centers—either cannot suffer large attenuations relative to the center of the area or must contribute only negligibly to the total effect. It is a reasonable guess that the principal effect of a high tone extends to about twice the separation of the centers of our effective electrode areas, which would give 1 db as the effective total attenuation.

We apply much the same reasoning to phase. For the high tones we measure phase differences up to 35° for 10,000~. For a single traverse this difference reduces to 17.5°. Portions of the responding system remote from the effective electrode center may suffer larger phase changes, but if they do so they must assume relatively insignificant parts in the total action. Let us take 35° as the extent of the phase variation for the bulk of the response wave at 10,000~. This variation of 35° is only one-tenth of a wave length, an amount that is wholly inadequate to support a traveling wave theory. For lower tones down to 1000~ for which the results indicate still smaller phase changes, and for still lower tones for which no phase change can be measured, the response can cover only an insignificant portion of a wave length. Even if we admit an experimental error of 5° for these tones, which is liberal for the lower frequencies, we still may declare that the active region can represent no more than 1⁄72 of a wave length.

We see from these results that the speed of sound conduction through the cochlea is so great that only a fraction of a wave of sound pressure can exist in the cochlea at any instant, and we see also that the basilar membrane responds to this pressure by movements that are practically in phase over the most active region of the membrane.

These are the conclusions that we arrive at from our study of the living ear of the cat. As we have seen, Békésy has drawn very different conclusions from his examination of artificial models and of dead human ears. We can of course point to possible differences in behavior of living and dead tissues. Also, as is obvious, we are dealing with two different species. Yet the hearing of the cat is so much like man's and its cochlear structure is so similar in general that we can attribute little of the theoretical disparity to this cause. There is a scale difference to be sure; the cat's cochlea is smaller than man's by a ratio of 2:3, and correspondingly the sensitivity of this animal is shifted upward along the frequency scale compared with man's to an extent of about half an octave. Yet these variations are not such as to justify one theory of hearing for man and a different theory for the cat.

In this connection it is important to emphasize what was said earlier in the presentation of Békésy's theory, that his observation of traveling waves was made in a mechanical model and not in the actual cochlea. A model, of course, may be made to do many things, and nearly every proposer of an auditory theory has sought to fortify his position by constructing a model that invariably is found to perform according to his views. Békésy only inferred the existence of traveling waves in the actual cochlea from observations of variations in amplitude and phase as a function of cochlear position and stimulus frequency. These variations he found to be consistent with the predictions of a traveling wave theory. The recent observations of Tasaki, Davis, and Legouix, carried out in the guinea pig, also show variations in the magnitude and phase of responses in the cochlea that are consistent with this theory.

Such consistency, of course, affords a measure of presumption in favor of a traveling wave theory, but it does not amount to a proof of such a theory. The indicated form of action could arise simply as an expression of the local characteristics of the basilar membrane. Particular regions when exposed to the vibratory energy communicated to them by the cochlear fluid may respond with amplitudes and phases determined by their own impedances and time constants. If different regions of the basilar membrane vary systematically in these characteristics their behavior may give the appearance of a traveling wave, in that the region of greatest activity seems to move along the cochlea. This is not a traveling wave in fact, however, and the distinction is more than one of descriptive terms. The real question has to do with the way in which a given region of the basilar membrane obtains the energy that makes it vibrate. The traveling wave theory states that this energy reaches the region wholly or in the main by way of neighboring regions of the membrane. What happens at one point of the membrane is thus determined by previous actions in other regions of the membrane through which the energy has passed. Our evidence indicates, on the contrary, that each region responds simply to energy communicated to it immediately from the cochlear fluid.

In the understanding of this distinction it is well to recall our analogy of the rope that is caused to execute traveling waves by movements imparted to one end. The energy is transferred through the rope itself. If someone should grasp the rope somewhere along its course and hold it firmly the traveling wave would proceed to this point and no farther. Similarly, if there are traveling waves in the cochlea, a restraint of motion at one place along the basilar membrane must cause the more remote regions of the membrane to remain inactive. Then, presumably, high tones would still produce patterns much as before and would still be heard, but low tones would give waves that failed to reach their proper regions in the cochlea and their perception would be altered or prevented. A fortunate observation made by Guild (3) is worthy of attention in this connection. He found in the Johns Hopkins collection of serial sections of ears an instance in which an aberrant blood vessel was present in the lower apical region of the scala tympani. This vessel connected the midportion of the basilar membrane to both the bony spiral lamina and the bony septum between apical and middle turns of the cochlea. This connection must have caused an appreciable damping of movements in this region of the basilar membrane, yet tests made of this ear before the person's death showed that the hearing for low tones was within normal limits.

Another anatomical variation needs to be mentioned in this relation. It is frequently observed in the study of serial sections of the ear that the extreme basal end of the basilar membrane for a distance of 2 mm or more is greatly thickened and often heavily calcified. Mayer observed this condition in the ears of old persons, whereas Crowe, Guild, and Polvogt found it sometimes in young persons as well. Such a basilar membrane must be rigid at the basal end, and incapable of taking up the vibratory movements of the cochlear fluid and transmitting them apicalward in the manner demanded by a traveling wave theory. And yet the evidence shows that these persons had normal hearing for all tones at least up to 8192~. These anatomical variations therefore present a serious obstacle to the traveling wave theories.

The objections that have been raised to the traveling wave theories by our experimental observations apply in a simple manner to all the regular forms of these theories in which sound is transmitted from the stapes to the fluid and then to the basilar membrane either by itself or with participation by the fluid. Such theories are those of Békésy, Ranke, Reboul, and Zwislocki, as well as some of the older theories. Certain of the later theories have taken more complex forms and require further discussion.

DeRosa's theory, it will be recalled, assumes two traveling waves, one in the fluid and the other in the basilar membrane, both progressing at relatively slow speeds. Our observations are contrary to his assumption that the speed of the fluid wave is around 30 meters per second for most frequencies and even slower for the low frequencies. Also, our observations do not support his assumption of large viscosity effects and consequent rapid attenuation of the fluid wave.

The Peterson and Bogert theory postulates a pressure wave in the scala vestibuli and another pressure wave in the scala tympani, each moving from base to apex at a rapid rate, at the speed of sound through water. However, according to this theory the stimulation of the basilar membrane is not caused directly by these waves but by their pressure difference, and the wave of pressure difference travels at a slower speed, between 350 and 7 meters per second for low tones. Sounds introduced into the apical end of the cochlea, as in our experiment, then should produce positive waves and a pressure difference wave progressing in a direction contrary to the normal one. The stimulating effects of sounds applied at the two ends of the cochlea as in our experiment would be just as in the other forms of traveling wave theory. Our results, therefore, are inconsistent with this theory. The same statement holds for the Huggins theory, which derives the pressure difference wave in the same way.

Fletcher's theory requires more searching consideration. Initially it is a direct-action theory. It supposes that pressure waves are propagated through the cochlear fluid with the velocity of sound through water, and the individual "elements" of the membrane—little regions that at this stage are considered as working in essential isolation from one another—respond to this pressure according to their own resonant characteristics. So much of this theory is in full accord with our observations.

The theory assumes further, however, that the pressure wave is subjected to considerable attenuation, and for the high tones this attenuation is sufficient to extinguish the wave after it has traveled only a short distance in the basal region of the cochlea. Our results reveal little attenuation and show that the pressure waves even for the highest tones extend all the way through the cochlea.

In Fletcher's theory, when once the elements of the membrane begin to respond to the pressures exerted upon them by the fluid, we enter into a second stage in which a traveling wave appears. Now the elements of the membrane are no longer considered as isolated, but they communicate their motions from one to another in a particular manner. This communication is from a given element to one farther up the cochlea. A wave thus starts at the basal end of the cochlea and travels apically, gathering energy and rising in amplitude as it travels, until it attains a maximum at some characteristic place. The speed of this wave is relatively slow. Its presence profoundly modifies the pattern of action initially produced, and presumably it largely determines the form of nerve excitation.

On first examination this theory seems to be consistent with our results, apart from its assumption of a rapid decrement for the pressure wave. A stimulus delivered to the apex of the cochlea should quickly traverse the cochlea by way of the fluid and set up a traveling wave running from the basal end apicalward just like a stimulus presented by way of the stapes. Simultaneous stimuli at the two ends of the cochlea as in our experiment should simply combine their effects, as we find to be the case.

Further consideration of this theory brings forth a few questions, however. It is difficult to understand why the initial actions of the elements are not stimulating, and how the stimulation process can be delayed until the traveling wave emerges and reaches its maximum. The cochlear potentials show no time lags commensurate with expectations on this hypothesis. It is also difficult to understand why an element should be able to transfer its motion to its neighbor on the apical side and not to the one on its basal side. Neighboring elements can hardly be presumed to vary in physical characteristics so profoundly that this variation amounts to a one-way valve for energy flow. Yet if any appreciable back flow is possible our apical stimulus should be able to set up a basally directed wave, with results unfortunate for the theory.

To us this second stage of Fletcher's theory seems superfluous. The first stage has already achieved a patterning of action over the basilar membrane that varies with frequency in the manner that the place principle requires. The derivation of a traveling wave is a secondary complication that serves no purpose that we can justify. The intended purpose is, of course, to give agreement with observations like Békésy's that seem to indicate the presence of a traveling wave, but, as we have pointed out, the observations do not require so elaborate an interpretation. A simpler theory, developed along the lines laid down in the first part of Fletcher's discussion, is fully adequate to explain the evidence before us and to support the place principle to the extent that the facts of hearing demand.

All of the objections that have been raised against the traveling wave theories hold equally well for the standing wave theories. On the basis of their assumptions, our experimental situation—with tones applied at the two ends of the cochlea—should produce a standing wave pattern extending throughout the cochlea. Our electrode when placed at various points along the cochlear capsule then should reveal variations of potential corresponding to the nodes and loops within. Nothing of the kind can be found.

14

Cochlear Patterns and the Stimulation Process

We have traced the passage of sounds through the peripheral conductive mechanism and into the cochlea, and then along the cochlea itself. We have still to consider in detail the pattern of action on the basilar membrane and the nature of the final processes of stimulation.

THE PATTERNS OF STIMULATION IN THE COCHLEA

The foregoing discussions have brought out two fundamental features of the action patterns of sounds in the cochlea. These features are the pervasiveness of the action of all tones and the variations in the form of the action patterns as a function of frequency. We now consider these features in detail.

THE RANGE OF TONAL ACTION

Numerous lines of evidence support the view that every stimulating tone reaches all parts of the cochlea and affects every hair cell, though not necessarily in equal degree. Perhaps the simplest evidence comes from the stimulation deafness experiments.

Stimulation Deafness Experiments. When the ear is stimulated with a pure tone and the intensity is progressively raised the cochlear potentials increase at first as a linear function of the sound pressure, as already described, and then at a high intensity level they depart from this form, reach a maximum, and thereafter decline. At these extreme intensities the ear is endangered, and if the rise of intensity is continued to the point where the responses fall off markedly an injury results. After such injury the original intensity function can no longer be obtained, and the same stimuli will now produce a new function of a lower magnitude than the first, as Fig. 114 shows.

What is of particular interest in this connection is the fact that

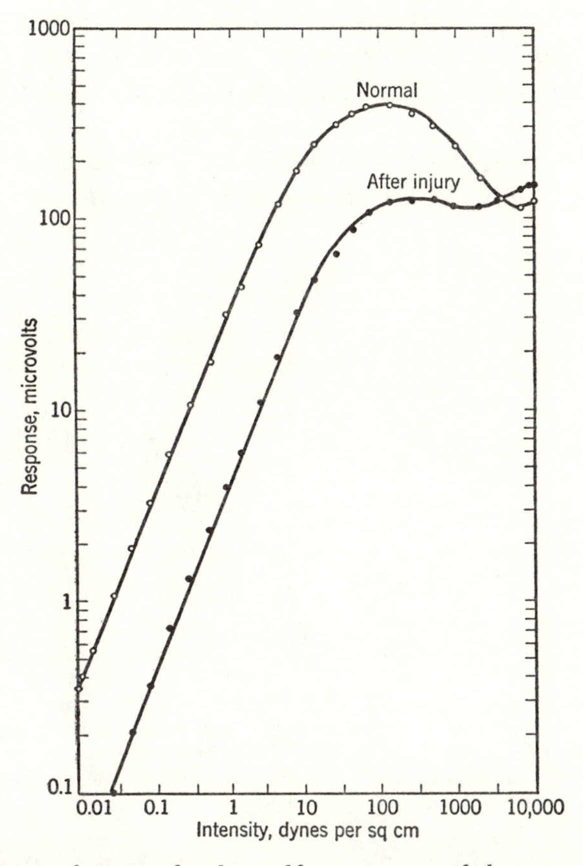

Fig. 114. Intensity functions for the cochlear responses of the cat on stimulation with a 1000~ tone, before and after injury by overstimulation with this tone.

after injury has been produced by overstimulating with one pure tone the cochlear potentials are impaired not only for this tone but for every tone throughout the audible range. This fact is shown in Fig. 115, where four curves are given, representing the losses of sensitivity incurred by four different groups of guinea pig ears after stimulation with tones of 300, 1000, 5000, and 10,000~. This figure brings out the further fact that for all but the highest stimulating tones the impairments bear no systematic relation to the stimulating frequency. The impairments are no greater in the region of the stimulating tone than elsewhere, but rather are general over the frequency range. For 10,000~, however, a regional variation is evident; this tone produces the most serious impairments at the high-frequency end of the range and progressively smaller impairments for the lower tones. Even this high tone, however, gives effects throughout the whole range studied.

A histological examination of ears after this kind of stimulation reveals no impairment of the middle ear mechanism. Indeed, our experiments, described in Chapter 9, have shown that this mechanism is able to withstand even greater levels of stimulation than any used here. The impairments occur in the inner ear as changes in the delicate structures of the organ of Corti.

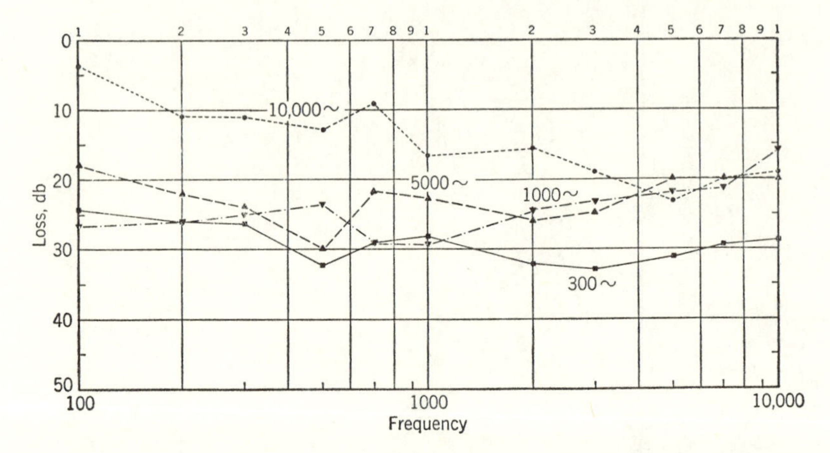

Fig. 115. Stimulation deafness as a function of the exposure frequency. In different groups of guinea pigs four exposure tones were presented, each at an intensity of 1000 dynes per sq cm and for a period of 4 minutes. The resulting impairments of cochlear potentials are shown in decibels below the original level. From Smith and Wever [*Journal of Experimental Psychology*].

The structural damage varies in degree according to the intensity and duration of the stimulation, and it varies regionally over the basilar membrane. It varies also according to the susceptibility of the individual ear, for some ears are particularly fragile and others are resistant to this treatment. The earliest stages of damage that are observable microscopically consist merely of alterations of staining qualities of the hair cells and a swelling of these cells. There is no doubt, however, that even milder degrees of injury occur, for a diminution of cochlear potentials is found in many ears in which a subsequent histological study reveals a cochlea that is altogether normal in appearance. The modifications that have been produced in these cochleas by the stimulation are of a subtle kind, not disclosed to visual examination. No doubt they consist of biochemical changes, perhaps something like the loss of nucleic acids and other proteins that Hamberger and Hydén by special methods have disclosed in the cochlear ganglion cells after overstimulation. From these mild forms the damage proceeds by degrees to the extreme stage in which the organ of Corti is completely disrupted and only a bare basilar membrane remains.

Under the conditions described for Fig. 115, when both the intensity and duration of stimulation are kept constant, the locus and range of damage produced in the cochlea bear a systematic relation to the frequency of the stimulating tone. Low tones produce damage in the apical region of the cochlea, usually extending over about two-thirds of the basilar membrane altogether, whereas high tones produce more restricted damage in middle and basal regions. These differences are shown in Fig. 116. When the intensities are extreme and the stimulation is prolonged the damage is even more extensive than that shown here. It is clear from these results that every tone at high intensities extends its effects throughout the cochlea.

We can also consider this problem in another way, by showing that limited regions of the cochlea will give responses to all tones. Figure 117 represents the ear of an animal that was first tested for normal responses and then was strongly stimulated with 300~, and finally was examined after a period of three weeks. The histological result as shown was the complete absence of the organ of Corti except for small portions at the two ends. Yet it was possible to record cochlear potentials to tones for all frequencies, when the stimulation was raised to high levels.

Form of the Intensity Function. That the stimulating effects of tones are extensive at low as well as at high intensities is shown in a consideration of the intensity function for the cochlear potentials. As we have seen, the cochlear potentials vary in magnitude as a linear function of sound pressure over an extensive range, from the lowest level at which they are measurable under present conditions (below 0.1 microvolts) to as much as 300 microvolts for certain tones, a range of more than 70 db. These potentials are the product of numerous elements working in parallel and adding their individual effects.

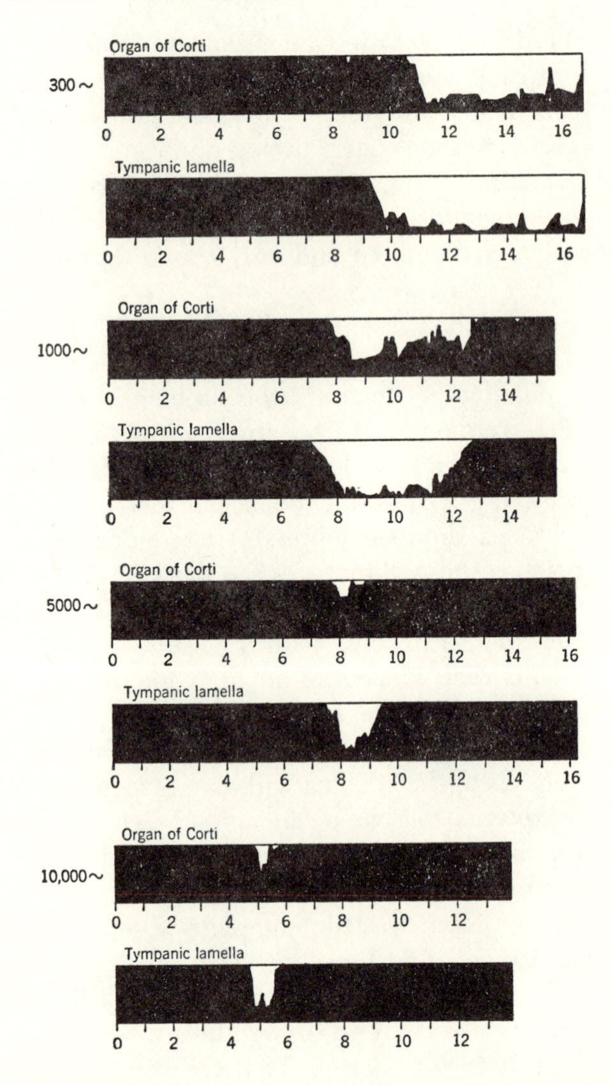

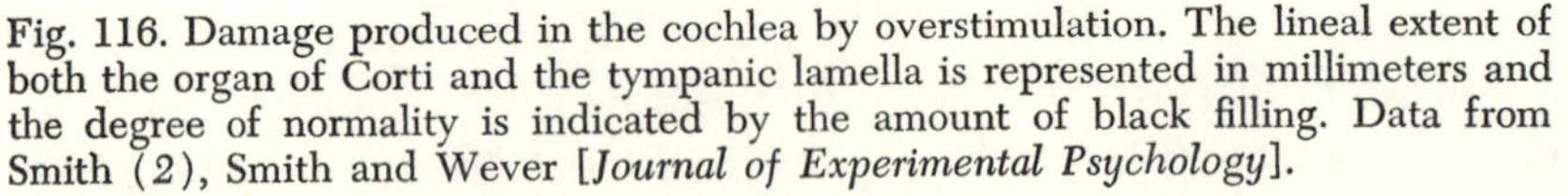

Fig. 116. Damage produced in the cochlea by overstimulation. The lineal extent of both the organ of Corti and the tympanic lamella is represented in millimeters and the degree of normality is indicated by the amount of black filling. Data from Smith (2), Smith and Wever [*Journal of Experimental Psychology*].

The evidence that the recorded potential is contributed to by numerous hair cells comes from the observation that the potentials are reduced in magnitude for all tones after damage to a portion of the cells. Most convincing are the results of chronic experiments, in which a delay of some weeks is interposed between the

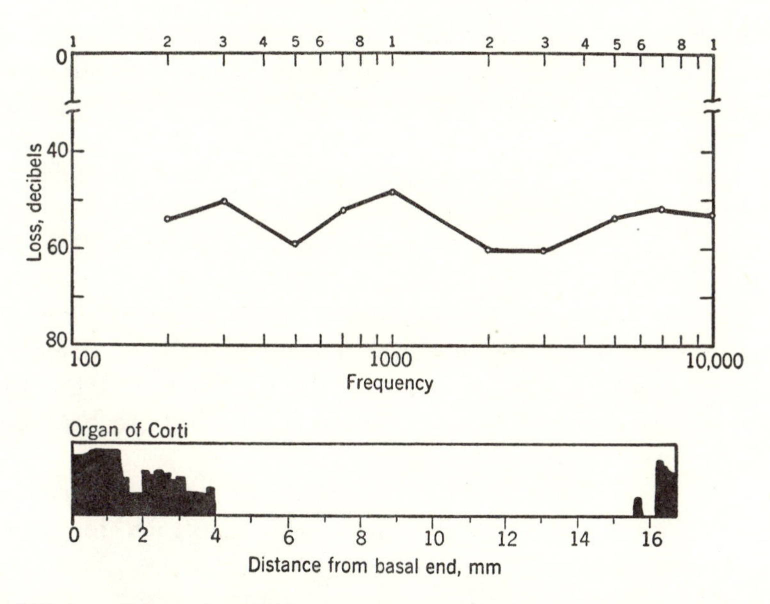

Fig. 117. Impairment of cochlear potentials and cochlear damage produced by an intense tone of 300~. A period of three weeks elapsed between the injury and the final study. From unpublished observations of I. E. Alexander and F. J. Githler.

overstimulation and the measurement of cochlear potentials, for then the pathological changes initiated by the stimulation have become stabilized. In such experiments there is a close correlation between the number of hair cells and the magnitude of cochlear potentials produced by a given stimulus. Alexander and Githler (2) in a study of the chronic effects of jet engine noise on the ear found a correlation of 0.9 between the over-all sensitivity as measured by cochlear potentials and the number and condition of the hair cells in the cochlea.

Potential generators in parallel combine their effects additively whenever the internal impedance of the generators is large relative to the impedance of the external circuit. From the results as described we conclude that the hair cells have a high internal impedance relative to the field in which their potentials are radiated and picked up by our electrodes. The same conclusion has been drawn in the study of nerve fibers, which likewise add their individual potentials.

As has been pointed out elsewhere (Wever, 5), the linear form of the intensity function for the cochlear potentials over a range as extensive as 70 db can only be accounted for on two assumptions, that the voltage output of the individual hair cell bears a linear relation to the sound pressure and that the number of cells in action does not change with intensity. Any other assumptions are unreasonable. If the number of active hair cells should increase with intensity the resulting function would not be linear but accelerated, unless it happened that the output of the individual cell were decelerated in exactly the same degree. For this fortuitous combination of functions to hold true over a range as extensive as 70 db is practically unthinkable. It becomes altogether unreasonable when we add the evidence that the function remains linear after many of the hair cells have been damaged, as Fig. 114 has shown.

From these considerations we conclude that the extensive spread of cochlear stimulation that was demonstrated for tones at high intensities holds likewise at low intensities. Every tone spreads its effects throughout the cochlea in a characteristic pattern whose form does not change with intensity but only varies in magnitude over a range extending from below threshold intensity to the beginning of overloading. After overloading sets in, the pattern becomes even broader than before.

THE FORMS OF THE ACTION PATTERNS

We wish to know the relative amplitudes of movement produced by a tone from point to point along the cochlea and how this pattern of movement varies according to the frequency of the tone. We have already discussed a number of results pertaining to this problem. We have seen in Fig. 98 the results reported by Békésy from his attempts at direct visual observation of the movements. Though these observations were difficult to make and were carried out in dead specimens rather than in the living ear, and any one view included only limited portions of the basilar membrane, we have no doubt about their general indications. They show that all tones affect broad regions of the cochlea and present systematic variations in form as a function of frequency, and with the low tones giving broader and more gently sloping curves than the high tones do.

For further information on this problem we have to depend upon results obtained by other methods, methods that are less direct but which have other compensating advantages, especially the advantage of dealing with the living, functioning ear. We have already referred to some of these results. The stimulation deafness experiments show clearly the variations in the forms of the patterns for tones of different frequencies. When sound pressure is kept constant the high tones reach an injurious level of action only over a limited area of the basilar membrane, whereas the low tones reach this level over a large area. The areal difference is about tenfold from 300 to 10,000~ under the conditions of that experiment, when the tones are presented to the guinea pig ear for 4 minutes at an intensity of 1000 dynes per sq cm, as Fig. 116 has shown.

We have seen the same variation of form reflected in the results of Fig. 106, where a difference of sensitivity is found between the two ends of the cochlea. As the foregoing discussion has brought out, this difference arises from the changing forms of the response curves as we pass from low to high tones.

Further Study of Intensity Functions. A comparison of the forms of the intensity functions for the cochlear potentials produced by different tones gives further information on this problem. The intensity function of a tone, as illustrated in Fig. 48, shows two features of significance in this relation. One is the level of potential at which the curve first departs from a linear form and the other is the maximum of potential attained.

As will be seen, the curve departs only gradually from linearity, and the designation of a limit is somewhat a matter of judgment. In this example the departure seems to be at a response level of 320 microvolts, but we must allow a variation of a few per cent about this figure. A degree of uncertainty is unavoidable in view of irregularities in the readings and the gradual change in form of the curve. It is somewhat easier to determine some particular amount of departure from linearity, such as a departure of 1 db.

The maximum response is more easily determined, though here too there are difficulties. At the extreme intensities necessary in this region the ear is unstable and particularly subject to injury. It is necessary to exercise more than ordinary care to avoid either temporary or permanent injuries that make further readings uncertain. Even a temporary injury, though fully recovered from in a few seconds or minutes, will make the ear more variable and more susceptible to additional damage. In these experiments we have explored the region of the maximum with particular care, applying each sound only for a second or so as needed to obtain a reading, raising the intensity by small steps as the maximum is approached, and stopping the series once a decline of the response shows that the maximum has been passed. Some ears will not withstand the stimulation involved in even this cautious procedure, but are injured. We have found a particular susceptibility to injury in ears that are below normal in sensitivity and especially in ones in which the presence of thickened membranes and scar tissue gives evidence of a former middle ear infection.

When cochlear responses are carefully measured at the round window for a number of tones it immediately becomes obvious that the functions for the low tones are much alike whereas those for the high tones show systematic differences. This is true for the guinea pig and cat alike.

Series 1. The first experiments to be described were carried out on guinea pigs, with the recording electrode on the round window membrane (Wever and Lawrence, 3). Some of the results are shown in Fig. 118. This is a composite graph in which variations in sensitivity for the different tones have been compensated for and the functions have all been made to coincide along their linear portions. In order to avoid confusion in this figure we have omitted functions for 50, 100, 200, and 500~, all of which belong in the region occupied by 300, 700, and 1000~. These functions for tones between 50 and 1000~ exhibit no marked variations from one another. Those for the high tones, on the contrary, distribute themselves in an orderly fashion as illustrated. As the frequency rises these curves show departures from linearity at progressively lower levels and attain maximums of progressively smaller amounts.

Some of the curves of this figure do not actually pass through their maximums because in this instance we were more interested in the lower portions of the functions and wished to avoid all risk of injury. Nevertheless, the individual courses of the curves are sufficiently well defined for our purpose.

A third feature is a variation in the forms of the curves in their departures from linearity. For the low tones the curves bend only gradually away from the linear form. For the high tones they break away somewhat more abruptly.

The systematic variation of the maximums for the high tones is disturbed beyond 10,000~. The curve given for 15,000~ does not

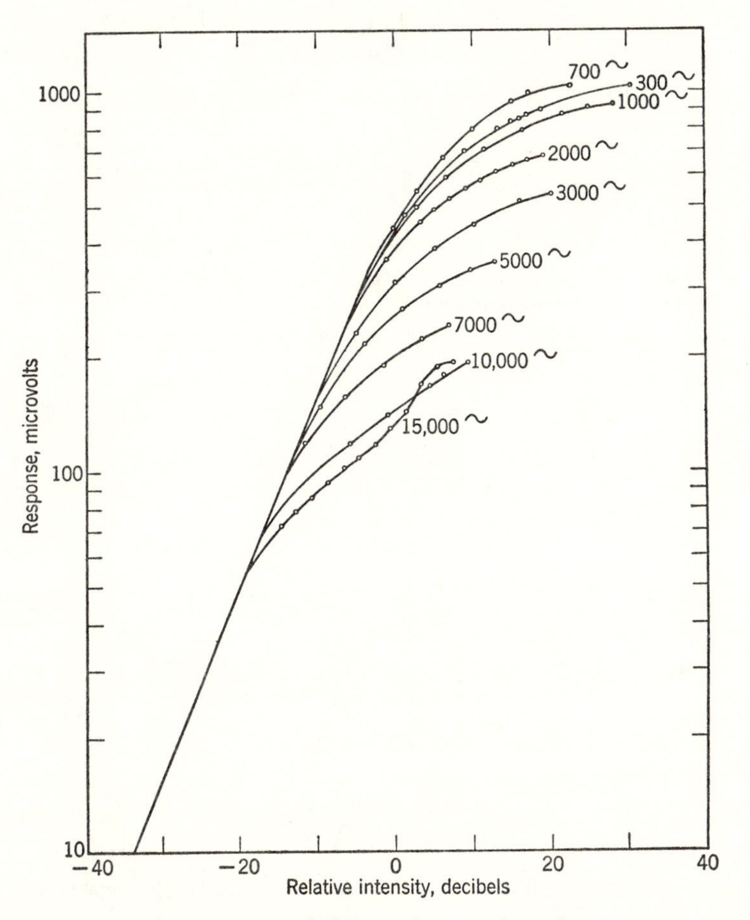

Fig. 118. Intensity functions recorded from the round window in guinea pigs for various tones, plotted after adjustment for differences in sensitivity so that the linear portions of the curves coincide. From Wever and Lawrence (3).

show a maximum in a true sense. The measurements at this frequency were made after all the others had been completed, and the intensity was raised beyond the point of safety as determined by earlier tests, with consequences as shown. The curve at first took a course well below that for 10,000~, as expected, and thereafter, at a high level, it quickly curled upward and then bent over sharply. At this point the ear was found to be damaged, and former readings could not be repeated. In only one among 21 guinea pig ears studied in this way was it possible to obtain a true maximum at 15,000~.

A clearer picture of the relations shown here is obtained by representing the results in a frequency plot as in Fig. 119. Here the magnitude of response is given on a linear scale along the

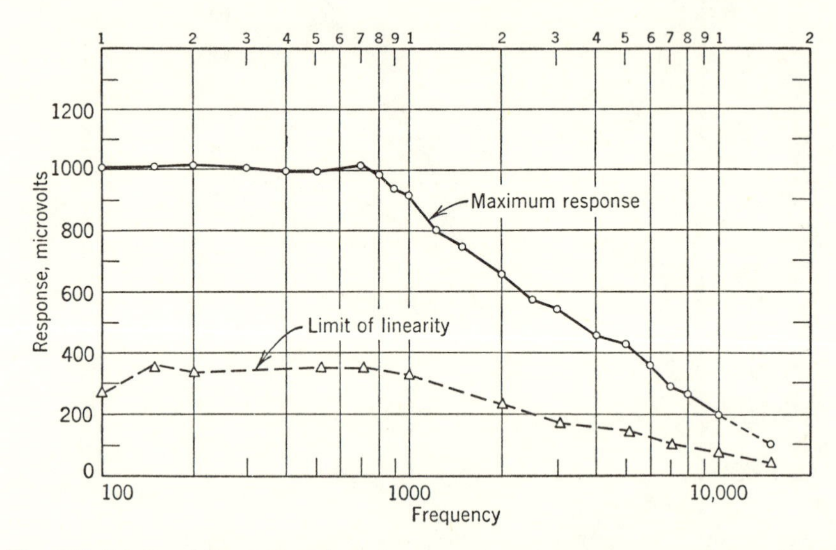

Fig. 119. Limits of linearity and maximum responses in a guinea pig's ear. These data were obtained from intensity functions like those shown in the preceding figure. From Wever, *Theory of Hearing*, 1949, John Wiley and Sons.

ordinate, with frequency shown logarithmically along the abscissa. The lower curve represents the responses at the points of initial departure from linearity and the upper curve represents the maximum responses.

Let us first consider the lower curve. For the low tones, up to 1000~, the limit of linearity is reached when the response has a magnitude around 350 microvolts, and then for the higher tones this magnitude falls off steadily to a value of 40 microvolts at 15,000~. The differences shown among the low tones are probably due to experimental variations, for an average of observations on many ears gives a fairly uniform curve in this region.

Consider now the upper curve of this figure. Over the lower portion of the frequency range this curve is nearly flat; for all tones from 100 to 700~ the maximums have a uniform value of about 1000 microvolts. Above 700~ the curve breaks sharply and then falls progressively to a value of 200 microvolts at 10,000~. In many of the ears this break occurred at 1000~ instead of at 700~. In such ears the decline of the maximums was a little more rapid, so that at 10,000~ the curve reached approximately the same level as shown here.

We now turn to an interpretation of these results and the development of our conception of the relations between the response function of a tone and its mode of action on the basilar membrane. As already brought out, we conceive of the measured response as a summation of potentials generated by all the hair cells throughout the cochlea. However, these cells do not enter equally into the action, and even if they did our electrode would not afford an equal representation of them. Some of the cells are stimulated vigorously, others more mildly. The cells immediately adjacent to the electrode show their effects strongly, and others at more remote positions do so only after suffering attenuation in the paths through the cochlear fluids. Results already described and presented in Fig. 110 indicate that this attenuation is about 1 db per mm. Consequently the response does not give a direct picture of the action, but one that is biased by the positional factor.

The mechanical action of a tone upon some one narrow region of the basilar membrane can be thought of in terms of the displacement of the membrane in this region, or alternatively as the velocity of the motion at this place. In either case we can represent the action as an amplitude. Then if we consider the action on all the elements over the membrane we can speak of an operating area for this tone.

A potential is generated by each hair cell in proportion to the amplitude (or velocity) of action at its location. This potential, attenuated according to the distance from the electrode, is combined with the potentials from all other cells at the recording electrode. As the stimulus intensity is increased the recorded potential increases proportionately as long as the action in every cell is linear. At some stimulus level, however, the operating amplitude becomes excessive and in certain of the cells the action becomes nonlinear. Naturally this nonlinearity first appears in the cells at the peak of the curve of action over the basilar membrane. Other cells away from the peak continue in linear activity until the stimulus is raised to still higher levels.

Let us now consider the form of the response in an individual cell. Because we regard the response as a whole as a composite representing many cells, we must infer a function of the same general type for the individual cell. However, if we consider the stimulus as applied unevenly, involving certain cells more vigorously than others, it follows that the individual function must bend more sharply than the composite one and must pass through its maximum more quickly. The composite curve is gentler in its changes because of the presence of many cells that are still operating in the linear portions of their functions.

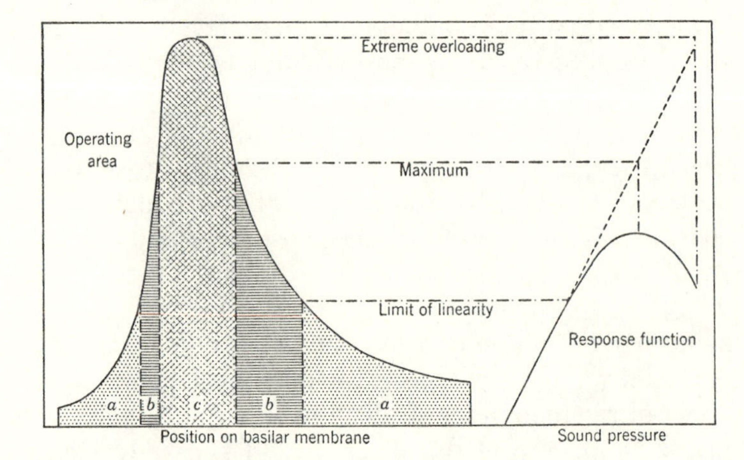

Fig. 120. A schema representing the relations between the operating area of a tone and the response functions of the sensory cells. Regions of the operating area are distinguished according to their efficiency of contribution to the total response, as described in the text.

Figure 120 will aid in picturing the situation that is conceived here. Shown on the left is the operating area of a tone as spread over the basilar membrane and on the right is a response function as conceived for any one sensory cell. Different regions are distinguished by stippling according to the contributions of their cells to the total response. Cells in the regions marked *a* are working in the linear part of their functions; as the sound pressure is raised their contributions to the response will rise in strict proportion. Cells in the regions marked *b* are all operating nonlinearly; their contributions are less than proportional to the magnitude of the stimulus, and some of them, the ones stimulated the most strongly, are making the maximum contributions of which they are capable. An increase in the level of stimulation will push these cells still farther into nonlinearity, and the ones most strongly stimulated will give a diminishing output. Finally, the cells in the peak area *c* are all in a stage of grave nonlinearity, and their contributions, already below proportionality, will be diminished further as the sound intensity is raised. Moreover, the cells occupying the highest peak are suffering damage as a result of the stimulation.

It is reasonable to suppose that we perceive the initial departure from linearity when the contributions to the recorded response made by the most strongly stimulated and overloaded cells comprise some fairly constant fraction of the total response. Because the factor of positional bias is involved, this departure from linearity will be the more readily perceived the closer the peak of the operating area to the recording electrode. When, as in these experiments, the electrode is on the round window membrane, this form of distortion will be the more evident the closer the peak to the basal end of the cochlea.

Now let us make the assumption that the distortion processes are similar for the various regions of the cochlea, which means that for any element the distortion enters at some particular level of action at its site. On the basis of this assumption we can say that at the points of departure from linearity the operating areas of all tones have the same peak amplitudes. The locally generated potentials likewise have the same amplitudes, though they must be seen by the electrode from the point at which it is placed and only after modification by the positional bias. Let us leave this bias out of consideration for a moment in order to make our conceptualization easier. Then we may interpret the observed differences in magnitude as reflecting differences in the form of spread of the operating areas over the basilar membrane. We can say that the patterns for the low tones are broad and flat, involving a great many elements in vigorous action. The patterns for the high tones, by comparison, are sharply peaked and increasingly so in regular progression as the frequency rises. For these tones there are relatively few elements in vigorous action.

Let us now bring in the factor of positional bias. In this connection it is important to bear in mind the evidence already presented that the peak regions for the high tones are located toward the basal end of the cochlea, and the more so the higher the frequency. The effect of the bias factor is then in general to increase the magnitude of the recorded responses for the high tones relative to those for the low tones. We have to conclude that the true operating areas for these tones are more sharply peaked relative to those for the low tones than the data of the lower curve of Fig. 119 would directly indicate.

Our conception of the cochlear activity is developed further as we consider the response functions beyond the initial stage of distortion. The fact that the curves for the low tones bend only gradually away from linearity, whereas those for the high tones depart rather abruptly, reflects the difference in the proportionate areas covered by the peak portions of the curves. For the low tones this proportionate area is small. Their curves have broad and rounded central portions and long, sloping ends. Distortion enters at the peak and then spreads progressively to the neighboring regions. For the high tones this proportionate area is large. The peak is sharp and narrow, the sides of the curve are steep, and the area outside the peak is small. Therefore the distortion enters almost at the same time for the bulk of the elements in the response area.

As we continue to raise the stimulus intensity the responses of those cells at the peak of the operating area will eventually pass their individual maximums and bend downward. These cells are now giving smaller responses as the stimulus intensity is increased. We can express the matter otherwise by saying that as the stimulus is increased these cells contribute negative increments to the total response. The response as a whole continues to rise as long as the positive increments contributed by other cells—those on either side of the peak—are paramount. Eventually as we raise the stimulus intensity a point is reached where the negative and positive increments are in balance. This is the point of the maximum. As we go still farther the negative increments supervene and the response as a whole bends downward.

Our further consideration of the relative magnitudes of the maximums confirms what we have already inferred about the form of the action patterns for different tones. For a large value of the maximum it is plainly desirable to have a broad, flat area over which the action is relatively uniform, so that many elements will continue to make positive contributions before those passing into their negative phase become too numerous. The low tones evidently satisfy this requirement.

The break in the maximum function around 700 to 1000~ reflects two changes in the tonal patterns as we go up the frequency scale. One is the progressive narrowing of the operating areas and the other is the movement of the peaks of these areas toward the basal end of the cochlea. Because of the bias factor this second change tends to counteract the first. It augments the responses as the frequency is raised while the other factor is diminishing them. Over the low-tone range a balance of these factors is maintained, and the maximums have a uniform value. Then in the high-tone range the narrowing of the peak areas gets the upper hand and the maximums fall rapidly.

A significant feature is the discontinuity in the maximum response function beyond 10,000~. As already pointed out, the potentials for the highest tones continue to rise until injury occurs, without attaining a maximum. This fact seems to mean that for these tones the area comprised by the peak portion of the pattern is particularly small with respect to the outlying area, so that the continuing rise of potentials from the outlying area is predominant. We have suggested that these tones may be beyond the limit of proper tuning of the ear, and operate only by forcing the response of the terminal elements at the basal end of the basilar membrane. We have then only the tail of a resonance curve, with the strongest action at the extreme basal end and a rapid falling off of action in elements farther up the cochlea. This form satisfies the requirement of an area of vigorous action that is particularly small.

A further factor probably enters into this situation. On theoretical grounds we should expect an asymmetry of the forms of the operating areas. A resonator is more readily set in action by tones below its tuned frequency than by tones above this frequency, and this difference is the more prominent the greater the damping. The skewed form of action as illustrated in Fig. 120 is therefore the one to be expected. The effect is to emphasize that part of the operating area to the basal side of its center.

In these observations the relatively strong participation by the basal end of the cochlea is further emphasized because the potentials are recorded from the round window. The result is to reduce the differences shown in the maximums. It follows that the high tones have narrower action patterns than the data would show otherwise.

Series 2. The indications of the above experiment are confirmed and extended by the one next to be described, which was carried out on cats by recording potentials both at the round window and at the apex of the cochlea. The round window electrode was a platinum foil applied in the usual way and the apical electrode was another foil applied sometimes to the bony surface at the apex and sometimes placed in a depression drilled in the bone at this point. The depth of the depression varied somewhat. The intention was to drill almost but not quite through the bone, so as to reduce the local resistance as much as possible without causing any mechanical damage to the cochlear structures. The drilling was done slowly by hand while observing with a dissection microscope, and it was usually possible to stop short of actual penetration into the interior. An accelerated oozing of fluid from the deeper layers of the bone was the sign that the drilling had gone far enough. Sometimes, in ears in which the bone was particularly thin, the drilling was carried too far and a hole was opened into the cochlea. The responses then were impaired.

This drilling into the cochlear capsule reduces the resistance that the potentials must pass through in reaching the electrode and so it gives increased sensitivity. On the other hand it produces greater variability of the results. It seems that some mechanical disturbance is caused by the drilling even when the bone is not penetrated. We have obtained results both with this drilling and without, and in general regard the recording from the intact cochlea as preferable.

When the apical electrode is simply placed on the bony surface the sensitivity, determined as the stimulus intensity required to produce some standard response such as 1 microvolt, is of the same order of magnitude as at the round window for the low tones, and then as the frequency rises the sensitivity here becomes progressively poorer relative to that at the round window. These relations are shown by the solid-lined curve of Fig. 121 and are

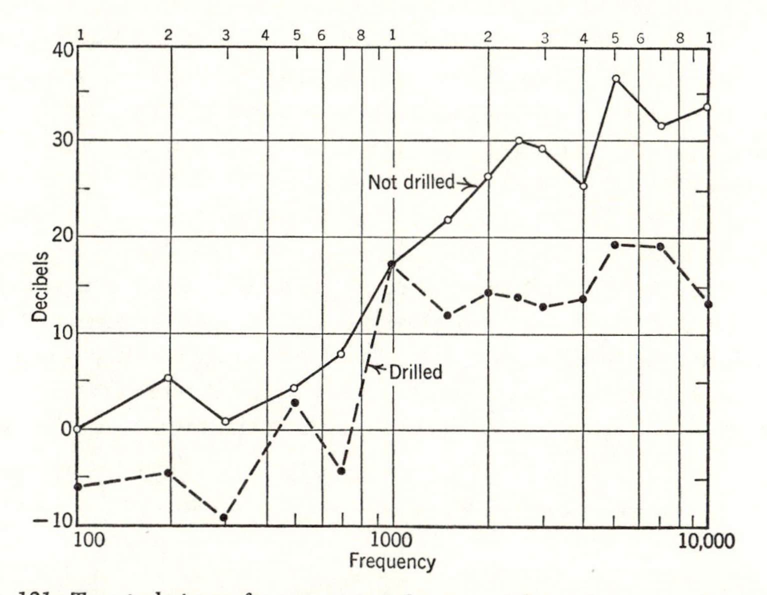

Fig. 121. Two techniques for comparing the potentials at the two ends of the cat's cochlea. For the solid curve the cochlear capsule was intact and for the broken curve a hole was drilled part way through the bone at the apex. The curves show, for these two conditions, the increase in stimulation necessary when recording at the apex to produce the same responses as when recording at the round window. The sensitivity is generally improved by the drilling but the function becomes more irregular.

like those already seen in Fig. 106. When the electrode was placed in a depression drilled in the apex this same ear gave the results shown by the dashed curve of Fig. 121. The sensitivity was improved by the drilling at all frequencies except at 1000~ where it remained the same. At the lowest frequencies the improvement was such as to make the apical position the more sensitive one. At the high frequencies the sensitivity was considerably greater than before but still was 15 to 20 db poorer than at the round window.

These differences in sensitivity at the two ends of the cochlea have two causes, already referred to earlier in this chapter. The factor of local resistance evidently still operates in favor of the round window position even after the cochlear wall is partially drilled through. Of most importance is the varying effectiveness with which the two electrodes are able to survey the operating area of the tone. Because for the low tones the sensitivity is as good or better at the apical position it appears that these tones produce the largest amplitudes of action at this end of the cochlea. Because the high tones show much reduced sensitivity here it appears that their action in the apical region is only slight.

As the stimulus intensity is raised the apical potentials increase linearly or nearly so and then bend over just as the basal potentials do. However, the departures from linearity and the maximums appear at much lower levels of response. These levels doubtless reflect the unfavorable local conditions, which reduce the responses in general, and therefore it is a little hazardous to take the absolute magnitudes very seriously. However, we can safely consider the forms of the curves in their variations with frequency.

Figure 122 gives a comparison of the maximum responses ob-

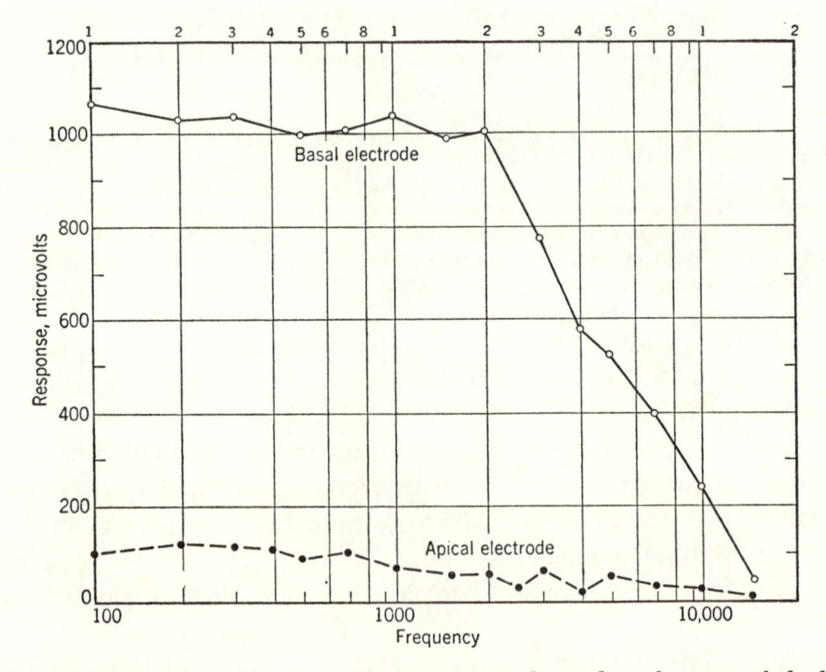

Fig. 122. Maximum values of cochlear responses observed in the cat with both basal and apical electrodes.

tained at the two electrode positions. The responses from the round window are much like those already described for the guinea pig. The maximum values are fairly uniform, around 1000 microvolts, for all tones up to 2000~, and then they fall off rapidly as the frequency rises further. Some cat ears show the break at 1000~ instead of at 2000~ as indicated here.

The responses obtained from the apex are fairly uniform in magnitude up to 700~ and then fall off gradually for the higher frequencies. There is no sharp break as seen in the round window function.

Curves were drawn also for the initial departures from linearity, but are not reproduced here. These curves follow much the same courses as the maximum curves, though with greater irregularities.

These results are more readily comprehended when we present them in relative terms, as differences in decibels between the readings at the two electrode positions. This has been done in Fig. 123 both for the maximums and for the magnitudes attained

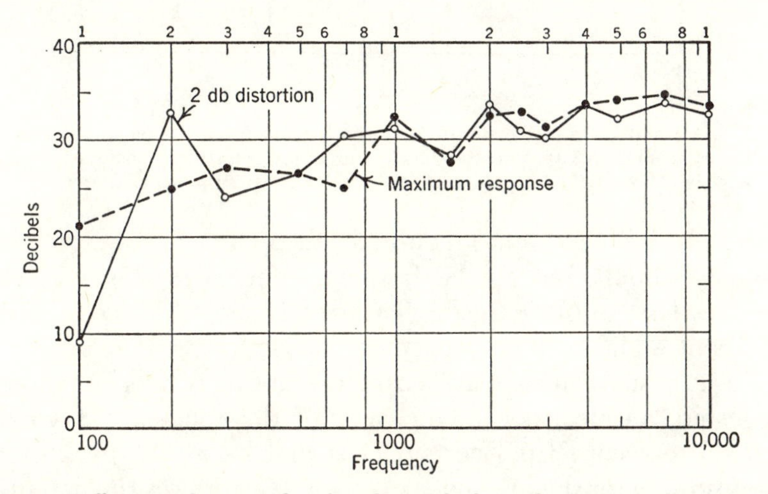

Fig. 123. Differences between basal and apical electrode positions for maximum responses and for 2 db departures from linearity. For all conditions the responses were greater at the round window position.

when the function departs from linearity by 2 db. The 2 db departure points are used because they are more reliable than the points of initial departure. The two curves are much alike, and it is clear that the differences found for the two electrode positions grow progressively greater as we go up the frequency scale.

Of interest in connection with the above data is a comparison of the stimulus values required for the attainment of maximums at the two cochlear positions. These values are shown in Fig. 124, again in comparative terms, as the differences in decibels between stimuli required at the round window relative to those required at the apex. It is evident that at the round window position the

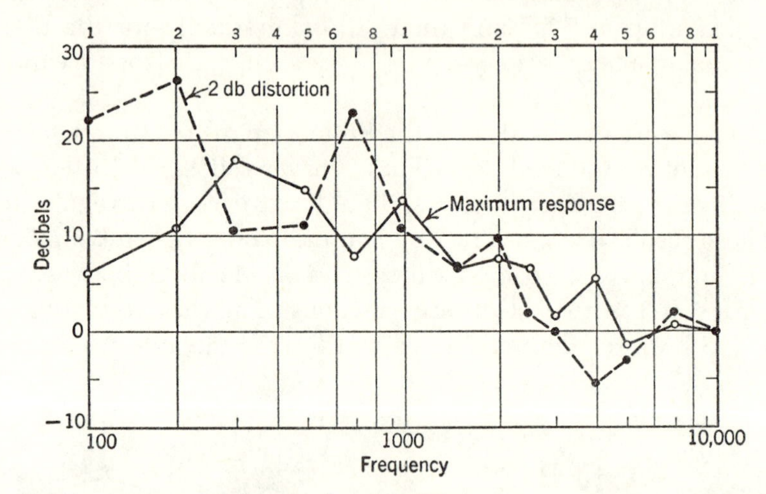

Fig. 124. A comparison of the stimuli required for maximum responses and for 2 db departures from linearity, for the basal and apical electrode positions. Positive decibels signify that the stimuli were greater for the round window position.

potentials for low tones continue their linear rise for much stronger stimuli, but as the frequency increases this difference becomes less, until at the highest frequencies the two positions are about equal in this respect.

We now seek an explanation of the above results. As already suggested, an electrode at the apex of the cochlea is favorably situated to record from certain cells that lie at the peak of the response area for the low tones. As the intensity rises the potentials rise also, but they soon attain a maximum. This maximum represents the point at which the cells in the peak region have suffered serious distortion and are making negative contributions that are balanced by positive contributions from other cells more remotely located. These other cells extend all the way to the basal end, but their effects are reduced by the resistance encountered in their paths.

As we go up the frequency scale the values of the maximums as measured from the apex continually decline for two reasons. They decline because the operating areas are moving away from the recording position and also because these areas are becoming more peaked and therefore smaller. An apical position is less advantageous for recording high-tone responses than a basal position is for recording low-tone responses because of the asymmetry already referred to in the forms of the response patterns. The low tones are able to spread broadly into the basal regions but the high tones do not involve the apical regions in anything like the same degree.

The form of cochlear localization indicated by these results departs widely from the classical notion that every tone has its individual place. These indications are also out of harmony with an intermediate position taken by Tasaki and Fernández on the basis of some observations of theirs. They inserted microelectrodes through holes drilled in the wall of the cochlear capsule at several places and recorded the cochlear potentials after various procedures designed to bring about a local inhibition of electrical activity. They used these electrodes as what they termed "differential electrodes," by which they meant that there was no indifferent position but both recording electrodes were inserted in the cochlea, usually at adjacent points.

These investigators reported that after potassium chloride solution was injected into the scalae at the basal end of the cochlea, or a direct current was passed through the tissues here, the activity of the basal end of the cochlea could be completely inhibited as measured by a pair of electrodes in this region. At the same time, a pair of electrodes in the apical region showed a reduction of potentials to low tones and a complete absence of potentials to high tones. From these results they concluded that the basal end of the cochlea is responsive to both high and low tones, whereas the apical end of the cochlea is responsive to low tones exclusively.

A careful consideration will show that these conclusions of Tasaki and Fernández do not follow from their results. What needs to be borne in mind is just what any pair of electrodes reveals about the local electrical activity. A pair of electrodes always represents the differences of potential existing from moment to moment in the tissues with which the two contacts are made. All electrodes thus are differential electrodes. If the two electrodes are in contact with cells that are undergoing different sorts of changes they will report differences and a deflection will be seen on the recording instrument, and if they are in contact with cells that are undergoing exactly the same changes they will report an absence of difference and nothing at all will be seen.

The results of Tasaki and Fernández thus show that their two electrodes in the apical region of the cochlea are recording the activity of cells that are behaving similarly in response to high tones. Of course one form of similarity is an absence of activity, but to conclude, as these authors did, that only inactivity is possible as an interpretation of the results is unwarranted. Our conclusion that the apical end of the cochlea participates in the response to high tones is fully consistent with these results as well as with the other results described earlier. These data of Tasaki and Fernández merely show that there is little differentiation of the apical end of the cochlea in response to high tones, or in other words that the response curve here is practically flat. This form is just what we have been led to expect from the other results mentioned above.

The further indication of the Tasaki and Fernández observations that the basal end of the cochlea responds to both high and low tones is also in agreement with these other results. The fact that these investigators by their method were able to demonstrate a spread of low tones but failed to do so for high tones is only a consequence of the limitations of their method and the systematic differences that exist in the degree of spread of high and low tones.

THE LOCAL STIMULATION PROCESS

We have still to consider the final steps of the stimulation process, in which the vibratory motions of the cochlear fluid act upon the hair cells and lead to excitation of the auditory nerve fibers. These steps include the communication of mechanical energy to the hair cells, the changes produced in these cells, and the action upon the nerve terminations.

THE COMMUNICATION OF ENERGY TO THE HAIR CELLS

There are two general theories regarding the transfer of vibratory energy to the hair cells. One is the theory advanced by Eduard Weber in 1841, referred to at several points in the preceding chapters, which states that the sound pressures exerted by the stapes produce mass movements of the cochlear fluid and that the basilar membrane lying athwart the path between the stapes and round window is moved in the process. The basilar membrane movements then lead to stimulation of the hair cells. The action upon the hair cells is thus indirect, involving first an action upon the basilar membrane.

A second theory regards the stimulation process as a direct transfer of acoustic energy from the cochlear fluid to the hair cells. The basilar membrane and its related structures are considered to play no part in this action except to provide a suitable support for the hair cells. This theory was advanced by Ayers in 1892 as the "sensory hair theory." He believed that the cilia of the hair cells were very long and extended far out into the endolymph of the cochlear duct where they were waved about by sound vibrations. The movements of the cilia were then transmitted through their roots to the body of the hair cell, where stimulation occurred. Bryant in 1909 espoused a similar theory. He referred to the cilia as being moved by the passage of sound much as the stalks in a field of grain are wafted back and forth by gusts of wind. Borgheson has recently revived this theory. Pohlman upheld the theory in another form, in which the action is directly upon the body of the hair cell rather than through its cilia. He specifically denied any "shuttle" movement of the cochlear fluid such as Weber had postulated.

The Ayers theory was based upon an erroneous conception of the cilia of the hair cells. These cilia are short and do not extend freely into the endolymph. They form a heavy tuft that seems to be firmly attached to the under surface of the tectorial membrane. From these relations it is difficult to conceive of their being acted upon directly by a sound and it seems more reasonable to assume that stimulation occurs through relative motion between the basilar membrane and the tectorial membrane.

The Pohlman theory of a direct action upon the hair cell by sound pressure is definitely denied by our results. We have already described our experiment in which a tone was delivered simultaneously to both oval and round windows and a cancellation of responses was obtained for a suitable adjustment of the magnitudes and phase relations of the two stimuli. The condition for cancellation is one in which the alternating pressures are balanced on the two sides of the basilar membrane and consequently this membrane is at rest. This is the condition that produces a maximum of pressure about the hair cells—double the pressure that one pathway by itself would cause—and yet we find that the hair cells remain unstimulated. The evidence as a whole gives strong support to the view that the basilar membrane plays an essential part in the process of hair cell stimulation.

In further support of this conclusion we can point to evidence on otosclerosis and its treatment by fenestration, as presented more fully later, in Chapter 16. This evidence shows that mechanical fixation of the stapes, which is a consequence of this disease, produces a loss of sensitivity of varying degree, according to the firmness of fixation, up to 80 db or more. Such extreme losses of sensitivity could not arise from this cause if the hair cells were directly stimulable by sounds, for sounds in penetrating the head and reaching the hair cells can hardly suffer attenuations as great as 80 db. It is easy to show by a calculation in terms of acoustic resistances that aerial sounds on striking a mass of solid bone will enter it with an attenuation of only about 35 db. It is evident that the mere presence of sound energy in the cochlea is not enough for stimulation and that to be effective this energy must be channeled through the cochlea so as to mobilize the cochlear fluid and displace the basilar membrane. This theory has received explicit formulation in Lempert's explanation of the results of the fenestration operation as a treatment for otosclerosis.

It is clear that in our modern acceptance of Weber's principle we need to permit certain modifications. The mass movement of the cochlear fluid is not perfectly free, but is constrained to a certain extent by the local properties of the basilar membrane. The form of this constraint varies according to frequency, as the results of the preceding sections have shown.

THE MECHANICAL ALTERATIONS OF THE HAIR CELLS

We are next concerned with the manner in which the movements of the basilar membrane affect the hair cells. In the early discussions of this problem it was assumed, without much consideration of alternatives, that the displacement of the basilar membrane acts directly in hair cell stimulation. In recent years there has been serious consideration of other aspects of the basilar membrane motion, such as the velocity or the spatial gradient of the motion, or other still less direct effects of the displacement, as possible stimulative agencies. For the most part these alternative suggestions for hair cell stimulation have been offered with a view to obtaining a range of action narrower than that of the displacement itself.

The Direct Displacement Hypothesis. The earliest explanation of the stimulation process was one offered by Hensen (*1*), who suggested that the upward movements of the basilar membrane caused the cilia of the hair cells to be thrust against the under surface of the tectorial membrane. Helmholtz and most other writers who have been concerned with this problem have adopted Hensen's view without developing it further. Ter Kuile, however, suggested certain revisions of the conception that would bring it closer in line with the anatomical relations of the parts. He pointed out that the inner pillar of the arch of Corti has its foot resting on the fibrous lip of the spiral lamina. This pillar therefore is not free to move up and down but can only rotate about its foot as on a pivot. Indeed, the whole inner portion of the basilar membrane swings about this pivot as the membrane is displaced. At the same time the tectorial membrane swings about its own point of anchorage at the vestibular lip of the limbus. Because the basilar membrane and the tectorial membrane are rotating about different axes there is a relative motion between the two. This is a sliding of one over the other where they come in contact, which is at the hair cells. In this sliding motion, ter Kuile supposed, the cilia of the hair cells are bent back and forth and thereby these cells are stimulated.

There are other possibilities for the direct operation of basilar membrane displacement in stimulating the hair cells. It may be pointed out that even in a sliding contact between the cilia of the hair cells and the tectorial membrane as described by ter Kuile there will be a vertical component of motion tending to move the ciliary tuft alternately toward and away from the body of the hair cell. This motion will cause a deformation of the hair cell and may be the effective stimulus. Another possibility is a simple bending of the hair cell (Guild, *5*).

Fischer's Spacing Hypothesis. Hensen (*1*) implied and Fischer said explicitly that the cilia normally stand a little distance away from the tectorial membrane and make only a momentary contact with it as a result of vibratory movements. Fischer's hypothesis was advanced with the idea of affording a restriction of effective stimulation to only the peak of displacement of the membrane. The remainder of the pattern of motion, he thought, could be ignored altogether.

When we examine this hypothesis we see that it achieves its objective of "sharpening up" the action only for threshold stimulation—where there is no need of it. Stronger stimulation will bring in hair cells on either side of the peak just as it would without the spacing between cilia and tectorial membrane. Such spacing would only dull the sensitivity in general without altering the spread of action.

Such a spacing, causing only momentary contacts at low intensities, would have the further effect of producing an impulsive form of stimulation. The nature of the cochlear potentials is such as to deny this kind of transformation of the sound waves. These potentials reproduce fully the forms of the pressure variations in these waves. For simple sinusoidal waves the potential changes themselves are sinusoidal and for complex waves they are correspondingly complex. This can be true only if the communication of pressure to the hair cells is continuous and is a faithful rendition of the varying amplitude of the sound waves.

Furthermore, there is good anatomical evidence that the cilia of the hair cells maintain a constant contact with the tectorial membrane. Sometimes in histological preparations when the tectorial membrane is curled away from its normal position some hair cells may be seen adherent to its tip, obviously having been pulled out of their sockets by means of their ciliary attachments (Wever, *5*). DeVries (*2*) by a special method, in which fresh temporal bones were frozen in liquid air and then broken into fragments for study, obtained clear evidence of a connection between the tectorial membrane and the cilia of the hair cells.

Mechanical Actions of Higher Order. The stimulative effect upon the hair cells is not necessarily a direct action of the displacement of the basilar membrane but may be some other aspect of the motion. In fact, there are almost unlimited possibilities, some of which will now be considered.

The Velocity Hypothesis. Considerations of the form of the ear's sensitivity function have led to the suggestion that it is the velocity of the motion communicated to the hair cells that determines its stimulating effect (Wever, 5). By this assumption the relatively great sensitivity of the ear to high tones is more readily explained, for the velocity of the basilar membrane motion is proportional to the frequency as well as the displacement of the sound waves. For any given tone the velocity pattern on the basilar membrane has the same form as the displacement pattern.

The Eddy Hypothesis. As we found in Chapter 13, Békésy in his auditory theory assumed an indirect action of the basilar membrane motion. The traveling wave along the membrane is said to produce an eddy movement in the fluid of the cochlear scalae. This fluid movement is most prominent in the region where the displacement wave reaches its maximum, and it is this movement that is supposed to stimulate the hair cells of the organ of Corti.

Békésy, like Fischer, proposed his hypothesis with the idea of obtaining a more restricted sensory stimulation than that directly afforded by the membrane displacement itself. However, this objective is not achieved in the eddy hypothesis any more than in the spacing hypothesis as long as a direct, linear relation holds between the eddies and the primary movements.

Békésy's observations did not establish the form of this relation. Ranke (*1*) on the basis of his mathematical treatment considered the fluid streaming as simply proportional to the displacement of the basilar membrane. This means that the eddy will spread out just as the displacement does as the sound intensity is increased.

Zwislocki was unwilling to accord the eddies any important role in stimulation, but he reported that his calculations indicated that they should vary as the square of the amplitude of basilar membrane motion. If this is true a narrowing of the form of distribution of the eddies would result. Against this possibility, however, is the fact that the cochlear potentials vary directly with the sound pressure and thus with the amplitude of movement and not as its square.

The Gradient Hypothesis. An obvious possibility is that the effective action on the hair cells is the gradient of amplitude. Insofar as the displacement differs at the two sides of a hair cell this cell is exposed to a shearing force. Huggins and Licklider considered this possibility and rejected it because their calculations showed that the narrowing effect upon the stimulation by this means was only slight and, perhaps even more serious, the shearing forces are extraordinarily small.

The "Curvature" Hypothesis. The shearing action just considered represents the first space derivative of the basilar membrane movement. We may also consider the second space derivative, which is the form of variation of this gradient. Huggins and Licklider discussed this aspect of the basilar membrane motion, but again their calculations led them to reject it. The sharpening effect by this means is only slightly greater than that indicated under the gradient hypothesis and likewise the forces developed in this manner are so minute as to be of doubtful effectiveness.

The Beam Hypothesis. We have seen in our discussion of the Huggins theory in Chapter 13 that the tectorial membrane is regarded as having the properties of a rigid beam and as imposing these properties upon the basilar membrane and its hair cells. The form of stimulation postulated in this theory is that of the fourth derivative of the basilar membrane displacement. As worked out by Huggins and Licklider this form of action is fairly well localized, covering perhaps a third of the range of the primary movement. They indicated also that the speed of progression of the "beam" wave is slow, thereby presenting the possibility of a further sharpening of the excitation by a principle of temporal summation suggested by Ranke, and next to be considered.

Ranke's Temporal Hypothesis. Ranke, (*3,4*) saw no clear possibility of the ear's selectivity being achieved by purely mechanical means, but suggested for this purpose a peculiar sort of relation between the peripheral activity and the effect upon the neural elements. He assumed that the excitatory effect of the basilar membrane action is not simply a function of the amplitude of movement but depends also upon temporal conditions. He suggested that summation is needed for stimulation of the nerve fibers, and for proper effectiveness a number of sensory cells in a particular region must deliver their excitatory impulses in close temporal sequence. Hence it is advantageous that the speed of wave movement in this region be slow. His theory, as we have seen, provides this condition. The traveling wave is described as slowing down as it proceeds up the cochlea until a maximum amplitude is attained, after which it suddenly speeds up again. The effective stimulation is thus the ratio between the amplitude of movement and its speed, and Ranke sought to show that this ratio has a relatively sharp peak as considered along the membrane.

We have now reviewed the various suggestions as to the manner of stimulation of the hair cells. We must look with considerable caution upon those hypotheses that depend upon complex forms of action like the first, second, or higher derivatives of the space patterns or secondary effects such as eddy currents. Such complex functions of the displacement are subject to extraordinary modifications as a result of structural variations that would produce only moderate changes in the displacement itself. The functional stability of the ear under various kinds of stress tells against such hypotheses. Moreover, the evidence derived from a study of the cochlear potentials, as described in the first portion of this chapter, indicates that the stimulation really is extensive over the cochlea and not limited to any narrow place.

ORIGIN OF THE COCHLEAR POTENTIALS

Closely bound up with the problem of stimulation in the cochlea is the question of the origin of the cochlear potentials. One theory of these potentials is that they consist of modulations of a base polarization of the hair cells. These modulations are considered to be brought about by deformations of the hair cells, or more particularly as dependent upon the velocity of such deformations.

This theory explains the distortion that arises when stimulation is excessive as due in part to a kind of rectifier action in the electrodynamic transformations. When the movements are great they can still raise the potential well above the base level but they can depress it only down to zero. Hence an excessive movement gives an electrical wave that has its alternate half-cycles clipped off. The theory also accounts for the effects of overloading by supposing that with violent stimulation there is a reduction in the base potential. This reduction is temporary unless serious damage is done to the structures. It is usually found after strong stimulation that the cochlear potentials are reduced for all tones and at all stimulus intensities, but that partial or even complete recovery to normal occurs after a few minutes of rest. It has recently been shown by Alexander and Githler that full recovery can occur even after a profound reduction of the potentials. In some instances they obtained normal responses, after a recovery period of three weeks, when the stimulation had been severe enough to depress the potentials by 60 db.

For a polarization of the hair cell to be present we have only to assume that the contents of the cell differ in chemical character from the endolymph bounding the cell and that the cell membrane provides a selective barrier to ions within and outside the cell. Conditions such as these are common in animal and plant tissues. Chemical peculiarities of the hair cells are easily seen in their special affinities for histological stains. Also Békésy (22) was able to measure in the living guinea pig a potential difference between perilymph and endolymph of 80 millivolts (with the endolymph positive) and between perilymph and cells of the organ of Corti a potential difference of 40 millivolts (with the cell contents negative); and the difference between these cells and the endolymph thus is 120 millivolts (with the cell contents negative). It is also of interest that these resting potentials are found to diminish on extreme oxygen deprivation and death, which are physiological changes that are known to cause a profound loss of cochlear potentials.

A base potential in the hair cells as large as 120 millivolts, as just indicated, is entirely reasonable. Potentials of the order of 30 millivolts have been measured in nerves, and calculations based on the chemical composition of nerve tissues lead to the prediction of an actual potential of 118 millivolts. The cochlear potentials as measured at the round window or from the surface of the cochlear capsule are much smaller, of the order of 1 millivolt with maximal stimuli, but these locations are far from optimal for recording. They are remote from the actual sites of the potential changes at the hair cells, and the generated potentials suffer an attenuation of approximately 1 db per millimeter of distance. Also, and probably more important, the internal resistance of the hair cells probably amounts to several megohms and further reduces the recorded potential.

Some other studies of Békésy (*21*) are of interest in this connection. He drilled holes at intervals along the cochlea of the guinea pig and inserted electrodes to measure the resting potential of the perilymph along the two scalae in reference to a fixed electrode placed in the perilymph of the vestibule. His results showed an increase in the magnitude of this potential in going from the basal to the apical ends of the scala vestibuli and then a fall from the apical to the basal ends of the scala tympani. Békésy did not explain this potential as due to differences in the concentration of ions in the perilymph. He argued that the chemical character of the perilymph must be the same throughout both scalae because the helicotrema permits free circulation of fluid and sound vibrations are always producing a streaming of the perilymph. Therefore, he said, the potentials as measured must be due to some active process in the cochlear duct.

Békésy also observed that the direct potential between round window and vestibule is suddenly diminished when a sound is applied to the ear, and this lowered potential continues as long as the sound is present. The diminution is proportional to the sound intensity until loud sounds are used that begin to overload the ear. For such overloading sounds the initial drop of potential is followed by a further gradual decline as the sound is maintained, and after the sound is removed the return of potential is only partial at first and continues over several seconds. The cochlear potentials (by which we mean as usual the alternating responses to sounds) exhibit a rise and fall immediately at the presentation and withdrawal of the sound. The direct potentials and the cochlear potentials thus have different courses during and after strong stimulation. Anoxia and death of the animal were found to have different effects upon these two types of potentials. Békésy therefore concluded that the two potentials are independent "to a certain degree."

In a further experiment Békésy (*22*) inserted a thin glass needle through the round window so as to press upon the basilar membrane, and he then found that this pressure changed the direct potential between an electrode in the perilymph of the vestibule and another in the perilymph of the scala tympani near the round window. Pressing the basilar membrane toward the scala vestibuli reduced the interelectrode potential. When the pressure was steadily maintained the observed potential was likewise steady. With a steady pressure no energy is being supplied to the basilar membrane and yet the potential difference continues; hence it follows that this potential difference does not represent a direct conversion of mechanical into electrical energy.

Békésy explored this point further by a consideration of the amounts of energy involved in the mechanical and electrical processes. He calculated the mechanical energy imparted to Reissner's membrane by pushing against it with a needle, and also calculated the electrical energy involved in the observed change of potential between the perilymph of the vestibule and the perilymph of the helicotrema near the point where the needle was applied. He found the electrical energy to be 19 per cent more than the mechanical energy. If this difference is real it obviously means that the electrical energy is not wholly derived from the mechanical energy.

This conclusion regarding the source of direct potentials was extended by Békésy to the cochlear potentials as well, for he said that they too do not represent a direct conversion of mechanical into electrical energy. His evidence was obtained in the following experiment. He applied a mechanical impulse to Reissner's membrane by suddenly pulling away by means of an electromagnet a small iron ball that had been placed so as to rest on this membrane. He then observed the resulting cochlear potentials, which showed that Reissner's membrane undergoes a damped oscillation under these conditions. The form of oscillation for a region of the membrane near the helicotrema indicated a natural frequency of $150^{\sim}$ and a damping factor (ratio of two consecutive amplitudes in one direction) of 3. This form did not change despite the change in over-all magnitude when anoxia reduced the cochlear potentials to a fourth of their normal value. Békésy therefore concluded that the cochlear potentials do not represent a direct conversion of mechanical into electrical energy. The reasoning is as follows. If there is a direct mechanoelectrical conversion, then a given amount of energy imparted to the moving system can be used up in only two ways, through frictional forces by which it is converted into heat and by the mechanoelectrical conversion. The conversion process as seen in the cochlear processes then lasts a certain time. Now if something is done, like introducing anoxia, to reduce the rate of the mechanoelectrical conversion, and the friction presumably remains the same, the conversion process must be slower and must last longer. Such a prolongation of the process was not found.

From the above evidence it might be concluded that both the direct and cochlear potentials are derived from a general pool of energy. However, the degree of independence of the two types of potential leads us to suppose that different pools of energy are involved. The basic metabolic processes by which these energy pools are supplied are naturally similar. These considerations favor the opinion already expressed that the energy source for the cochlear potentials lies in a polarization of the hair cells. That for the direct potentials may be a polarization of other cells of the organ of Corti or of outlying cells. It is possible that the hair cells contribute to the polarization from which the direct potentials arise, but there is no necessity that they do so.

The exclusion of a direct mechanoelectrical conversion of energy, as indicated by this evidence of Békésy's, seems to rule out certain of the simpler hypotheses regarding the generation of the cochlear potentials.

Among these simpler hypotheses is the membrane hypothesis, according to which movements of a polarized membrane are said to produce the observed effects. Reissner's membrane has usually been preferred for this role, though the tectorial membrane and the basilar membrane have been mentioned as possibilities (Adrian; Hallpike and Rawdon-Smith). Another hypothesis is the piezoelectric hypothesis, which supposes that the hair cells or other cells act like piezoelectric crystals to generate potential changes when they are altered in shape. Other theories depend upon the assumption of electrodynamic and electrostatic transformations in the auditory cells. These theories have never seemed very reasonable in view of the limitations of physiological processes. The evidence of Békésy's experiments leads away from these theories and toward the polarized cell theory as described.

EXCITATION OF THE AUDITORY NERVE FIBERS

A final problem concerns the manner of excitation of the auditory nerve fibers. Here we have three theories that regard the process respectively as mechanical, chemical, or electrical in nature.

It has been suggested that the process of stimulation of the auditory nerve fibers may be mechanical, involving merely the deformation of the naked axis cylinders. Against this possibility we have the evidence on certain albinotic cats. These animals have normal middle ears but their inner ears are malformed. There is a complete absence of hair cells and varying atrophy of the supporting cells. The nerve supply of the cochlea, however, is often only moderately affected and sometimes both nerve fibers and ganglion cells are present in nearly normal amounts. These animals are deaf: they are completely unresponsive to sounds and never give any cochlear potentials (Howe and Guild, Howe). Therefore it appears that when hair cells are absent the cochlear nerve fibers are unaffected by sounds. The transmission of mechanical changes directly to these fibers does not suffice for their stimulation, but a participation by the hair cells is essential.

The second hypothesis is that the hair cell in response to mechanical deformation secretes a chemical substance that acts as a mediator and leads to nerve excitation. This hypothesis is a part of the general assertion that chemical mediators are always required for the transmission of responses from one cell to another, whether the transmission be from sensory cell to neuron, across the synapse between neurons, or across a neuromyal junction. The evidence for this assertion comes mainly from observations of chemical activity in ganglia of the autonomic nervous system. In these ganglia it has been shown that stimulation results in the liberation of a substance, identified as acetylcholine, which is capable of exciting the activity of postganglionic neurons and muscle fibers. This substance is rapidly destroyed by an enzyme, cholinesterase, of widespread distribution in tissues throughout the body as well as in the blood. That acetylcholine and similar chemical substances are somehow involved in the activities of the autonomic nervous system can hardly be disputed, but that they are the actual stimulating agencies here is still a matter of theory.

The theory becomes particularly tenuous when it is extended beyond the autonomic nervous system, to excitatory processes in other nerves, in sense organs, and in skeletal muscles. When this extension is made it is also necessary to assume that the acetylcholine is produced with flashlike suddenness and is as promptly removed, so that the whole process can occur within the limits of the refractory period of the tissues. For the ear, which is one of the most rapid-acting of organs, this temporal requirement is particularly severe.

Derbyshire and Davis accepted the chemical mediator theory for the ear because they measured latencies between the cochlear potentials and auditory nerve impulses of the order of 0.5 to 0.6 milliseconds. They argued that these times were too long for a direct action of the cochlear potential upon the nerve fibers. It must be pointed out, however, that their measurements were made by observing the electrical effects produced at the round window by click stimuli. The reported latencies depended upon the assumption that by examining the complex wave of potentials from this stimulus it was possible to decide what part was due to the cochlear action and what part was due to nerve action. This assumption is open to question.

Recent evidence obtained by Martini and by Gisselsson are considered as favoring the chemical mediator theory as applied to the ear. Martini reported that the perilymph of pigeons contains cholinesterase, or something similar, with the property of neutralizing acetylcholine, and Gisselsson verified this observation. Gisselsson found this enzyme also in the endolymph of the cod and of the cat. The fact is not surprising in view of the wide distribution of this substance, but we fail to see how its presence gives any presumption of chemical mediation. The concentration was found to be even less in both perilymph and endolymph than in blood.

Martini went further and claimed to have demonstrated the presence of acetylcholine in the perilymph of pigeons as a result of tonal stimulation. This observation Gisselsson was unable to confirm, though he repeated the experiments exactly and besides pigeons tried rabbits, guinea pigs, cats, and human subjects.

In further experiments Gisselsson was able to show, in guinea pigs and cats, that the injection of a number of substances, including physostigmine and neostigmine by themselves and both hexastigmine and fluostigmine when followed by acetylcholine, results in an alteration of the latency of the cochlear potentials. From these observations he argued that a chemical mediator is required for the production of the cochlear potentials. Also, on this basis, he came to the conclusion that the cochlear potentials were of nervous origin.

These conclusions from Gisselsson's results are not compelling. In the first place, the conditions of the experiment did not exclude the possibility that the changes of latencies were due to changes in the mechanical properties of the middle ear. The drugs mentioned are convulsants and would be expected to affect the middle ear muscles along with the other muscles of the body. Indeed, it was noted in the experiments with physostigmine that the changes of latency were coincident with the appearance of general muscular twitching. As the results of Chapter 8 have shown, the middle ear apparatus introduces a considerable phase shift into a conducted sound, and an alteration of this apparatus affects the phase relations.

Gisselsson considered this possibility of accounting for his results but rejected it because he believed that the animals were anesthetized so deeply that the middle ear muscles were rendered inactive. Our own experience casts doubt on this assumption. We have observed repeatedly that with progressive anesthesia these muscles continue in activity for some time after general bodily movements have ceased. We have further observed that in animals treated with strychnine these muscles take part in the general spasms. Hence we must seriously entertain the possibility that the observed latency changes had a mechanical origin.

Gisselsson's further conclusion that the cochlear potentials are of nervous origin must be set aside on the basis of the fact, now well established, that these potentials continue in full strength after the auditory nerve fibers have been severed and have fully degenerated.

The third theory of cochlear nerve stimulation, which we have held to since the electrical potentials of the ear became known, is that these fibers are stimulated by the cochlear potentials. The fibers form a complex dendritic network about the bases of the hair cells and thus are at the immediate site of the potential changes. The potential across them must amount to many millivolts even for a moderate sound, and the resulting current must be considerable. We know that nerve fibers in general are excited by electric currents, and if anyone is to deny an excitation here he must present a convincing argument that the nerve fibers are insulated from the cochlear potentials. The observation of latency in the nerve response is not a proof of another, slower form of energy transmission. The transmission may be rapid and the latency may be in the action set up in the nerve fiber.

An acceptance of the electrical theory of excitation does not preclude an acceptance of a role of chemical products in hair cell activity. Indeed, our modern theory of cell action in general is that it is electrochemical. Unquestionably the electrical activity of the hair cell is based upon chemical states and the energy is derived ultimately from metabolic processes. We can even go further and accept the idea of a "flash" of chemical change at the moment of deformation of the hair cell. But even if this chemical "flash" were itself stimulating it could neither act as rapidly nor extend as far from the hair cell as the electrical change. The actual nerve excitation thus must be electrical.

It is more reasonable to suppose that the chemical changes in the hair cell, instead of determining the immediate excitatory response itself, are related to the general state of the cell and its efficiency of electrical output in response to mechanical changes. We can then explain the temporary impairment of hair cell activity that is shown as a reduction of electrical potentials after overstimulation.

PART VI

Clinical Applications

15

The Forms of Conductive Deafness

In the foregoing chapters our primary concern has been the study of the reception and transmission of sounds by the ear under normal conditions. Though at times it has been helpful to make use of evidence obtained from abnormal ears and ears subjected to unusual stresses, this has been done for comparison with the normal and to bring out more clearly the principles involved in the usual behavior of the system. Now, in this part of the book, we extend our interest to the pathological ear. We consider here the various forms of derangement of the conductive mechanism, as caused by accident or disease or deliberate surgical effort, and seek to explain their functional effects. In this discussion we hope to discover to what extent the principles that we have seen at work in the normal ear are still in operation in the defective ear.

TYPES OF DEAFNESS AND THEIR DIFFERENTIATION

Our concern is with the conductive types of deafness, which arise from impairments of parts of the mechanical system. We must distinguish these from sensory and neural types, which arise from impairments of the electrophysiological systems of the inner ear and the nervous pathways.

The determination of the type of deafness in any particular case depends upon objective and subjective signs and upon the history of the disorder. Such things as closure of the external auditory meatus and perforation of the drum membrane are obvious to otoscopic examination. Infectious diseases involving the middle ear usually have a typical history of pain referred to the ear and of a discharge from the meatus, and often leave visible changes in the drum membrane. Certain other diseases, like mumps, measles, and scarlet fever, when by ill chance they leave deafness as an after-effect, are known to produce their main damage in the cochlear and neural structures. Likewise, the lasting changes of hearing that follow upon an exposure to extraordinarily powerful sounds, such as gunfire, can be attributed to damage to the inner ear and its nerve supply. Often, however, the organic basis of an auditory impairment is not immediately apparent and a diagnosis must be made on the basis of the whole clinical picture, with particular consideration given to the nature of the hearing.

It is well known that the primary basis of this diagnosis is a comparison of sensitivity as measured by air conduction with that by bone conduction. In conductive deafness there is an impairment of air conduction but usually only slight variations from the normal in bone conduction. In sensory and neural deafness, on the other hand, a loss of hearing is found for air and bone conduction alike.

Of further importance is the fact that a conductive loss can be fully compensated for by raising the intensity of stimulation. The conductively deafened person can hear and understand speech about as well as a normal person if the speech is sufficiently loud. He is only handicapped a little because loud speech is not as well enunciated as ordinary speech. The person with sensory and neural deafness, on the other hand, is helped little if at all by an increase in the stimulus intensity. Ordinary sounds reach his cochlea with sufficient strength but either the hair cells or their nerve fibers are defective and these elements are incapable of representing the sounds adequately. Therefore the discrimination of complex sounds, such as speech, becomes a trying task. Increasing the intensity beyond a moderate level only overloads the responding elements and the resulting distortion adds to the task of understanding. It is commonly observed that the person with this kind of deafness is unable to use a hearing aid to advantage. Also such a person finds himself severely handicapped in the presence of a background of noise. His problem of discrimination is then particularly difficult. The conductively deafened, by contrast, is usually less handicapped in the presence of noise because normal persons unconsciously raise their voices above this background.

The diagnostic problem is somewhat complicated by the fact that in many persons conductive and sensory-neural impairments are simultaneously present. It then becomes difficult to determine precisely what the organic changes are and to what extent the functional losses are due to the one or the other type of defect.

CLINICAL TESTS OF HEARING

A great many tests have been devised to provide the basis for clinical diagnosis. Of these the oldest and most numerous are the tuning fork tests, mostly identified by the names of their inventors. Such are the Weber test, Rinne test, Schwabach test, Bing test, and absolute bone-conduction test. These tests and others have been fully described in treatments on audiometry and need no detailed discussion here. The first two have proved to be the most valuable for routine use.

The Weber Test. This test is based upon observations made by E. H. Weber in 1834. If, with a person of normal hearing, the stem of a vibrating tuning fork is pressed against the head at the vertex, or elsewhere in the midline, the tone will be heard as located inside the head or perhaps as coming from the point of contact. If now one ear is covered with the hand or, better, is closed by inserting the finger into the meatus or by pressing the tragus inward, and the fork is still held in position at the midline, the tone is perceived as stronger in the closed ear. The closure of the meatus is considered as producing an artificial conductive deafness on one side and thereby causing a lateralization of the tone to that side. Therefore, it is concluded, a person with true conductive deafness in one ear should lateralize a sound to this ear when the stimulation is applied to the midline.

The practical test consists of applying the fork to the midline and asking the person to report in which ear the tone is heard as louder. Incidentally, it is necessary to caution the person to base his report upon his subjective experience and not upon the conclusion that he will naturally draw that he ought to hear the sound better in his good ear.

When the sound is lateralized to the ear that further testing shows to be the poorer one the conclusion is that this ear has a conductive impairment. On the contrary, when the sound is lateralized to the better ear, a sensory or neural impairment is indicated as present in the other. This test of course fails to be diagnostic when the two ears are equally affected. Its results are difficult to interpret also when a mixed type of deafness is present.

There has been much discussion as to the reason for the lateralization of a bone-conduction tone to the side of an obstructive lesion. The major experimental effort has been expended in extending Weber's original observations on the effects of an artificial obstruction. Pohlman and Kranz showed that occluding the external meatus with stiff wax caused changes of bone-conduction acuity that varied as a function of frequency. For low tones the acuity was increased by 18 db or so, and as the frequency was raised this effect was diminished until it disappeared around 2600~. Békésy (*3*) carried out a similar experiment by use of a rubber plug in the meatus. He took the precaution to provide an air leak in the plug to prevent any raising of the air pressure in the meatus on the insertion of the plug. As a result of plugging the meatus in this way he obtained an increase of acuity for all tones below 2500~, with the greatest effects, around 20 db, in the region from 800 to 2100~.

Many hypotheses have been advanced in explanation of these effects of occluding the meatus. Rinne suggested that the effects are due to an altered resonance of the meatal cavity. However, closing the outer end of this cavity should double its normal resonance frequency and give the most improvement around 8000~, which decidedly is not the case.

Mach (*1*) appealed to a reciprocity principle and said that if aerial sounds could affect the cochlea then the cochlea when vibrated by bone conduction should radiate a part of its energy as sound into the air. An occlusion of the meatus prevents this radiation and thereby conserves the stimulating energy. This hypothesis can hardly be evaluated as it stands because it fails to specify the particular routes of energy flow for effective stimulation. An outflow through the meatus may or may not be detrimental depending on the circumstances of stimulation. In the translatory form of bone conduction it is definitely advantageous to have freedom of vibratory movement along the ossicular chain.

Békésy tested Mach's outflow theory by attaching a large mass of wax to the mastoid process. Such a mass ought to form an avenue of outflow in Mach's sense, but there was no appreciable effect. Békésy suggested that the vibration of the skull in bone conduction causes a periodic change in the dimensions of the meatus through movements of the condyle of the mandible. The alternating air pressures that these movements produce in the meatus are greater when its opening is occluded.

Bouman suggested that closing the meatus on one side reduces the masking effects of external noises and thereby makes the sound seem louder in this ear. He said that the lateralization usually found in the Weber test was greatly reduced when the test was carried out in a soundproof room, whereas it was exaggerated when carried out in the presence of a loud noise. However, the existence of a lateralization effect in silence, even though it is reduced, remains unexplained by this theory.

An important fact, well established in clinical experience, is that the obstruction to conduction giving the lateralization effect need not be in the meatus, as it is in the artificial case with which the experimental studies have been concerned. It may be anywhere in the path of conduction to the cochlea. The theories that have been devised to account for obstruction of the meatus are less serviceable in relation to other forms of conductive impairment. Thus Rinne's and Békésy's theories do not apply to a fixation of the ossicular chain as a whole or an immobilization of the stapes.

For an explanation of the effects of ossicular fixations as seen in the Weber phenomenon we must consider more particularly the nature of the stimulation process in bone conduction. Our earlier discussion of this process has shown that there are two main possibilities, known as the translatory and compressional modes of stimulation. It is clear that an ossicular fixation will reduce the translatory mode but will considerably enhance the compressional mode. In general, any change that affects the motional impedance of the peripheral mechanism will affect one or both of these modes of stimulation by bone-conducted stimuli. However, from our present knowledge it is difficult to predict for the human ear what the net effect on hearing will be. For the cat the results already presented in Fig. 91 show that an ossicular fixation definitely reduces sound transmission, especially for the middle frequencies.

The Rinne Test. This test, first described in 1855, is a comparison of hearing by bone conduction with that by air conduction for the same tuning fork. The stem of the fork is held against the mastoid bone on one side until it just ceases to be audible, and then it is removed and its prongs are held opposite the meatus on the same side. When the ear is normal the tone becomes easily audible after the change to air conduction and continues to be heard for several seconds. When the ear has a conductive impairment the tone remains inaudible after the change to air conduction. The usual relation—air conduction better than bone conduction—is known as the positive Rinne response, and the other, characteristic of conductive deafness, is the negative Rinne response.

It is obvious that the degree as well as the direction of the difference between air and bone conduction is of importance in diagnosis, and this consideration has led to many variations and refinements of the simple procedure just described. It will also be obvious that the observed relations depend upon the physical properties of the tuning fork itself: upon the relative effectiveness of stem and prongs as sound sources when used in the prescribed manner, and the rate of decrement of the vibrations of these parts. These considerations must be borne in mind if any attempt is made to treat the results quantitatively or to compare results obtained by different examiners. They do not present any serious obstacle to the use of the test by a given examiner as long as he maintains a standard procedure for striking and applying a particular fork.

The comment just made applies to all tuning fork tests, and indeed to all clinical tests in general. There are many procedures that yield useful information about the condition of the ear, and the important thing is not so much the particular choice of tests or procedures as the regularity and care with which the examiner applies the ones that he happens to choose, and the body of experience that he has from which to interpret his results. A few simple tests carefully made and interpreted with insight are better by far than elaborate procedures carried on indifferently. Generally speaking, the two tests described here, together with audiometric determinations of acuity by air and bone conduction, will suffice to distinguish conductive deafness from other types. The further specification of the disorder will involve the otoscopic examination and the patient's past history of disease.

BLOCKING OF THE EXTERNAL AUDITORY MEATUS

The simplest form of impairment of the transmission of aerial sounds to the ear is a blocking of the external auditory meatus.

Such blocking is often produced by a plug of cerumen. The accumulation of cerumen causes no very noticeable effects upon hearing until the meatus is almost completely closed. The effects then increase rapidly as the occlusion grows complete, and especially if the mass comes to lie in close contact with the tympanic membrane.

Dishoeck and DeWit studied the effects of various forms of closure of the meatus. Filling with wax, without contact with the tympanic membrane, produced a loss to all tones of about 40 db. Bunch reported six trials in as many ears in which the meatus was packed with petrolatum-soaked cotton. The average results, as shown in Fig. 125, represent a loss of 30 to 40 db for all tones.

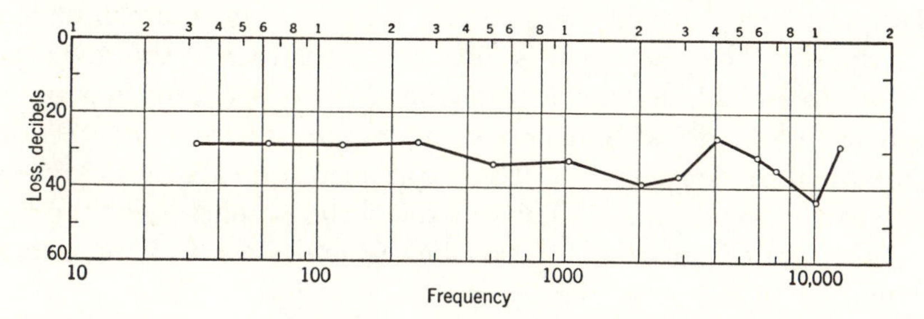

Fig. 125. Effects of blocking the external auditory meatus in man. Data from Bunch.

Occasionally a developmental malformation of the ear occurs in which the external auditory meatus is either occluded or this canal is missing altogether. The auditory effects are much like those of packing the meatus with wax or cotton as just shown, if the deformity is limited to the meatus. Usually this is not the case, but malformations are present in the ossicular mechanism as well. The hearing then is more seriously affected, and is often reduced as much as 60 db below normal for all tones (Woodman).

PERFORATIONS OF THE TYMPANIC MEMBRANE

Another simple form of conductive lesion is a perforation of the tympanic membrane. Politzer (5) noted that such perforations vary in location according to their causes. Those produced by penetrating objects are more often in the posterior part of the membrane than elsewhere, because of the curved form of the meatus. Those caused by sudden pressure changes, as by a blow on the side of the head or an explosion, are most often in the anterior-inferior quadrant. Those caused by the slow pressure of a suppurating middle ear are most frequently in the anterior-inferior or the posterior-superior quadrants, and somewhat less often in the pars flaccida.

A number of writers have expressed the view that a simple perforation of the tympanic membrane, where the ossicular chain is unimpaired, has little or no effect upon hearing. Others have said that the low tones only are affected. Most of the recent experimental studies, however, have indicated that as a result of a perforation the hearing is impaired for all tones.

Bordley and Hardy reported audiograms for two persons who had sustained a traumatic rupture, without secondary infection, in one ear. The hearing was diminished relative to the good ear for all tones included in the test, from 256 to 4096~, by an average amount of 12 db without any clear relation to frequency. These investigators also carried out an experiment on cats by the use of the cochlear potentials. They measured the potentials for the five octave tones from 256 to 4096~ before and after making several kinds of incisions in the tympanic membrane. These incisions caused impairments, for all the tones tested, in amounts varying usually from 5 to 20 db. The losses varied somewhat depending upon the site of the incision, but in any location were fairly uniform over the frequency range.

Békésy (*10*), on the other hand, observed changes only for the low tones. He measured the movements of the ossicles in cadaver specimens both when the tympanic membrane was intact and after making a perforation of about 1 sq mm, and observed losses only for tones below 400~. The reduction of ossicular motion was 48 db at 10~, 29 db at 50~, and 12 db at 100 and 200~. Békésy explained this result by saying that the elasticity of the tympanic membrane is great and a small opening has no very significant effect upon it.

Payne and Githler made a thoroughgoing study of this problem, with systematic variation of the size and locus of the perforations. They worked on anesthetized cats and used the cochlear potentials to indicate the effects upon transmission over the range from 100 to 10,000~. After an initial set of readings with the membrane intact, they made a small perforation with an electrocautery and repeated the readings. This perforation was then enlarged step by step until the entire membrane was removed, with further readings at each step. The site of the initial perforation and the manner of its enlargement took three forms. (1) An "anterior" or "*A*-type" perforation was begun in the superior margin on the anterior side of the manubrium, as shown in Fig. 126, and en-

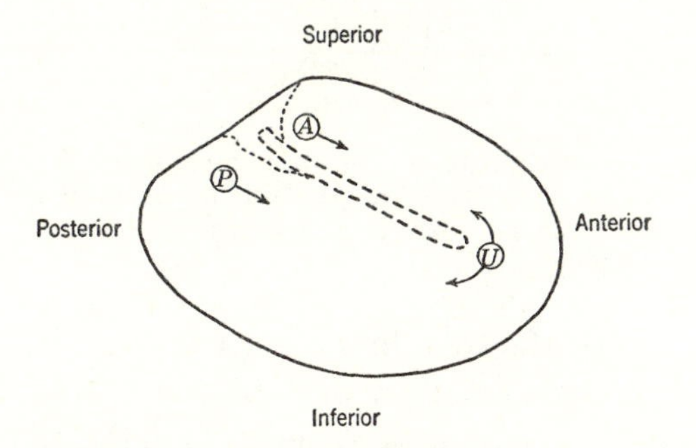

Fig. 126. Sketch of the cat's tympanic membrance showing the three types of perforation procedure used in the Payne and Githler experiments.

larged in the anterior-inferior direction as far as the umbo and then superiorly and posteriorly until all the membrane was removed. (2) A "posterior" or "*P*-type" perforation was begun at the posterior-superior margin and enlarged by going anteriorly and inferiorly, around the umbo, and then superiorly. (3) An "umbo" or "*U*-type" perforation was begun directly below the tip of the manubrium and enlarged by working alternately on anterior and posterior sides. Determination of the site and form of each perforation was made by observing the membrane through a microscope with a gridded reticle in one lens and transferring the pattern to graph paper. Finally the areas were measured with a planimeter and expressed as a percentage of the total area of the membrane.

All types of perforations caused a loss of conduction throughout the frequency range. As shown in Fig. 127, in which the results for all three types of perforations have been averaged, this loss is fairly uniform up to 2000 or 3000~ and then becomes less at the higher frequencies for the small and moderate sizes of perforations. For the large perforations this relationship is reversed. As

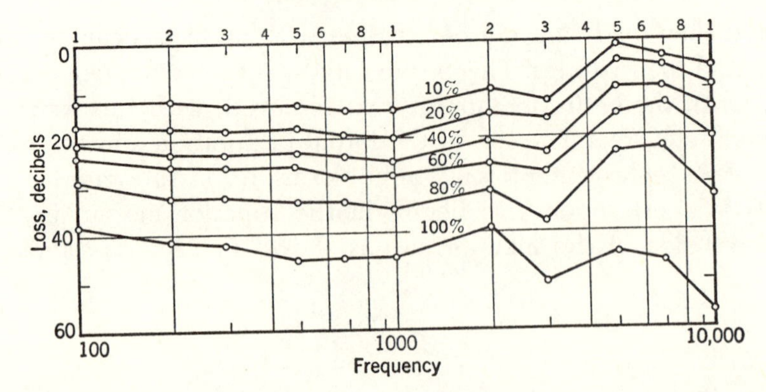

Fig. 127. Effects of perforations of the tympanic membrane of the cat, for various frequencies. Results from three types of perforation were averaged. The numbers above the curves show the percentage of membrane removed. From Payne and Githler [*Archives of Otolaryngology*].

the last remnants of the membrane are removed the transmission of high tones diminishes rapidly until finally, when the membrane is wholly absent, these tones are the most seriously affected.

The effect of the site of the perforation is most clearly shown in Fig. 128. Here each curve is for a given type of perforation and

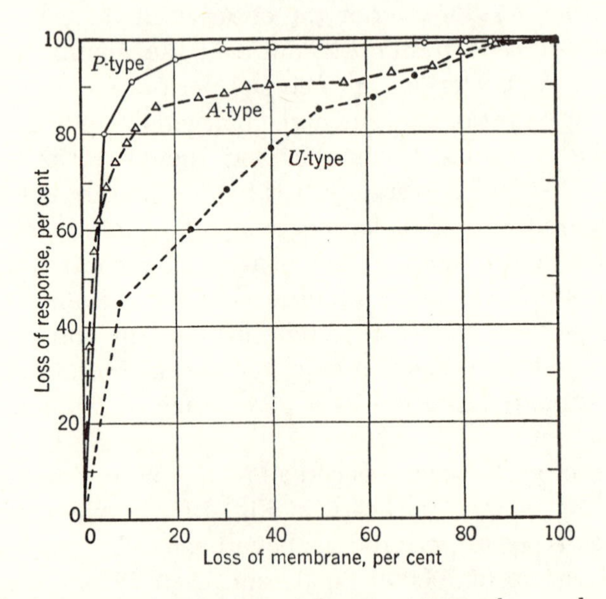

Fig. 128. Effects of size of perforation of the tympanic membrane, observed in the cat. Curves are given for three types of perforation, each representing an average of results for tones from 100 to 2000~. Data from Payne and Githler.

represents the average loss for all tones up to 2000~. For all three types, but especially for *A* and *P* types, the loss is particularly great for the small perforations. The results obtained by Payne and Githler for the high tones, in the region of 3000 to 7000~, showed curious variations. For *A* and *P* types of perforations the curves did not rise as rapidly at first as they did for other tones. Then oftentimes there was a reversal of direction, and the transmission was made actually better than normal. This improvement was usually only 1 or 2 db, but in one instance it amounted to 7 db. As the perforation was enlarged the function quickly reversed its direction, and thereafter showed increasing impairments. The results for 10,000~ were more regular, but like those for the other high tones the changes were rather gradual as a function of size of perforation.

These results prove that the classical view about the effects of perforations is correct as long as the perforation is small: then the hearing is seriously impaired for the low tones and only slightly affected for the high tones. When the perforation is large, however, the most serious effects are found for the high tones.

Because the effects of perforations of the tympanic membrane vary with the size and locus of the perforations and also with the frequency of stimulation, it is clear that the conductive system is being altered in very complex ways. We know too little at present about the mechanical details of this system to say with certainty what these alterations are, and the following suggestions are offered only tentatively to provide a basis for further study.

(1) The primary effect of a perforation of the tympanic membrane is a reduction of the surface on which the sound pressure is exerted. This effect is simply proportional to the size of the perforation.

There is also a secondary effect. A perforation permits the sound to pass through the opening and to exert pressure on the inner surface of the membrane, and this back pressure acts in opposition to the primary pressure. The amount of sound passing through the opening and thus opposing the primary action depends upon both the size of the perforation and the sound frequency. If the perforation is small its edges will present a certain amount of friction to the passage of sound, and this friction increases with the sound frequency. Thus we can account for the fact that the small perforations have the greater effects upon the responses to low tones.

(2) Because this secondary effect is added to the primary effect it follows that the loss of response is out of proportion to the size of the perforation. The loss is very large even for small perforations and comes close to the maximum amount even before a half of the membrane is lost. This rapid initial rise is most noticeable for the low and middle tones. It is less prominent for the high tones because for these tones the secondary effect is reduced by the edge friction just mentioned. Eventually, as the perforation is made larger, the frictional effect disappears and the losses for the high tones come to equal and finally even to exceed those for the low tones.

This last stage, in which the responses to high tones are impaired more than those to low tones, appears only as the last remnants of the membrane are removed. Then the transformer action of the ossicular system is lost and the sound must act directly upon the two cochlear windows, a condition that is particularly unfavorable to the high tones.

(3) Because the tympanic membrane is effective only insofar as it communicates its motions through its attachment to the manubrium, it follows that a perforation is particularly serious when it is in the vicinity of this attachment.

(4) The tympanic membrane also is a part of the tuned mechanism of the ear. It supplies a large part of the stiffness of this system. A perforation that reduces this stiffness therefore modifies the vibratory characteristics of the system as a whole. This effect will be greater when the perforation is in the more tense regions of the membrane. Thus we account for the observation that small perforations in the superior portion of the membrane cause greater changes than perforations near the umbo. Moreover, this modification of vibratory characteristics will be most serious for tones in the region of mechanical resonance of the system. We know from results presented earlier that one of the two principal resonances of the cat's middle ear is in the region of 5000 to 7000~. Thus we can account for the curious variations in sensitivity to these tones that are observed as the result of perforations.

THE EFFECTS OF MIDDLE EAR INFECTIONS

Infection of the middle ear is the commonest of auditory disorders and produces the greatest variety of conductive changes. An infection involving swelling of the membranous lining of the Eustachian tube will close its duct, and then the absorption of the air in the middle ear cavity produces a negative pressure. Inflammation of the mucous membranes of the middle ear cavity may cause an accumulation of fluid and a partial or complete filling of this cavity. If the Eustachian tube is closed this accumulation of fluid may exert a positive pressure, even to the point of rupturing the drum membrane. The fluid will also damp the movements of the drum membrane and ossicles. Long-continued suppurative processes in the middle ear may cause various tissue changes. New connective tissue may be formed, and finally may be changed into fibrous tissue or in rare cases may even become calcified into bone. Finally these suppurative processes may erode away a part or all of the ossicular mechanism.

The auditory effects of these changes are somewhat difficult to determine clinically because usually a number of such changes occur at the same time. Often as a result of infection there are cochlear and neural involvements as well.

Polvogt and Bordley carried out a histological study of ears for which hearing tests had been made before death, and in their selection of cases they attempted to avoid inner ear complications by requiring a negative Rinne response. For the purposes of their study they divided the middle ear into ten regions, and they found that the 20 ears studied had lesions in nearly all of these regions. Audiograms for two of these ears are shown in Fig. 129. In one (dashed line) the middle ear changes were of moderate severity, and consisted mainly of a marked thickening and a minute perforation of the tympanic membrane and heavy adhesions between the head of the malleus and the roof of the epitympanum. In the other ear (solid line) the tympanic membrane was almost wholly destroyed, the outer ossicles were separated from the stapes, and both oval and round windows were covered with scar tissue. As we see, the hearing was profoundly affected in both of these ears. The losses averaged 59 db in one and 64 db in the other for the range up to 4096~, and above that point nothing was heard with the test intensities available.

Because of the complex nature of the changes resulting from middle ear infections it is necessary to turn to experimental studies, especially on animals, for a further analysis of their effects. We have already seen the effects of perforations of the tympanic membrane. Also, in Chapter 11, we have seen the effects

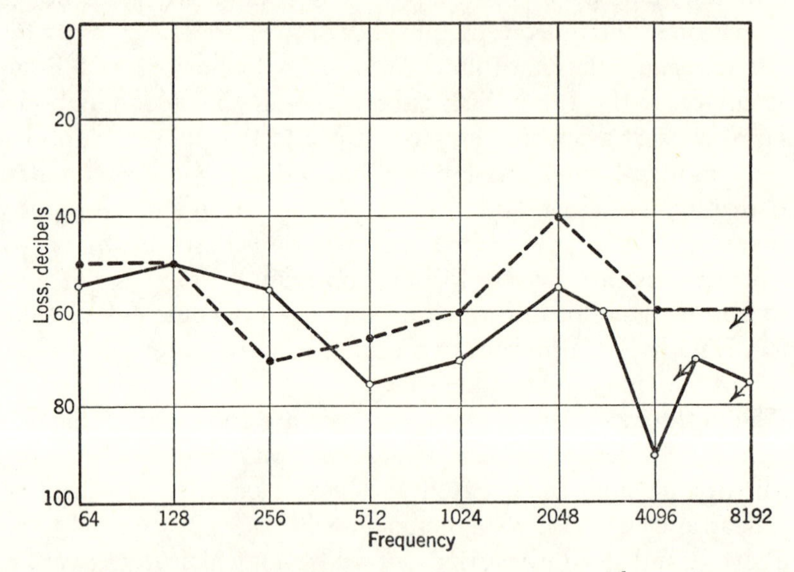

Fig. 129. Effects of middle ear infections on hearing in man. The arrows at certain points indicate that these tones were not heard at the highest intensities produced by the audiometer. After Guild (7).

of pressure changes in the middle ear cavity. A number of studies have been concerned with the effects of fluid loading of the tympanic membrane.

THE EFFECTS OF LOADING OF THE TYMPANIC MEMBRANE

Fluid in the middle ear may obstruct the movements both of the drum membrane and of the ossicles, but there is little doubt that the main effect is upon the drum membrane. Some students have investigated this problem by loading the membrane from the outside. This method does not correspond exactly to the clinical situation, but its obvious convenience recommends it. Luescher (2,3) in his experiments on human subjects followed two procedures. In one he had his subjects lie on the side, with the meatus directly upward, and placed small weights on a little stand whose lower end was resting on the umbo. He found that weights up to 6.66 grams could be tolerated, but a load of 10 grams was definitely painful. Such loadings caused a general reduction of hearing, most prominent for the low tones, for which it amounted to 3 or 4 db.

In another procedure Luescher placed drops of mercury on the drum membrane. These tended to gravitate to the anterior-inferior quadrant, thus primarily loading this region of the pars tensa. These loadings, applied in amounts up to 2 grams, also caused a general loss of hearing but much larger than that caused by corresponding loadings of the umbo, and with a progressively greater effect the higher the frequency.

Dishoeck and DeWit carried out procedures similar to the two used by Luescher, with similar results. Loading the umbo reduced the acuity for low tones in amounts from 5 to 15 db and improved the acuity for high tones up to 8 db. Covering the tympanic membrane with mercury caused reductions that rose progressively from 4 db at 128~ to 19 db at 9747~. Filling the meatus with mercury gave larger effects, especially at the low frequencies. Covering the tympanic membrane with a solution of collodion and letting it evaporate to form a closely adherent layer gave no change at 256~ and a progressive impairment for high tones that amounted to 40 db at 8192~.

These results suggest that the application of a weight to the drum membrane by means of a prod actually constitutes a fixation and thus gives an increase in the stiffness of the ossicular mechanism. Therefore its effect is mainly to reduce the conduction of low tones. A loading of the membrane with a liquid, on the other hand, mainly adds to the mass of the moving system and has its main effect upon the high tones. Completely filling the meatus not only adds mass but also forms a substantial barrier which reflects the sound entering the ear. This reflection is general for all tones. From these observations we can expect that fixations of the ossicles caused by adhesions in the middle ear will chiefly affect the conduction of low tones, and loading of the tympanic membrane by fluid in the middle ear cavity will have more general effects but will most seriously reduce conduction for the high tones.

CAVITY EFFECTS

The question has often been raised as to the effect of the middle ear cavity upon the transmission of sounds. Any enclosed volume of air that is exposed to sounds acts as an elasticity because the alternating pressures must compress and expand this air. The elasticity is inversely proportional to the volume of the air.

Békésy (*8,10*) expressed the opinion that this volume elasticity is considerably greater than the elasticity of the drum membrane. He measured the elasticity of the ear as a whole as presented at a point in front of the drum membrane and obtained the same value as that characteristic of a volume of air of 2 cubic cm, which, he said, corresponds closely to the actual volume of the middle ear cavity. On the basis of this observation he explained the absence of nonlinearity in the action of the drum membrane and also the failure (as he thought) of a small perforation to have any substantial effect upon hearing.

If we grant that the volume elasticity of the middle ear cavity is a large factor in the operation of the ossicular mechanism, then we should expect a considerable effect upon hearing when the middle ear cavity is partially filled with exudate or other substance. The increase of elasticity from the reduction of the effective volume would be expected to produce an alteration of conduction, and mainly a reduction for the low tones.

Our study of the cavity effect in the cat fails to give support to the view that the elasticity from this source is of any importance in relation to the total elasticity of the ear. Opening the middle ear cavity and thus effectively eliminating its volume elasticity causes only minor changes in transmission as shown in the cochlear potentials. Figure 130 gives some of our results. The differences between open and closed bulla are always small, attaining a value of 5 db at only one frequency. From these observations it appears that the elastic reaction of the tympanic cavity aids the reception of the high tones by 1 or 2 db and hinders the reception of low tones by 2 to 5 db.

EFFECTS OF BREAKING THE OSSICULAR CHAIN

A break in the ossicular chain may come from the accidental intrusion of an object, from an excessive pressure change, or, as already mentioned, from erosion of the ossicles. In man the particular nature of these changes is usually difficult to determine. Often also the purely mechanical changes are followed by other changes due to infection. Therefore the effects of the interruption of ossicular continuity are best determined in animal experiments.

The following experiments were carried out on the cat by the

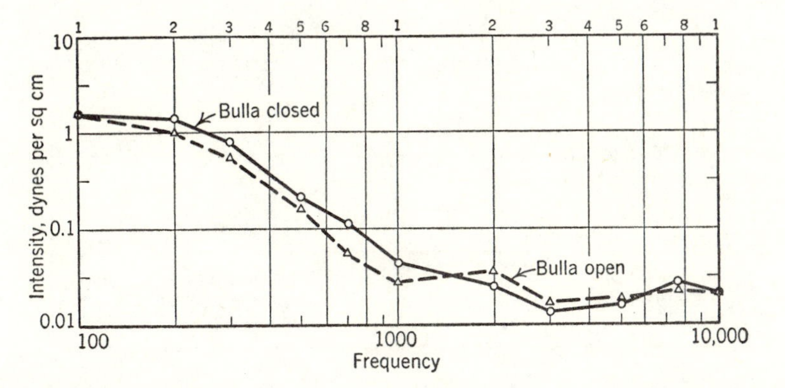

Fig. 130. Sensitivity of the cat's ear when the auditory bulla is open and closed. The curves show the sound intensity necessary to produce a standard response of 10 microvolts. From Wever, Lawrence, and Smith (*1*) [*Archives of Otolaryngology*].

cochlear potential method. In the first series the ossicular chain was broken at the incudostapedial joint, with care to leave all other working parts undisturbed. The drum membrane and outer ossicles remained in position and the stimulating sounds were delivered at the meatus as usual. Readings of the cochlear potentials were taken before and after the breaking of the chain and the differences in decibels are shown by the lowermost curve of Fig. 131.

We see that the interruption of the ossicular chain causes a profound loss of transmission. This loss is about 34 db at 100~ and increases to 60 db in the middle range, after which it diminishes to about 42 db for the highest tones. The average loss is 52 db.

The loss just described is considerably greater than we should expect simply from the elimination of the transformer action of the middle ear mechanism, which we have found to average about 28 db. There are two principal reasons for this heightened effect.

(1) One reason is that the drum membrane and outer ossicles are now worse than useless. They stand in the path of the incom-

ing sound and cause both reflection and absorption of this sound to no good purpose. This fact is proved by the following experiment. The normal sensitivity was determined as before, the incudostapedial joint was broken, and then the drum membrane, malleus, and incus were removed. The tube through which the stimulating sounds were introduced remained in its usual position

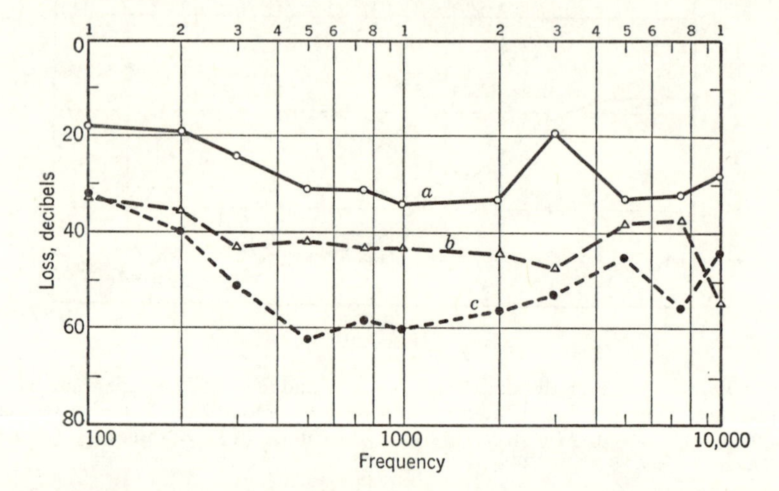

Fig. 131. Losses of sensitivity as shown in the cochlear potentials of the cat, under three conditions of nonfunctioning of the ossicular chain: *a*, when the sounds were delivered directly to the oval window; *b*, when the sounds were presented at the meatus but drum membrane, malleus, and incus were removed; *c*, when the sounds were presented at the meatus with all parts of the middle ear mechanism in position except for a separation of the incudostapedial joint. Data from Wever, Lawrence, and Smith (*1*).

at the entrance to the meatus. Measurements of cochlear potentials now showed better transmission than when the drum membrane and outer ossicles were present. The middle curve of Fig. 131 gives some of the results. It is evident that the removal of the useless parts has improved the transmission by about 15 db for tones of the middle range, and by somewhat less for the low and high tones.

We have in this second form of the experiments a modification of the cavity of the ear, inasmuch as the removal of the drum membrane joins the space of the meatus to that of the middle ear proper. However, in view of the results already presented on cavity effects, we do not attach any great importance to this change.

(2) The second reason for the large impairments resulting from an interruption of the ossicular chain is that now the sounds have more nearly equal access to both oval and round windows. Their stimulating effects upon the inner ear tend to be counteractive, as we have seen earlier (page 219). This cancelling effect can be eliminated by bringing the sound tube over only one window and sealing it there so as to prevent the entrance of sound through the other window. This was done in the third stage of our experiments on the cat. To facilitate the placing of the sound tube the crura of the stapes were broken off, leaving only the footplate in the oval window. The upper curve of Fig. 131 now shows the loss when the sound is thus presented only to the oval window. We see that the transmission is further improved. The simultaneous stimulation by way of both windows evidently has a cancelling effect of about 12 db.

Theoretically this cancelling effect might have been much greater, and indeed would have been complete if the sounds reached the two windows with equal intensities and with identical phases. This condition for complete cancellation is highly critical, however, as we have seen, and anything that alters the intensity or the phase relations between the two pathways will move the combined response away from the zero point. It is evident that some conditions are present for so altering the two ways of stimulation. The 12 db loss that we have been able to attribute to this cancellation effect allows us to form a general idea of the effectiveness of these conditions. To account for the observations we can assume that the sounds at the two windows have the same phase but differ by 14 db, or that they have the same amplitude but differ in phase by 15.5°, or that they differ in smaller amounts in both amplitude and phase.

Some observations made by Békésy (*15*) are relevant to this problem. In cadaver specimens he used a reciprocity method to determine the relative stimulation at oval and round windows. With the drum membrane and all three ossicles absent, he introduced separate sounds from sources located behind the window openings and adjusted the pressure and phase relations of these sounds to obtain a null at the meatus. Considered inversely, his observations tell us what the relative pressures and phases should be at the windows for a single sound applied through the meatus.

Unfortunately, his results are presented in a form that makes their interpretation difficult for our purpose, but if we make certain reasonable assumptions we can say that the pressures in front of the two windows should differ by about 0.2 db at 250~ and by increasing amounts up to 0.9 db at 3000~. At the same time, the phase differences increase from 1° at 250~ to 5.2° at 3000~.

These differences are much too small to account for our observations on the cat. With these differences the cancellation effects would be considerably greater than 12 db. However, we believe that Békésy's method eliminated the most important factor that is operating to produce intensity and phase differences between the two windows. These differences, according to our observations, are mainly due to the presence of the stapes in the oval window. We have already presented the evidence that various manipulations of the stapes cause variations in the phase for high tones. It is reasonable to suppose that this ossicle, or even the footplate by itself, is mainly responsible also for the intensity differences found between the oval window and round window pathways.

It is fortunate that this condition and others are present to prevent a complete cancellation effect when the ossicular mechanism is no longer functioning to give the oval window pathway an advantage. Otherwise many cases of conductive deafness might be much more severe than they are. It is evident that the hearing in such cases might be improved if something could be done to increase further the difference in stimulation of the two windows. If the sound were prevented altogether from reaching one window, as in our third experiment, the hearing should be raised to the level imposed by the absence of the transformer effect. It would be even better, of course, if the sound were given equal access to both windows, but in a contrary phase relation, for then we should have a summation that is 6 db better than action upon one window alone.

FIXATION OF THE OSSICULAR CHAIN

A conductive impairment of particular importance is a fixation of the ossicular mechanism. At least three forms are recognized.

Fixation by Fluid. One form of fixation is found in the acute stage of middle ear infection as a result of the accumulation of fluid and semifluid substances in the tympanic cavity. As already suggested, these substances add mass and friction to the moving structures, and when they begin to fill the spaces they add stiffness as well. The result is an impairment of sensitivity that mainly concerns the high tones at first and then in the advanced stage includes all tones.

Simple Mechanical Fixation. We have also made note of the tendency of infective processes to produce new tissue growth and to cause adhesions between different parts of the moving mechanism and between these parts and the walls of the tympanic cavity. These adhesions add friction to the motions of the mechanism and impair the transmission for all tones. If they become heavily fibrosed they may also add stiffness and cause a particular reduction for low tones.

Some of the animal experiments show how serious a mechanical fixation may be, and reveal further the form of the impairment when the fixation adds greatly to the stiffness of the system. In a series of experiments on cats by Smith, already referred to, a steel needle was introduced into the auditory bulla and anchored so that its midportion passed close by the head of the stapes but without touching it. A measurement of the ear's sensitivity in terms of the cochlear potentials was made at this point. Then a thread was passed about the stapes and tied firmly to the needle. As a result the transmission was sharply reduced, by amounts varying in different trials with the firmness of the anchorage, but sometimes by as much as 55 db for the low tones. A reduction of 55 db represents a physical fixation of 99.8 per cent: the amplitude of the conducted sound was reduced to 0.2 of 1 per cent of its former value.

The average results obtained in this experiment on stimulating with aerial sounds are shown in the lower curve of Fig. 91. A prominent feature is the progressive change in the effects as a function of frequency. The transmission is most seriously affected for the low tones and the effect becomes less as the frequency rises. This relation reflects the fact that the fixation adds a great amount of stiffness as well as frictional resistance to the moving system.

Otosclerosis. The most serious fixation of the ossicular mechanism is caused by the disease of otosclerosis. This condition and the surgical measures that have been developed for its treatment will be the subject of the next chapter.

16

Otosclerosis and the Fenestration Operation

OTOSCLEROSIS is a disease of the bone in the region of the oval window that leads to a partial or complete fixation of the stapes to the margin of this window. The result is a serious impairment of hearing. Strictly speaking, this is clinical otosclerosis, for the disease of the bone can be present, evidently over many years, without leading to stapes fixation or causing any noticeable auditory impairment.

The cause of this disease is unknown. There is evidence of a hereditary basis, for it often runs in families. It is more common in women than in men. Guild (*6*), from a study of 1161 pairs of serially sectioned temporal bones, found the incidence of histologic otosclerosis—bone lesions with or without stapes fixation—to be 6.5 per cent for white males and 12.3 per cent for white females. In comparison, this figure for negroes of both sexes is only 1.0 per cent. A study of the distribution of cases according to age showed the highest incidence among persons from 30 to 49 years of age.

Among the 81 ears of this study showing otosclerotic lesions of the bone only 10 showed fixation of the stapes. The results indicated that in all 10 of these ears there was impairment of hearing, though other conditions often complicated the picture. The evidence supports the conclusion that an otosclerotic area of the bone does not affect the hearing unless it produces a fixation of the stapes.

Most of the otosclerotic areas were located anterior to the oval window, and when they involved the stapes they most often fixed the anterior end of the footplate or its anterior crus.

The histological study of various otosclerotic lesions suggests a regular progression of events in the development of a clinical disorder. The study has been carried out on post-mortem material and, more extensively, on auditory ossicles and portions of the labyrinthine wall that were removed during an operation on the ear. According to Lempert and Wolff, the first stage is a change in the blood vessel walls, fragmentation and aggregation of red blood cells, and other vascular changes. Perhaps as a result of these vascular changes the bone begins to show evidences of chemical modifications. The walls of the Haversian canals in the affected areas stain an intense blue with hematoxylin. This pathological bone can even be discerned in the living ear on observation with low magnification, as Lempert has found. The affected region shows a pink color as compared with the normal yellowish bone.

The chemical changes in the bone evidently lead first to a partial decalcification, and then further changes occur that represent the attempt of the tissues to repair the damage that they have suffered. A scar is formed, which may be mainly fibrous or bony or a mixture of fiber and bone.

Unfortunately, the scar does not necessarily preserve the form of the original structure, but often builds up more abundantly than normal in some regions. When these changes take place in the region of the oval window a bony bridge may form across the annular ligament thus binding the stapes to the margin of the window. Sometimes a part or even the whole of this ligament is bridged over or entirely replaced by the new bone. Often several narrow bridges are formed with normal-appearing remnants of the ligament between them.

It is a reasonable assumption that with the progression of the disease these bridges slowly enlarge and finally fuse to form a continuous connection. As the bridges become more numerous and more continuous the fixation increases correspondingly. The extreme stage is that in which the annular ligament is fully replaced by bone and for practical purposes the oval window is obliterated. The clinical course of the disease supports this view. Typically, the disease first becomes detectable in symptoms of mild hearing loss during early adulthood and the loss gradually becomes greater over a period of several years.

There is reason to believe that in its earlier stages the otosclerotic fixation mainly impairs the hearing for the low tones, much as we have seen to be true for an artificial fixation. Later, as the fixation continues, the hearing grows worse for all tones but particularly so for the middle and high tones. Still later, in many instances, the high tones drop out altogether. Figure 132 gives examples of hearing loss due to otosclerosis.

The high-tone loss, when it occurs, is nearly always found for bone conduction as well as for air conduction, indicating that there is an involvement of the cochlea or the nerve as well as of the conductive mechanism. It is a matter of dispute whether this

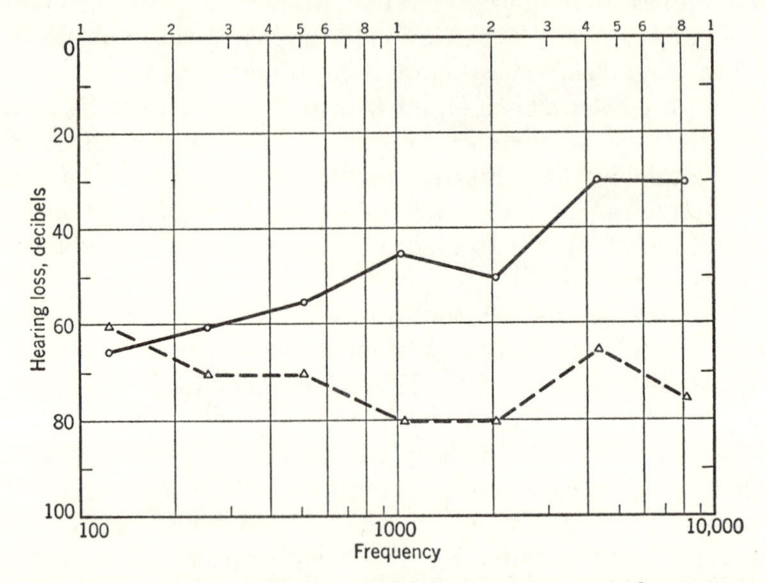

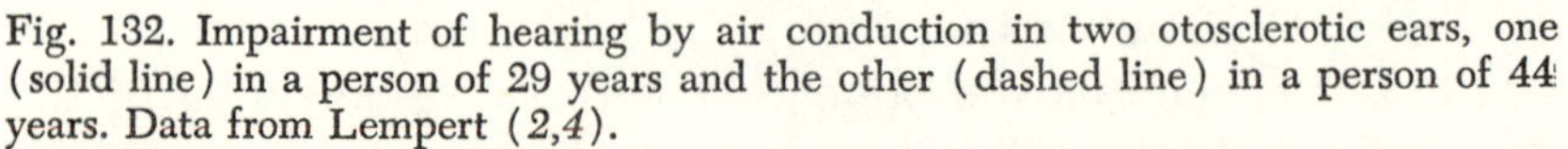
Fig. 132. Impairment of hearing by air conduction in two otosclerotic ears, one (solid line) in a person of 29 years and the other (dashed line) in a person of 44 years. Data from Lempert (2,*4*).

inner ear or nerve effect is a consequence of the otosclerotic lesion. It is plausibly suggested that the disease may produce toxic by-products that affect the sensory or neural cells. However, Guild (2) observed no more serious degeneration of the hair cells and cochlear nerve fibers in ears containing otosclerotic lesions than in ears of persons of comparable age not containing such lesions. It therefore appears likely that the high-tone losses that often occur in advanced otosclerosis merely represent an independent disorder that has been added to the otosclerosis.

THE FENESTRATION OPERATION

One of the outstanding accomplishments in the field of otology is the development of the fenestration operation for the relief of otosclerotic deafness. It is based upon the discovery that the hearing makes a dramatic return toward normal when an opening is made through the bone to the perilymphatic space at the vestibular end of the labyrinth. This discovery was made long before the knowledge could be put to any practical use (see Pierce, *et al.*).

Early surgical measures to relieve the effects of otosclerosis consisted of attempts to loosen the stapes, as practiced by Miot in 1890, or to pull it out of the oval window. Blake and Jack used the extraction procedure in several forms of disease of the ear, including otosclerosis, but Jack reported that often the footplate was so firmly fixed that it could not be dislodged. When the stapes was removed the observations showed an improvement of hearing, but this improvement could not be maintained either because the opening soon closed or because the continuing escape of perilymph caused the deterioration of the labyrinth.

Politzer (*4*) and later Siebenmann condemned this operation as both useless and dangerous, but many continued to experiment with it. New openings were tried both in the region of the oval window and in the promontory near the round window, but with little success. An important step, ascribed to R. Bárány, was the shifting of the site of the opening to a region more remote from the active otosclerotic process. He exposed and opened the posterior vertical semicircular canal, and reported favorable results immediately after the operation. Jenkins selected the horizontal canal as a more accessible site.

Holmgren experimented with the anterior vertical canal as a locus for the bony fistula, and addressed himself particularly to the problem of keeping it open. He covered it with a piece of the mucoperiosteum from the adjacent bony wall. However, such a flimsy covering fails itself to survive and in consequence the labyrinth dies. Sourdille placed the fistula in the horizontal canal and covered it with a skin flap taken from the external auditory meatus. This covering prevents the continued loss of perilymph and thereby preserves the labyrinth.

THE LEMPERT FENESTRA NOV-OVALIS OPERATION

To Lempert belongs the credit for having made the fenestration operation a practical procedure and for having continued over many years to improve it in every detail. He also vigorously attacked the basic theoretical problem of why this disease affects the hearing as it does and how fenestration gives improvement. He has called his procedure the nov-ovalis operation because it provides a new functional window in the place of the one that has been lost.

The Lempert operation makes an endaural approach to the inner ear region, that is, one through the entrance of the external auditory meatus. The passage is then posterior to the drum membrane and directly to the mastoid cortex covering the posterior portion of the otic capsule, including the three semicircular canals. The resulting exposure is shown in Plate 8. The incus is removed and the head of the malleus is cut off. An opening is then made in the lateral semicircular canal close to its ampulla. According to the most modern technique this opening is made by first thinning the bone over the canal by use of a dental burr and then cutting out a cap of bone. This method avoids the accidental introduction of bone dust or chips into the perilymph space. The perilymph space and the membranous canal are brought into view as shown in Plate 9. This opening is now covered with a skin flap which consists of a portion of the wall of the meatus together with the drum membrane with which this wall is continuous. The drum membrane is freed from its bony sulcus except at its inferior and anterior margin so that the whole tympanomeatal membrane can be diverted backward. The meatal portion covers the fenestra while the tympanic portion forms a kind of canopy over a portion of the middle ear cavity. This reduced middle ear cavity, which contains the round window, is thereby walled off from the meatus and the space about the fenestra.

The meatal part of the tympanomeatal membrane is pressed firmly upon the bone and also partly into the fenestra itself and is maintained there by means of surgical packing until attachment occurs. The extension of the skin into the opening serves to prevent the growth of bone from the fenestral rim and so keeps the fenestra open.

The improved hearing that follows this surgical procedure is shown in Fig. 133. The solid curve represents the usual result, in which the hearing is returned within 20 to 30 db of the normal, which is a level enabling a person to hear ordinary conversation and generally to conduct his affairs without particular handicap. Sometimes, in a small percentage of cases, depending upon various conditions and especially the condition of the inner ear and the auditory nerve, the improvement is less, as the dashed curve shows. Here the return is only to a level around 40 db below

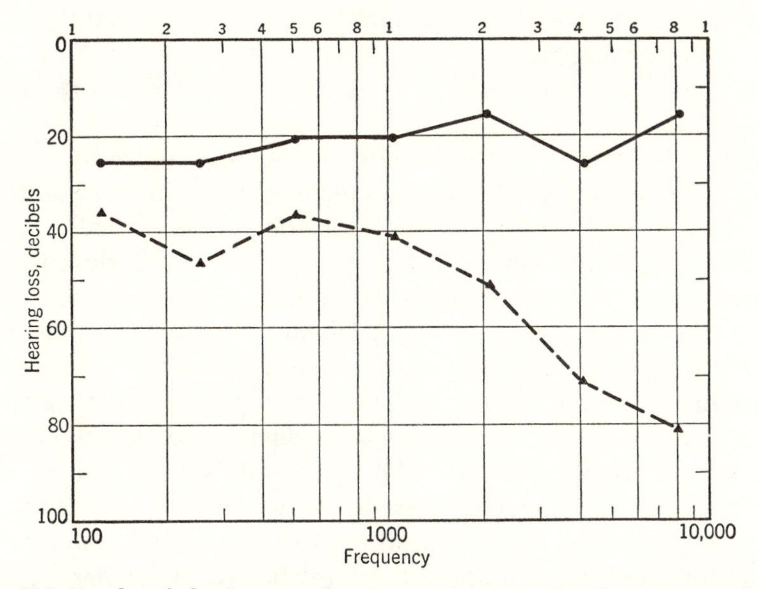

Fig. 133. Results of the Lempert fenestration operation for the two ears shown preoperatively in Fig. 132. Data from Lempert (*2,4*).

normal for the principal range, and this amount of sensitivity is not sufficient for serviceable hearing. In particularly fortunate cases the return is within a few decibels of normal.

This improvement of hearing is lost and the audiogram returns to its preoperative level if the fenestra is closed by fibrous and bony growth. The improvement is then restored if the fenestra is reopened.

Many variations of this operation are possible, and all are beneficial as long as they satisfy the condition that the opening must be into the perilymphatic space at the vestibular end of the labyrinth. Within the limits of surgical access the opening may be in any one of the semicircular canals and anywhere along the course of a canal or in the vestibule proper and, barring accidents, an improvement follows. The opening may be large or small, and there may be two or more of them. The surgical approach may vary and the middle ear structures may be variously handled. Thus Holmgren used the postauricular approach and left the middle ear mechanism undisturbed, whereas Lempert employs the endaural approach and greatly modifies the middle ear mechanism as described. Both of these procedures produce improvements, though to be sure not equally favorable ones. On the other hand, an opening into the scala tympani will not serve. In fact, such an opening may make the hearing worse than before (Lempert, 7).

It is possible to demonstrate the effects of fenestration in animals below man. Otosclerosis is exceedingly rare in these animals, if indeed it exists at all, but stapes fixation is occasionally found, probably as a developmental anomaly. Van Eyck described a pigeon in which the columella was grown fast to the margin of the oval window. Cochlear potentials in response to sounds were found to be feeble at first, but were made noticeably stronger by fenestration of the horizontal canal. Lempert, Meltzer, Wever, and Lawrence (2) produced an artificial otosclerosis in the monkey by breaking the ossicular chain and fixing the stapes with cement. The cochlear responses to sounds were greatly reduced by this procedure, but a fenestration of the lateral semicircular canal returned them to about the level before the stapes fixation.

We now have to explain why the fenestration operation improves the hearing. In so doing we shall also enlarge our understanding of the loss that occurs in otosclerosis.

THE FENESTRATION THEORIES

There are three theories of fenestration: the decompression theory of Holmgren, the "new tympanic system" theory of Sourdille, and the mobilization theory of Lempert.

The Decompression Theory. According to Holmgren, the loss of hearing in otosclerosis is due in large part to an excess of perilymphatic fluid pressure. Fenestration restores the hearing by relieving this excess of pressure.

The first weakness of this theory, as Lempert (3) has pointed out, is the lack of any evidence that in otosclerosis an excess of fluid pressure exists. When the fenestra is made there is never any outrushing of fluid as there should be if it is under pressure. No one who has studied the microscopic sections of otosclerotic ears, as far as we know, has reported as typical of the disease a pronounced bulging of the round window, as should exist in the presence of such pressure.

The second weakness is that after the fenestra is made it is immediately covered and tightly sealed, and indeed if it were not so covered the continuing fluid loss would cause the death of the labyrinth. The perilymph space is thus closed after the operation and the fluid pressure ought to rise to the same level as before, thus cancelling any improvement. It is well known that this does not happen.

A third point is that if decompression were truly effective a fistula into the scala tympani ought to give a favorable result, whereas it fails to do so (Lempert, 2).

Finally there is the evidence already referred to (page 196) that heightened pressure of the labyrinthine fluid does not affect the conduction of sounds through the cochlea. Lempert, Wever, Lawrence, and Meltzer made the following observations in the monkey. A special pressure needle was made by grinding flat the end of a No. 25 hypodermic needle and slipping over it a sleeve made of a piece of a larger needle and soldering the two parts together so as to leave the end of the smaller needle protruding about 0.5 mm. An opening was drilled through the bony wall of the ampullated end of the lateral semicircular canal into the perilymph space without injuring the endolymphatic labyrinth. This opening was then reamed to a diameter just fitting the end of the pressure needle, and this needle was inserted to make a tight connection. The pressure needle was connected to a gauge and a source of air pressure.

While a tone of constant intensity was presented to the ear the cochlear potentials were measured first for the zero pressure condition and then after various positive pressures were applied to the perilymph. Pressures of 10, 20, and 50 mm of mercury were used (i.e., pressures up to 66,600 dynes per sq cm), which caused an easily visible bulging of the round window membrane. In no instance was there any effect upon the potentials.

Békésy (*12*) had carried out a similar experiment on cadaver specimens. He measured the vibratory movements of the round window membrane in response to sounds and found them to remain constant while various static pressures were exerted upon the labyrinthine fluid up to a point just below 4×10^6 dynes per sq cm. At this enormous pressure something gave way with an audible click and the labyrinth became leaky, and thereafter sound conduction was found to be impaired by 10 to 20 per cent. These experiments show that the conduction of sounds through the fluid pathways of the cochlea is unaltered by tolerable increases in the fluid pressure. Our experiments, just referred to, go further in showing that this increase of fluid pressure does not impair the movements of the basilar membrane or the functioning of the hair cells of the organ of Corti.

If the fluid pressure has no effect upon the operation of the ear it is obvious that the decompression theory must be discarded.

Sourdille's Theory. The explanation of the effects of fenestration offered by Sourdille was based upon the idea of the creation of a new acoustic "amplifier" in the ear. He conceived that the tympanomeatal flap, acting in association with the incus, provided a path along which sounds would travel to the new window and be enhanced in the process. The supposed manner of operation of this "new tympanic system" remains obscure, but the incus is said to be an essential part of the structure, and he insisted that it be preserved in the operation. However, Lempert in developing his fenestra nov-ovalis operation showed that the incus could be removed without altering the beneficial effects of the fenestration. He pointed out also that the skin flap becomes firmly adherent to the underlying bone and hence can hardly serve as a free conductor of sound. Further, the flap may be completely severed between drum membrane and fenestra without any noticeable effect upon the hearing. This flap is sometimes accidentally torn in the course of the operation, yet its pieces may be laid down or it may be replaced altogether with a piece of skin taken from some other part of the body, and—always provided that the graft lives and grows in its new site—the functional results are as usual.

The Mobilization Theory. The theory of fluid mobilization was suggested by Lempert in his original presentation of the one-stage fenestration operation in 1938 and was more explicitly stated in his further reports of 1940 and 1941. He conceived that the opening of the labyrinth increases the effectiveness of aerial sound in setting up a reciprocating motion of the labyrinthine fluids. The new opening takes the place of the occluded oval window and once more permits a transfer of pressure between scala tympani and scala vestibuli and across the basilar membrane.

This theory is supported by many lines of evidence. It rests upon the secure foundation of Eduard Weber's conception of cochlear mechanics, according to which there must be a mass displacement of the cochlear fluid for movement of the basilar membrane and stimulation of its hair cells. As we have seen, this conception is firmly supported by the evidence of our experiments in which sounds are applied simultaneously at oval and round windows. The results of these experiments make it clear that in the normal operation of the ear, when the middle ear mechanism is present, the sounds are applied at the oval window and cause a fluid displacement between this point and the round window. Effective stimulation also occurs if the sounds are applied directly to the oval window, without the middle ear mechanism, though the efficiency is less after the loss of the transformer action that this mechanism provides. The stimulation is also effective when the sounds are applied to the round window and the oval window is left free as the point of pressure discharge. The reciprocating movement of the fluid, in whatever way it is established, moves the basilar membrane back and forth and stimulates its sensory cells.

When one window is closed this reciprocating motion of the fluid can no longer take place. We have seen in the animal experiments already described (page 241) that a rigid blocking of one window causes a serious impairment of sound conduction. Blocking by such materials as wax causes reductions of 20 to 30 db. It is obvious that the firmer blocking caused by tissue growth in the oval window in otosclerosis will immobilize the fluid even more completely.

When the cochlear capsule contains only one opening the sound energy entering by this opening spreads throughout the cochlea. Because the dimensions are small relative to the wave length of the sound the spread for all practical purposes is instantaneous and at any moment the pressure is the same at all points within the capsule. The basilar membrane is acted upon equally from all directions and is not displaced from its resting position.

Figure 134 will make these relations clear. Here the cochlea is represented schematically as made up of two chambers separated by the basilar membrane. The upper chamber is a combination of scala vestibuli, cochlear duct, and the free space of the vestibule, and the lower chamber is the scala tympani. The stapes is shown as ankylosed, and aerial sound, indicated by the large arrow, is considered as applied to the round window. The effect is a pressure distributed uniformly over the inner wall of the cochlear capsule, as represented by the small arrows. At any point on the

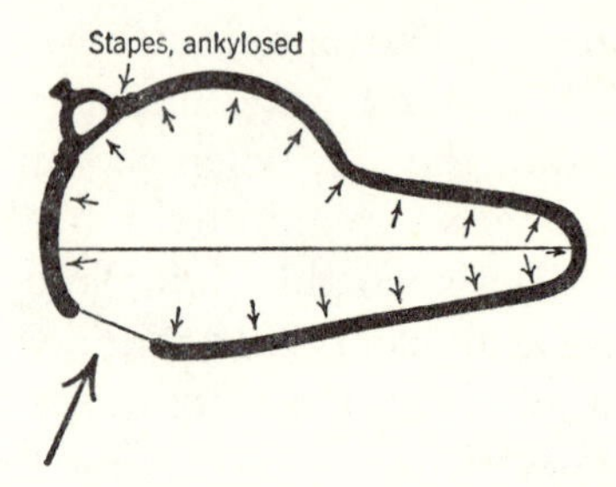

Fig. 134. Aerial stimulation in otosclerosis. Sound entering by way of the round window produces pressures that are uniform throughout the cochlear fluid and no movement of the basilar membrane results. From Wever (6) [*Annals of Otology, Rhinology, and Laryngology*].

basilar membrane the pressure is equal in all directions and this membrane is not displaced.

When a new window is surgically created to take the place of the lost oval window the cochlear fluid is remobilized. Now the sound pressure exerted at one opening displaces a quantity of fluid in the path leading to the other opening. It vibrates this fluid and moves the basilar membrane in doing so, thereby stimulating its sensory cells. This situation is represented in Fig. 135. Here a fenestra, covered by a skin flap, is shown in the vestibular region. Sound applied at the fenestra causes a pressure discharge through the cochlea. The arrows show the direction of discharge at the moment that a positive pressure is exerted at the fenestra. The basilar membrane is then displaced toward the round window. A moment later, as the pressure reverses its direction, this displacement will be away from the round window. The stimulation would be equally effective if the sounds were applied at the round window.

The mobilization theory explains why the fenestra must be located in the vestibular part of the labyrinth and not in the tympanic part. For a pressure discharge to traverse the basilar membrane, the two windows obviously must be on either side of it. A new opening in the cochlear promontory, into a part of the scala tympani, would merely permit a movement of fluid between this point and the round window without involving the basilar membrane.

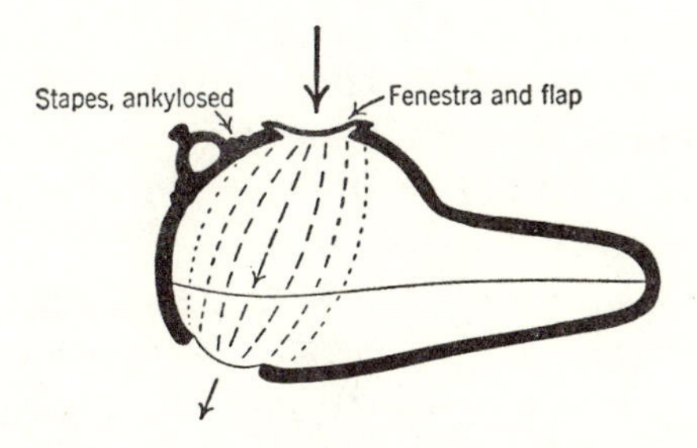

Fig. 135. Aerial stimulation in the otosclerotic ear after the fenestration operation. Sound applied to the fenestra causes a displacement of fluid toward the round window and involves the basilar membrane. From Wever (6) [*Annals of Otology, Rhinology, and Laryngology*].

This theory also explains why, as is generally reported from operations carried out under local anesthesia, the hearing is best at the moment that the new opening is made in the labyrinth and recedes somewhat when this opening is covered with the skin flap. It explains likewise the progressive loss of sensitivity that occurs in certain instances when the skin flap is made hard and rigid by fibrous growth and the fenestra is invaded and finally closed by fibrous tissue and bone, and it explains the return of sensitivity when this invading tissue is removed in a revision operation.

ACOUSTIC ROUTES TO THE FENESTRATED EAR

We have still to consider the problem of the pathway of entrance of sound to the cochlea after reconstruction of the ear in the fenestration operation. It is sometimes suggested that this pathway is the new fenestra and at other times that it is the round window. Let us first examine the physical situation.

In the Lempert operation as described the middle ear cavity is greatly modified. A portion of it remains beneath the tympanal portion of the tympanomeatal membrane. This cavity contains the round window and the now nonfunctional oval window.

A larger cavity is present as an extension of the meatus. It includes a small part of the original tympanic cavity, the epitympanic space from which the incus and the head of the malleus have been extracted, and a considerably enlarged mastoid space. Sounds have an unobstructed path from the outside through the meatus to the depth of this larger cavity where the fenestra lies with its transplanted covering. They may reach the round window only after traversing the drum membrane. Two series of observations bear on the question of the relative effectiveness of these two routes in the process of stimulation.

Effects of Perforations of the Drum Membrane. If at any time during the operation or afterwards a perforation is made in the drum membrane the hearing is immediately impaired. Later, when the perforation heals, the hearing is restored to the level obtaining before this injury. Figure 136 shows results obtained by

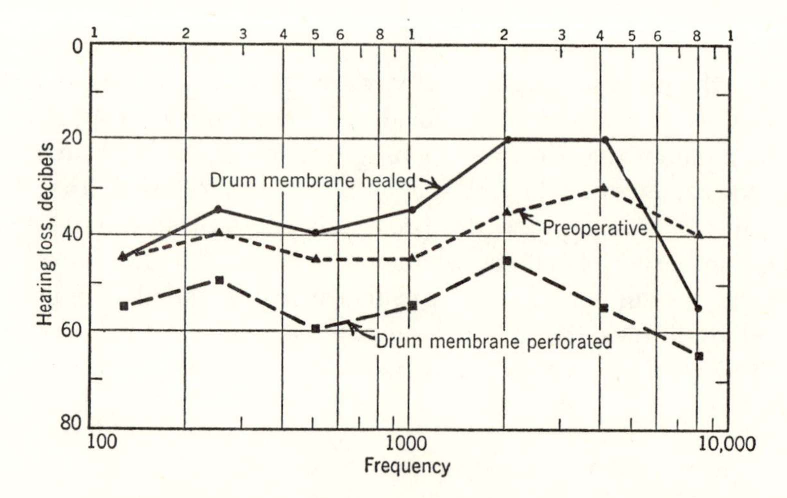

Fig. 136. Effects of perforation of the drum membrane on the sensitivity of the fenestrated ear. Lower curve, four weeks after operation, when an improvement of hearing is normally expected. A hole, accidentally torn in the drum membrane during the operation, was still present. Upper curve, ten weeks after operation, when the perforation had become healed. Zero on the ordinate represents the average sensitivity of the normal ear. From the case records of Dr. Julius Lempert.

Lempert. Here the middle curve represents the audiogram obtained before the operation. The drum membrane was accidentally torn in the course of the surgery, and the lowermost curve shows the audiogram obtained four weeks later, at a time when ordinarily the hearing would have been much improved. Instead it is 10 to 25 db worse than it was preoperatively. Still later, ten weeks after the operation, when the perforation in the drum membrane had become healed, the hearing was greatly improved, especially for the medium high frequencies, as shown in the uppermost curve.

Effects of Blocking the Drum Membrane in the Fenestrated Ear. The following test was carried out by Lempert on fenestrated ears after complete healing had taken place and a satisfactory recovery of hearing had been obtained. A quantity of petrolatum-soaked cotton was packed upon and around the tympanic membrane with the object of sealing this membrane from the exterior as thoroughly as possible and preventing the passage of sounds through it to the round window. Audiometer tests carried out before and after the packing and also after the packing was removed showed only a slight effect upon sensitivity. Some of the results are shown in Fig. 137. As will be seen, there was a

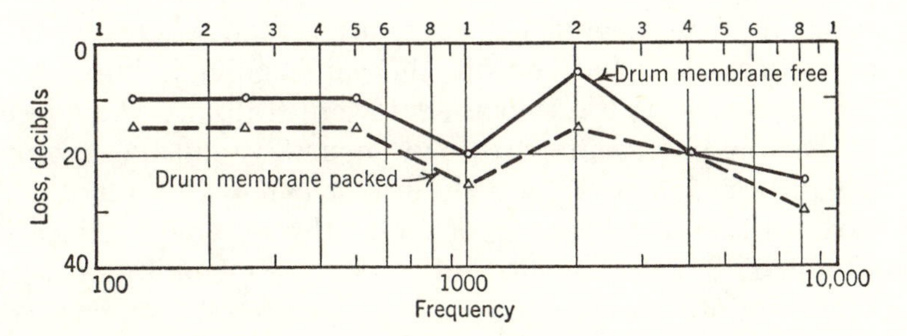

Fig. 137. Effects on the sensitivity of a fenestrated ear of a blocking of the drum membrane. Observations by Dr. Julius Lempert.

reduction of sensitivity of 5 db for most tones as a result of the packing.

The observations on the effects of perforations of the drum membrane show that the round window is involved in the transmission of sounds to the fenestrated ear, whereas the observations on the effects of packing the drum membrane show that the round window is not the only route. Evidently both the fenestral and round window routes are involved in the stimulation.

It is further evident that the sound does not enter by these two routes equally and in the same phase relation. If they did so, as earlier results have shown, there would be a cancellation of response. That such a cancellation does not occur means that in some manner there arise changes of intensity or phase, or both, in the fenestral and round window routes to the cochlea.

The hypothesis was advanced earlier (Wever, *6*) that these changes are mainly due to the presence of the drum membrane in the round window pathway. When this membrane is perforated, its phase-changing action is disturbed and a partial cancellation of the sounds is the result. Later, when the perforation heals, the phase-changing action is restored and the ear's performance is improved.

Our studies on the cat, described in Chapter 8, give support to this view. We have observed that the drum membrane in combination with the malleus and incus causes large phase changes in the conducted sounds, and a similar action is to be expected of this membrane and a part of the malleus.

These results seem to indicate that the phase changes produced in the sounds reaching the round window exceed 120° and even approach 180°, in view of the observation that, when the round window pathway is blocked off, the ear's acuity is impaired at most frequencies. If the impairment were 6 db we should conclude that the phase difference had been 180° and the two windows had been operating equally in conveying the sound to the cochlea, for then the blocking of one pathway would reduce the sound to half. Smaller impairments under these conditions signify that the reinforcement of the two pathways falls short of this optimum either because the phase difference does not attain 180° or because the intensities in the two pathways are not equal, or for both these reasons.

Further results on this problem were obtained by Skoog and Nilsson. Skoog reported two series of observations. One, carried out on 50 fenestrated ears at times varying from one to three years after the operation, consisted of blocking the drum membrane with paraffined cotton plugs in the manner already described in the experiments by Lempert. However, in Skoog's tests there was never a loss of sensitivity as a result of the blocking. In 55 per cent of the trials there was a marked improvement in the low and middle range, in 27 per cent a smaller improvement, and in the remaining 18 per cent no notable change. Evidently no tests were made for the high tones. Figure 138 gives the results of one of the experiments, representative of those in which marked improvements were obtained from the blocking.

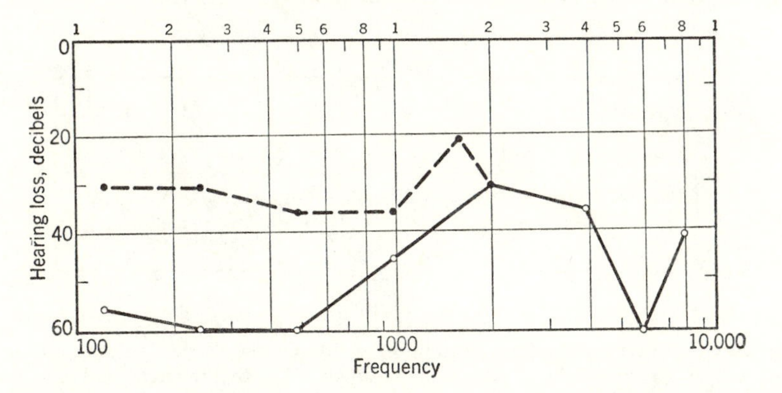

Fig. 138. Effects of blocking the drum membrane in the fenestrated ear, according to Skoog. The solid line represents the sensitivity of the ear (relative to the normal) before blocking and the broken line represents the sensitivity after blocking.

The second series of experiments by Skoog and Nilsson consisted of hearing tests made in the course of fenestration operations on patients under local anesthesia. The tests were carried out at different stages of the surgical procedure by the use of sounds from a loudspeaker held at a standard distance over the patient's head.

In these tests it was found that the maximum hearing was present after the fenestra was opened but was not yet covered by the flap, and the end of the flap was folded back so as to cover the drum membrane. An audiogram for this condition is shown in curve *a* of Fig. 139. Similar results were obtained when the fenestra was uncovered and paraffined cotton was packed both on the surface of the drum membrane and in the facial nerve region posterior to it, thus protecting the round window from the sound. Covering the fenestra with the flap when the drum membrane remained well covered gave results about 5 db worse than the above two conditions.

When the end of the flap was held up so that both the fenestra and the drum membrane were uncovered there was a loss of sensitivity, relative to the maximum condition, of 10 to 15 db for the low tones and of 5 db for the high tones. This condition is shown in curve *b* of Fig. 139. When the flap was placed over the fenestra and the drum membrane was left exposed—the usual condition of the ear when the fenestration operation is completed and all dressings are removed—the hearing was diminished by 10

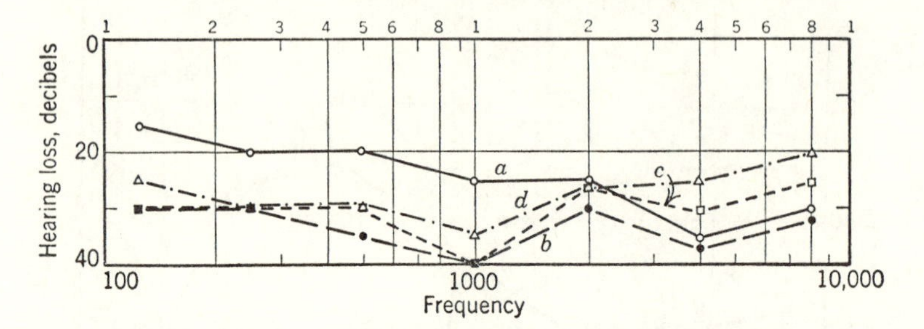

Fig. 139. Sensitivity after four procedures in the course of the fenestration operation, performed under local anesthesia. For curve *a*, the fenestra was uncovered and the drum membrane was covered; for curve *b*, both the fenestra and the drum membrane were uncovered; for curve *c*, the fenestra was covered with the skin flap and the drum membrane was uncovered; for curve *d*, the fenestra was covered with the flap and this region was also blocked off with paraffined cotton while the drum membrane was uncovered.

to 15 db for the low tones and improved by 5 db for the high tones, as shown in curve *c* of Fig. 139. When the fenestra was securely covered with paraffined cotton and the drum membrane was left exposed, thus permitting sounds to enter only by the round window route, the hearing relative to the maximum condition was made worse by 10 db for the low tones and improved by 10 db for the high tones, as shown in curve *d* of Fig. 139.

These results are easily summarized. The best hearing was found when one acoustic pathway was open and the other was closed. When both were open the hearing was worse, evidently because of unfavorable summation or "interference."

These results are in agreement with those obtained by Lempert in indicating that both fenestral and round window routes are involved in the stimulation of the fenestrated ear, but they are not in agreement as to the phase relation obtaining between these two pathways when the operation is completed. The Lempert results, as just pointed out, indicate a favorable form of summation, which according to the principles brought out earlier require that the sounds entering the two windows be in the region of phase opposition. The Skoog results, on the contrary, by showing an unfavorable form of summation, require that these entering sounds be in the region of phase agreement.

Now it well may happen, and indeed it must be predicted from theory, that the physical conditions determining the phase relations are highly critical. Seemingly slight variations of the surgical procedure might easily make the difference between a phase opposition and a phase agreement in the two pathways. Skoog reported that he was using the Lempert technique, but individual variations in technique are almost inevitable. In this connection it should be noted that Skoog's packing experiments were carried out on ears in which the fenestration was only moderately successful. In Fig. 138, taken from his report, the average hearing one year after the fenestration is 48 db below the normal. Lempert's experiments, on the contrary, were done on ears in which the operation was particularly successful; the average in Fig. 137 is only 14 db below the normal. The degree of success clearly depends upon the phase relation of the pathways, and if this relation is a favorable one any further manipulations, such as the blocking, may disturb this relation and make the hearing worse. By the same token, a poor result, obtained because the phase relation is unfavorable, stands a good chance of showing improvement as a result of these manipulations. The results obtained by Skoog and Nilsson during the fenestration operation, which consistently indicate unfavorable interactions between the two pathways, may also depend upon the particular techniques, but we cannot be more explicit on this point. The problem is a difficult one and requires further study. We feel sure, however, that the phase relation between the two pathways is the crux of the matter, and that in the surgical treatment of otosclerosis serious concern must be given by each individual operator toward the adoption and maintenance of a technique that will result in the most favorable phase relation.

THE STAGES OF OTOSCLEROSIS

In the course of this discussion we have identified four primary changes in the operation of the conductive mechanism as a result of the otosclerotic lesion, namely, mechanical fixation of the ossicular chain, loss of the transformer system, immobilization of the cochlear fluid, and phase interaction between the two pathways of entrance of sounds to the cochlea. We now consider how these physical conditions vary in the course of the disease and how they modify the ear's efficiency in sound conduction.

The progressive nature of otosclerosis has already been described. The disease is characterized by a gradual, even insidious form of onset, and for a long time in the early stage the depreciation of hearing will hardly be recognized. Eventually a clear awareness comes to the person that his hearing is impaired. Medical advice is not usually sought until this impairment amounts to 40 db or more, when the person begins to find himself definitely handicapped in his social relations. From a physical standpoint this is already a serious impairment, for a 40 db loss represents a reduction of the conduction of acoustic energy to one ten-thousandth of its normal value. The disease can then go on to far more devastating losses of 80 db or more, and an 80 db loss when stated in terms of energy is a conduction of one hundred-millionth of the normal amount.

We have already discussed the gradual formation and enlargement of a bony bridge across the annular ligament and its progressive fixation of the stapes. Let us now attempt an analysis of the changing physical conditions and their particular effects.

In the early stage, when the loss of transmission is no more than 20 db, the physical situation is relatively simple. We have only to deal with the mechanical fixation of the ossicular mechanism. The sound is still transmitted by way of this mechanism and the transformer action is retained. The loss of hearing results because the increased stiffness that is reflected to the drum membrane causes less sound energy to be taken up by this membrane and because much of the energy that is taken up is wasted in overcoming the increased frictional resistance and in producing vibrations of the bony capsule. In other words, the usual impedance match is now disturbed by an increase of stiffness and resistance.

As the fixation continues and approaches 30 db the picture grows more complicated. The fixation effect now cancels the advantage to conduction afforded by the transformer action, and the ossicular route to the cochlea is no better than the aerotympanic route through the round window. With the two routes operating with much the same efficiency we have to consider phase interactions between them. In general, because the middle ear mechanism retains its phase-changing characteristics, the interactions will be expected to have only slight effects over most of the auditory range, but at certain frequencies there may be serious cancellation effects. Also, because the phase action of the middle ear is somewhat unstable, subject to modification by changes in the structures, we can expect the hearing to show wide variations from time to time.

As the fixation exceeds 30 db the round window route to the cochlea gains in importance. The loss of hearing should now proceed relatively slowly for a time because the only progressive change is the increasing stiffness imparted to the drum membrane and its barrier effect toward the entrance of sound waves to the tympanic cavity.

Then, as the oval window finally is closed we have the ultimate and most grave effect, which is the immobilization of the cochlear fluid. The immobilization negates the effectiveness of the round window route and drops the hearing to profound depths.

The fenestration operation, as we have seen, effects a remobilization of the cochlear fluid. Also, as carried out by the Lempert method, it removes the barrier presented by the stiffened drum membrane and gives the sound waves freer access to the cochlear fluid. The transformer action formerly provided by the middle ear mechanism of course is permanently lost.

Because the transformer action is lost we must expect this operation to fall short of a complete restoration of hearing. As we have seen in experiments on the cat this action is responsible for a gain of efficiency of about 28 db. This gain is not necessarily the same in man, but is probably of the same order of magnitude. A study by Davis and Walsh showed an average acuity in fenestrated ears of 27 db below the normal.

Sometimes after fenestration the hearing is decidedly better than this. It may return within 20 or 10 db or even closer to the audiometric zero. This fortunate result does not represent an error of measurement, but is real: it is repeatedly shown by the hearing tests. Therefore it has to be explained in view of the loss of the ossicular mechanism.

It is possible to account for a part of this unexpected gain from a particularly favorable phase interaction, for if the phase relation between the fenestral and round window pathways is 180° and the intensities by these pathways are equal it is possible to obtain a gain of 6 db relative to conduction by one pathway alone. Yet we should still expect hearing something like 22 db below the average. Therefore we have to go further. We believe that the explanation lies simply in the nature of the audiometric zero. It is obtained by averaging results from many persons whose hearing lies within a fairly wide range of "normal." The ears that after fenestration show a particularly fortunate result are probably ones that had considerably better than average hearing before the onset of the disease. These persons, like the others in which a good surgical result is obtained, have been returned to a level perhaps 20 to 30 db below their former one.

BONE CONDUCTION IN THE OTOSCLEROTIC EAR

A further question concerns bone conduction in the otosclerotic ear before and after the fenestration operation. The general opinion is that bone-conduction sensitivity changes little if at all as a result of otosclerosis and that it is still unaltered after fenestration. A few writers, however, have reported small losses of bone conduction in otosclerosis and corresponding gains after fenestration. Juers in a study of 28 patients found that fenestration improved the bone conduction by about 1 db at 256~ and by increasing amounts up to 13 db at 2048~. Woods found a progressive loss of high-tone acuity by bone conduction as characterizing the early stages of otosclerosis, and this loss was recovered, at least in part, after fenestration. He reported an average recovery in the middle freqencies of 11.4 db. Nilsson and Skoog reported improvements of bone conduction varying from 9 to 15 db over the range of 250 to 3000~ as a result of fenestration. There were no further changes as a result of various manipulations during the operation.

In this connection we may recall that in Smith's experiments on the cat in which a mechanical fixation was made of the ossicular chain there was a moderate reduction in the cochlear responses to bone-conducted sounds. This effect was most marked in the region of 500 to 1000~, where it reached average values of 15 to 20 db.

This region includes the region of mechanical resonance of the cat's ear.

We have already recognized two principal modes of stimulation by bone-conducted sound. In the translatory mode of stimulation the cochlear capsule is effectively shaken back and forth by the sound, and the inertia of the fluid contents causes them to tend to lag behind. When in the normal ear the vibratory motion is in line with the oval and round windows the fluid can move alternately toward one window and the other. Also the ossicular chain because of its inertia tends to drive the stapes in and out of the oval window when vibration is along its axis of free motion, and the fluid is likewise propelled back and forth. The basilar membrane lies in this path of fluid movement and is displaced in the process.

In the compressional mode of stimulation the cochlear fluid is caused to surge in and out of the cochlear windows, but because the ossicular chain acts as a load on the oval window this displacement is mainly through the round window. Also, because of asymmetry of the fluid on the two sides of the basilar membrane a part of the displacement takes place across this membrane and stimulates its hair cells.

When in otosclerosis the oval window becomes occluded the translatory mode of stimulation largely fails for the same reason that aerial sound applied to the round window is then relatively ineffective in displacing the cochlear fluid. A translatory movement of the capsule carries all its contents along, without any relative displacement of either the fluid or the basilar membrane, and no stimulation occurs. On the other hand, an occlusion of the oval window is no handicap to the compressional mode of stimulation. Indeed, the displaced fluid now can find relief only by way of the round window, and this stimulation should be more effective than before. Because one of these modes of stimulation is reduced and the other remains or is even enhanced we should expect to find an altered picture of bone-conduction sensitivity. However, it is difficult to say what new form this sensitivity should take. We should expect a variation with frequency and with the particular techniques used in applying the stimulus.

When the otosclerotic ear is fenestrated the bone conduction is altered once more. The translatory mode of stimulation can now occur, but only through the inertia of the fluid and perhaps of the coverings of the fenestral and round window openings. The reactive complications formerly provided by the ossicular system are now absent. The compressional mode of stimulation is probably reduced from what it is in the otosclerotic ear because some of the cochlear fluid can now be displaced toward the fenestra. Yet because the skin flap covering this opening is less yielding than the round window membrane there is still a differential action much as in the normal ear.

PART VII

Further Theoretical Considerations

17

Analytical Treatment of the Middle Ear Mechanism

In Chapters 5 to 7 we studied the basic properties of the middle ear and reached the conclusion that this structure acts as a mechanical transformer, aiding the transfer of acoustic energy from the air to the cochlear fluid. We now consider this problem further, with special regard to the efficiency of the transformation process and the variations in this efficiency with frequency.

As the foregoing discussion brought out, the efficiency of transmission from air to cochlear fluid depends upon the accuracy of matching of the acoustic impedances of the two media. With a proper transformer ratio the inner ear impedance as presented at the entrance to the ear will appear the same as that of the outside air.

In an extended gaseous medium the acoustic impedance can be taken as a constant; it is practically unvarying within the audible range. For air under standard conditions the specific acoustic impedance has a value of 41.5 mechanical ohms per sq cm. In an extended liquid medium this impedance is likewise constant. However, our earlier analysis has shown that in the cochlea we are dealing with a fluid in only limited quantity and in a special situation where it is bounded and restrained by membranes. We are therefore unable to treat this medium as unlimited and to regard its impedance as constant. It is probable that the elastic forces exerted by the cochlear membranes are at least as important as the molecular forces of the fluid, and therefore the impedance will be subject to variation with frequency. We can further expect modifications of the transmission on account of properties of the middle ear incidental to its transformer action, and these also will be related to frequency.

THE TRANSFORMER RATIO OF THE MIDDLE EAR

Our earlier study of this problem has shown that the transformer function of the middle ear is carried out by a combination of two mechanical actions, one a lever action of the ossicular chain and the other a kind of hydraulic action of drum membrane and stapedial footplate. We were able to obtain measurements of these actions and thus of the over-all transformer ratio. For the cat's ear the lever ratio was found to be about 2.5 and the effective areal ratio between drum membrane and stapedial footplate was measured as 24.3. The over-all transformer ratio is then 60.7. For the human ear the lever ratio is represented as 1.31 and the effective areal ratio as 14.0, giving an over-all ratio of 18.3. The impedance transformation is then 3684 for the cat and 336 for man.

Some measurements carried out by Békésy (*11*) on the cadaver ear are of interest in this connection. After draining out the cochlear fluid he applied a sound to the inner surface of the stapedial footplate and adjusted its intensity and phase so as to cancel the effects of another sound of the same frequency introduced at the drum membrane. A capacitative probe brought near the surface of the stapes showed when the cancellation was complete. Under these conditions the sound at the stapes is exerting its pressure directly, whereas the sound applied to the drum membrane is operating through the middle ear transformer. The ratio of the two sound pressures that give cancellation at the stapes therefore represents the transformer ratio. Figure 140 shows the results.

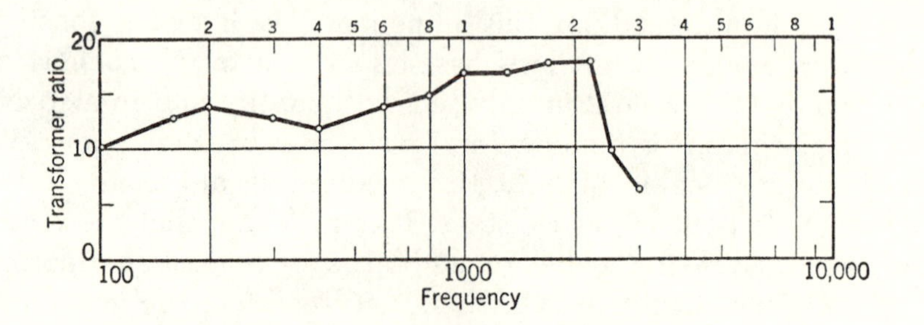

Fig. 140. The transformer ratio of the human (cadaver) ear according to Békésy (*11*).

The ratio as measured in this manner rises slowly with frequency from a value of 10 at 100~ to a value of 18 at 2200~, and then falls away. The mean value over this range is 13.6.

Now, the ossicular lever action ought not to change with frequency, and the areal ratio will not change either as long as the drum membrane continues to operate as a whole, which it should do at all the low frequencies. The variations obtained in Békésy's results therefore are due to disturbances of the pressure measurements by a reflection into them of incidental response properties of the ossicular system. The larger values observed for the transformer ratio probably are most representative of its true magnitude. Indeed, the mean value over the middle range from 600 to 2200~ is 16.5, in good agreement with the figure of 18.3 obtained from anatomical measurements.

Fletcher (*4*) in a recent treatment of the dynamics of the middle ear made use of certain other experimental data of Békésy's to determine the transformer ratio, with results very different from those just described. In these experiments, also carried out on cadaver specimens, Békésy (*12*) obtained measurements, for a number of conditions, of the volume displacement of the round window membrane as produced by acoustic stimulation. Again, as in the foregoing experiment, he used a cancellation method, balancing a tone applied by the oval window route with one introduced at the round window. The tone as heard through an auscultation tube leading to the observer's ear fell to zero when the cancellation was complete, and when this condition was obtained a reading of the current in the balancing receiver indicated the volume displacement of the round window membrane. The same displacement of course holds for the whole column of cochlear fluid.

The standard observations were made with the middle ear and the cochlear structures intact and with the stimulating sounds applied to the drum membrane. These observations showed a fairly uniform displacement of 3 to 8×10^{-10} cubic cm per unit of sound pressure (i.e., for 1 dyne per sq cm) over the range of measurements from 160 to 4000~. Békésy also measured the volume displacement after removal of the whole middle ear system, including the stapes and the round window membrane. The stimulating and cancelling sounds then acted directly on the cochlear fluid, which was restrained only by the internal cochlear membranes—Reissner's membrane and the basilar membrane. These results showed volume displacements considerably larger than before at low frequencies, but these displacements fell off progressively in magnitude as the frequency was raised until finally, at the higher frequencies, they became smaller than in the intact ear.

Fletcher divided each of the first set of values by corresponding values in the second set and regarded the quotients as representing the mechanical transformer action of the middle ear. The reasoning behind this treatment is that the first set of data represents the action of the ear as a whole and the second set represents the action of the inner ear, so that the computation gives results for the middle ear. These results are presented as curve *a* in Fig. 141.

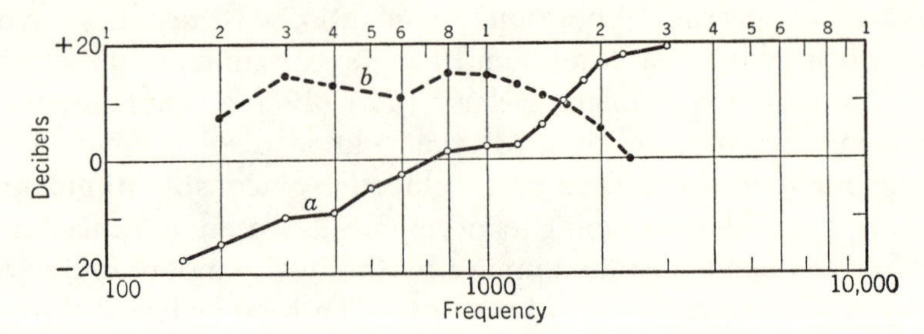

Fig. 141. The pressure amplification afforded by the middle ear mechanism, as derived from observations by Békésy (*12*). Curve *a* shows the results of Fletcher's computations and curve *b* the results of our own.

Fletcher extrapolated these data downward to 20~ and upward to 10,000~, and concluded that for all the low tones, up to 650~, there is no pressure amplification by the middle ear, but rather a diminution of pressure in the process of transmission. Indeed, the extrapolated curve indicates a loss of more than 300-fold or 50 db for the lowest frequencies. In this sort of action Fletcher saw an advantage to practical hearing in that the system discriminates against disturbing noises which are mostly of low frequency. At 650~ (or a little higher according to the curve shown) the transformer ratio is unity, and a given pressure in the stimulus sets up an equal pressure in the cochlear fluid and thus across the basilar membrane. For higher frequencies the transformer system gives a gain which rises to a maximum around 19 db and thus represents a pressure increase of about 9-fold.

Fletcher attributed the poor performance of the ear at low frequencies to a low mechanical impedance across the basilar membrane at these frequencies, and in his further theorizing brought this condition in relation with his conception of the general mechanics of the cochlea.

The picture presented here is surprising in relation to what has gone before and especially in view of the fact, well established in clinical observations, that loss of the middle ear always has a deleterious effect upon hearing for low tones as well as others (Békésy, 8). Results for the cat, given in Fig. 41, show that removal of the middle ear causes a general loss of sensitivity, and the loss is only a little less for the low tones than for the others. Hence we need to examine a little more critically the results on which Fletcher's conclusions are based.

We note that Békésy's second set of observations was made on a preparation lacking both stapes and round window membrane. A comparison with the normal ear therefore does not yield results attributable merely to the middle ear, but exhibits properties of the two cochlear windows also. Both the annular ligament holding the stapedial footplate in the oval window and the membrane of the round window provide stiffnesses that doubtless enter largely into the ear's performance and account in part for the form of the function shown.

To pursue this point we now turn to certain other results obtained by Békésy in this same series of experiments. In one of his tests he removed the drum membrane and outer ossicles but left the cochlea intact, with the stapes in the oval window and with the round window membrane undisturbed. If we divide the values of volume displacement observed in the intact ear by the values observed under this condition we obtain results that we consider to be more properly attributable to the operation of the middle ear transformer. Such results are represented by curve *b* of Fig. 141. Here we see a strikingly different picture. The transformer action is positive at all frequencies up to 2400~. It represents a fairly uniform gain of 10 to 14 db for all tones from 300 to 1500~, and a falling off for higher tones. These results are in reasonable agreement with those presented earlier, all of which testify to a useful transformer action for the low tones as well as others.

FURTHER DATA ON THE IMPEDANCE OF THE EAR

The results that have just been described are of further interest in relation to the problem of the ear's impedance and its variations as a function of frequency. If we consider the ear as a total system, without regard to pressure transformations within it, we can use the data on volume displacements of the cochlear fluid in a determination of the impedance. If the volume displacement per unit of sound pressure is multiplied by the angular velocity ($2\pi f$) we obtain the volume velocity per unit of pressure, and the reciprocal of this quantity is the acoustic impedance. Results of this calculation for the cadaver ear with its essential structures intact are shown by the uppermost curve of Fig. 142. The ordinate scale is

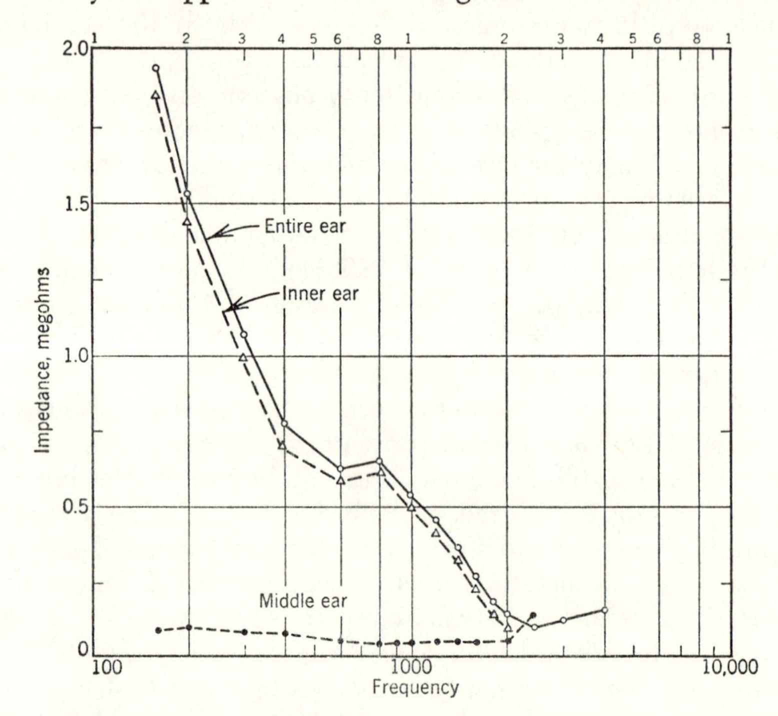

Fig. 142. The acoustic impedance of the cadaver ear according to observations by Békésy (*11,12*). One set of observations gives values for the ear as a whole, another set gives values for the middle ear, and the differences represent the inner ear.

in millions of acoustical ohms. Similar computations were carried out on the basis of other measurements made by Békésy (*11*) in which the volume displacement of the stapes was measured in an ear in which the cochlear fluid had been removed. We thereby obtain values for the impedance of the middle ear, and these values are given by the lowermost curve of Fig. 142.

Now, the impedances represented by the upper and lower curves of Fig. 142 are really vectorial quantities, as were the volume displacements on which they were based, but though Békésy seems to have measured the phase angles in at least some of his tests he did not report the results quantitatively. Still this is no serious handicap to our further use of the data in finding the impedance of the inner ear. We can take the differences between these two curves and say that these differences represent minimum values that the inner ear impedance must have. This method gives uncertain results only beyond 2000~ where the upper and lower curves come together.

These results show very large values of impedance, especially for the ear as a whole. Even for the middle ear these values exceed what we should expect for a system that ought to be fairly well matched to the impedance of the air. It seems likely that post-mortem changes have added greatly to the resistance and stiffness of the structures, or that other conditions of the experiment have affected the results. Therefore the absolute magnitudes as shown here are not to be taken very seriously. The feature of importance is the indication of much larger impedances for the inner ear than for the middle ear, and we regard this relation as acceptably demonstrated in spite of the experimental and analytical difficulties of this approach to the problem.

The last two points on the curve for the intact ear suggest that for the higher frequencies the impedance rises progressively, but the data hardly extend far enough to establish this point with certainty. Such a trend is to be expected theoretically, for at the higher frequencies we look for the moving masses of both the fluids and the membranes to have an increasing influence in raising the impedance.

Let us now turn to the impedance data obtained for the human ear by more direct acoustic measurements, already described in Chapter 4 and summarized in Fig. 16. These data represent the impedance of the ear as a whole as seen from the external auditory meatus. They are given in terms of transmission loss in decibels, but by means of a decibel table the values are easily converted to express the fractional transmission (T). To obtain the impedance we have only to apply the transmission formula, already given as $T = 4r/(r+1)^2$, where r is R_2/R_1 and R_1 is the specific acoustic impedance for air, which is 41.5 mechanical ohms per sq cm. The formula is transposed and solutions are obtained for R_2 at frequencies for which T is known. Results of such calculations are shown in Table 3. Here the first column shows frequency,

TABLE 3. THE IMPEDANCE OF THE EAR

Frequency (cycles)	Transmission loss (decibels)	Fractional transmission	R_2, apparent impedance (ohms per sq cm)	R_i, inner ear impedance (ohms per sq cm)
100	14.6	0.035	4700	1,580,000
200	14.4	0.036	4500	1,510,000
300	13.0	0.050	3240	1,090,000
500	10.4	0.090	1740	585,000
700	8.1	0.155	984	330,000
1000	5.5	0.282	505	169,000
1500	4.2	0.380	347	117,000
1600	4.3	0.371	359	121,000

the second column the transmission loss in decibels, the third column the fractional transmission, and the fourth column the values of R_2, the ear's apparent impedance, in mechanical ohms per square centimeter.

Now if we neglect the contribution of the middle ear to the total impedance, which Fig. 142 has shown to be small at most frequencies relative to the inner ear impedance, and if we accept a constant value for the middle ear transformer ratio of 18.3, we can use these data to derive the inner ear impedance. It is only necessary to multiply the values of R_2 given in column 4 by the square of 18.3, which is 336, the impedance ratio. The result is given in column 5 as R_i in mechanical ohms per square centimeter.

It is of interest to compare these values of specific acoustic impedance with that for sea water, which is 161,000 mechanical ohms per sq cm. We see that for the low tones the inner ear impedance is much larger than the impedance of a free fluid, but this impedance falls off as the frequency rises. Around 1050~ we have approximate equality with sea water, and above that frequency the inner ear impedance is the smaller.

As we see in Table 3, the human ear is most efficient in the region of 1500~. Here the transmission loss is 4.2 db, representing a transmission efficiency of 38 per cent. At 100~ this efficiency is only 3.5 per cent, corresponding to a loss of 14.6 db. If the transformer ratio of the middle ear were 53, or greater by 2.9-fold than it is, the transmission at 1500~ would be perfect but that for the low frequencies would still be deficient by about 6 db. If the human ear had a transformer ratio as large as we have measured in the cat (60.7), the efficiency at 1500~ would be 98 per cent and that for 100~ would be about 32 per cent or down by only 5 db. If the inner ear impedance for the cat's ear is of the same order of magnitude as the data indicate for man, then the efficiency figures just cited will hold for the cat, and its ear will have to be regarded as slightly superior to man's. The results given in Fig. 39 for the general sensitivity of cat and human ears do not indicate any consistent difference over the frequency range in consideration here, and we shall have to wait for further data before we can form any final conclusion on this point.

ACOUSTICAL ANALYSIS OF THE EAR'S PERFORMANCE

In our treatment so far the emphasis has been placed on the function of the middle ear apparatus as a mechanical transformer, and both its variations of efficiency in this respect and other peculiarities of performance as a transmitter of sound have been regarded as secondary and indeed as disturbing aspects of this operation. Now we need to turn our attention to the mechanical peculiarities themselves. In this approach we consider the middle ear simply as a sound-recording instrument and seek to understand its characteristics on the basis of established acoustic principles, and especially the principles already known to apply to man-made devices for the recording of sounds.

Ideally, this analytical treatment of the problem might consist merely of setting up equations of motion for the middle ear as a vibrating system under external forcing and then of solving these equations. Mach in 1863 was perhaps the first to make a serious attempt of this sort. Frank in more modern times offered the most elaborate treatment and attempted to supplement the mathematical analysis by empirical observations. Further treatments that are mainly mathematical and deductive have been presented recently by Esser and Onchi.

Frank (*1,2*) worked on a background of long experience with recording apparatus, and after writing the differential equations of motion for the middle ear he sought particular solutions that would be relevant to the practical conditions. Initially this procedure is only a matter of mathematical derivation; the equations are worked out to show relations among the variables that are expressed quantitatively by the included constants. Then comes the important and difficult phase of the development, which is to make a final specification of variables that may have been indicated only vaguely at first, and to identify these variables and the constants as dynamical and anatomical realities.

This latter process is largely inferential. From anatomical considerations Frank decided what parts of the middle ear were to be taken as unitary masses and what ones were so loosely coupled as to be dynamically separate. He considered the three ossicles as operating with a good deal of independence and hence he included three masses in his treatment. He also designated separately the elasticities of the suspensory ligaments, the damping coefficients involved in the motions, and other mechanical characteristics that depend upon the geometrical relations of the parts. He gave particular attention to the drum membrane which he regarded as a stretched membrane.

Frank dealt with a series of conceptual models, the final one of which he believed to be so nearly equivalent to the ear that it could serve the purposes of hearing equally well. By consideration of the response properties of this model he arrived at a symbolic solution of the problem. A practical solution (assuming that the underlying assumptions are correct) will then depend upon an actual determination of the values of the constants.

The masses of the more obvious parts, such as the ossicles, can be measured directly. The inertias of these masses in motion can either be measured or calculated from the anatomical dimensions, provided that the axes and modes of motion are sufficiently well known. The other features present greater difficulties. With Broemser's help, Frank set out upon an empirical determination of these characteristics in cadaver specimens.

As had already been done by Dahmann and others, Frank and Broemser placed small mirrors on the moving parts of the middle ear and attempted to measure the vibration during stimulation by sounds. Frank reported that the results obtained in these studies were fully in accord with the conclusions that he was able to draw from his mathematical analysis. However, he gave no detailed description of these results and made only a summary report on three characteristics, namely, the natural frequency, the moment of inertia, and the damping coefficient of the ossicular chain.

(1) Frank and Broemser measured the natural frequency of the ossicles as a whole in three different ears as 1092, 1110, and 1340~, for which the mean value is 1181~. These figures are in fairly good agreement with those for the point of resonance of the living ear as indicated in impedance measurements, described earlier. Békésy (5) by two methods obtained values between 800 and 1500~.

(2) Frank and Broemser found the inertia of the ossicles in rotating about the anterior ligament of the malleus to be 2.5×10^{-3} grams per sq cm. This value was obtained both by calculation and by measurement. For the measurements the ossicular chain was attached to a thin rod oriented along the line of the axial ligament and was made to oscillate about this rod. The inertia was determined from the frequency of oscillation and the moment of torsion of the rod.

(3) They reported the damping coefficient of the ossicular mechanism to be 0.3. This figure evidently was obtained by delivering a sudden force to the mechanism and observing the rate of decline of amplitude from one oscillatory wave to the next. Two or three waves could always be observed clearly before the oscillation was damped out. Békésy (5) reported values between 0.25 and 0.33.

We must suppose that measurements on other characteristics if attempted gave unsatisfactory results or at any rate failed to yield any usable values of the constants.

Lack of knowledge about these other characteristics, especially about the elastic coefficients, made it impossible for Frank to set forth any single prediction concerning the dynamic behavior of the middle ear. Therefore he formulated a number of alternative characterizations according to the assumptions that could reasonably be made as to the relative magnitudes of the constants. These possibilities need not be treated in detail here. They amount to saying that the quality of performance of the middle ear depends upon (1) a large effective area of the drum membrane, (2) some optimum value for the ossicular lever ratio, (3) a low mass for malleus and incus, and (4) strong elastic connections between the ossicles themselves relative to their connections with the tympanic walls. These conditions are rather obvious and their formal expression fails to add appreciably to our understanding of the behavior of the system. It would be different, of course, if in connection with the mathematical statements we were able to specify actual properties and dimensions for these anatomical relations.

Not only are we lacking in these specific data, but the very framework of Frank's development of the problem is in doubt. Perhaps most serious is the dependence of this development upon the assumption that the drum membrane in man is a stretched membrane, for this assumption is not borne out by recent evidence. Likewise doubtful is the assumption of loose coupling between the malleus and incus.

Esser dealt with the middle ear apparatus in two papers, one on what he called the "two-piston problem" and the other on the drum-lever problem as raised by Helmholtz. By methods much like those used by Frank he developed an equation expressing the efficiency of energy transmission by the middle ear. The constants in this equation represent the masses and elasticities of the moving parts, the densities and elasticities of air and cochlear fluid, the geometrical proportions of the lever arms, and the relative surfaces of drum membrane and stapedial footplate.

Because the elastic coefficients are unknown, Esser's formula can be used only at a single frequency, the resonance frequency, at which the effects of the masses and elasticities are counteractive. The formula then reduces to one that expresses the relation between the resistive properties of the two media (air and cochlear fluid) and the mechanical factors operating to increase the effective pressure of the transmitted sound. This reduced formula is not in agreement with what has been said in Chapters 6 and 7 and accepted elsewhere in our discussion. It does not state, as we have supposed, that the pressure is increased by the amount of the areal ratio between drum membrane and stapedial footplate but rather that this increase is equal to the square root of the ratio. Accordingly, Esser's formulation makes the impedance matching depend principally upon a lever action in both the ossicular chain and the drum membrane. When we consider that the drum-lever hypothesis is unsupported by the evidence and that the ossicular lever has only a small ratio, it is clear that this formulation is inadequate to account for the impedance matching produced by the middle ear.

Onchi likewise set out to deal with this problem by writing equations for the motions of the middle ear that included terms for the mass, stiffness, and friction of all the anatomical elements. Then, however, he found that his equations were too complicated to handle mathematically. He thereupon proceeded to "deduce" the dynamic properties of the system, which means simply that he stated the principles and assumptions that obviously or implicitly were included in the equations in the first place. His treatment therefore fails to lead to any new insights about the middle ear. Indeed, like the others, it contains a number of assumptions that are not borne out by the experimental evidence, such as the assumption of a high degree of nonlinearity in all the ear's actions.

It is now apparent that the analytical treatment of this problem cannot proceed any further without a broader base of empirical information. Some additional data on the cat's ear are now available in the observations reported in the preceding pages. Of particular interest in this connection are the results on sensitivity and phase relations.

As already described, our observations on the ear were made in two stages because of limitations in experimental method. The actions of the drum membrane, malleus, and incus were determined first because these parts are subject to surgical removal. The actions of the stapes then were studied by comparing the effects of stimulating the cochlea by way of the oval window and by way of the round window, and with the assumption that these two paths of stimulation differ principally by reason of the presence of the stapes in the oval window.

Let us review this evidence. We find in Fig. 43 that the peripheral portions of the middle ear, including the drum membrane, malleus, and incus, are responsible for the presence of two principal regions of resonance and one region of antiresonance in the response curve. A study of these results in detail shows that for all the low tones, up to the first resonance point near 1000~, the middle ear's behavior is mainly determined by the stiffness of its suspensory system. For these tones the vibratory motions show a phase lead over the driving force of the applied sounds. In the region of 1000~ the effects of this stiffness are just balanced by the effects of the mass of the moving parts and only friction places a limit upon the action. Here the motion is in phase with the acting force. Then near 3000~, for a typical ear, the phase curve changes rapidly from lagging to leading and correspondingly the sensitivity shows a narrow region of loss. Here is a point of antiresonance in the system. In the region of 4000 to 5000~ a further rapid change occurs. The phase alters from leading to lagging and the sensitivity again passes through a maximum that represents a region of resonance. Thereafter, as the very high frequencies are passed through, the phase is generally lagging, signifying that the mass effect mostly prevails, but there are many rapid variations in the curves and many differences among individual ears.

The response characteristics of the stapes have already been seen in Figs. 40 and 45. This ossicle has but minor effects upon both sensitivity and phase over most of the frequency range. Its presence in the oval window gives a small improvement of sensitivity, especially for the higher frequencies. Its phase effects are slight over the low and intermediate ranges, and significant changes appear only above 5000~. In this frequency region a phase lag appears and rapidly increases to a maximum around 6500~. Here the function is discontinuous, indicating a point of antiresonance. The function then represents an advance of phase which rapidly changes as the curve crosses the zero line at 12,000~, where resonance appears.

The results obtained in these two stages of our study have been combined in Figs. 41 and 47, which now represent the performance of the entire middle ear. The complete functions show three regions of resonance alternating with two regions of antiresonance.

ELECTRICAL AND MECHANICAL MODELS OF THE MIDDLE EAR

We have pursued this problem further by the use of electric circuit analogies and then by a consideration of their mechanical counterparts. The advantage of using electrical analogies is that nowadays the experience with complex electric circuits is considerably more extensive than with mechanical systems. Therefore it is easier, when given the problem of determining the system required to produce a certain form of action, to work it out first in terms of electrical elements and then, by use of well-known relationships, to translate this system into the mechanical form.

Actually there are any number of electrical and mechanical models that will represent the form of the middle ear's characteristics. Our aim has been to discover the simplest analogical arrangements that will give a reasonable approximation to this function. Thereby we reduce the problem to its lowest terms and facilitate its further handling by analytic methods. Indeed, such a simplification is about the only advantage to be expected from this sort of treatment.

An electric circuit representative of the action of the more peripheral portion of the middle ear—the drum membrane, malleus, and incus—is shown in Fig. 143. This circuit consists of three

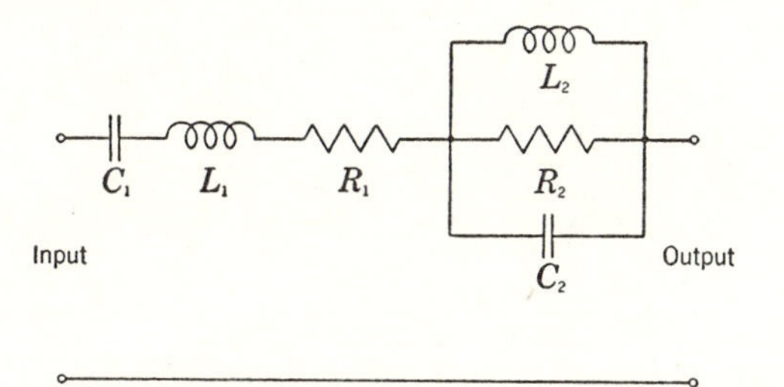

Fig. 143. An electric circuit analogy of the peripheral portion of the middle ear. C_1, C_2 are capacitances, L_1, L_2 are inductances, and R_1, R_2 are resistances.

elements in series, a capacitance C_1, an inductance L_1, and a resistance R_1, followed by three other elements C_2, L_2, and R_2 in parallel, all in the active or high side of the circuit. The mechanical counterpart of this circuit is shown in Fig. 144. Here we have two masses M_1 and M_2, two stiffnesses S_1 and S_2, and two frictional elements *a* and *b* interconnected as shown. The friction in bearing *b* dissipates some of the energy communicated to M_1 and this dissipation is designated as D_1. Similarly, the friction in *a* dissipates some of the energy to M_2, and this dissipation is designated as D_2. The division of M_2, S_1, and S_2 into symmetrical halves is merely to give an appearance of dynamical balance to the structure.

An electrical analogy for the stapes is shown in Fig. 145. It consists of an inductance L_3 followed by a parallel combination of

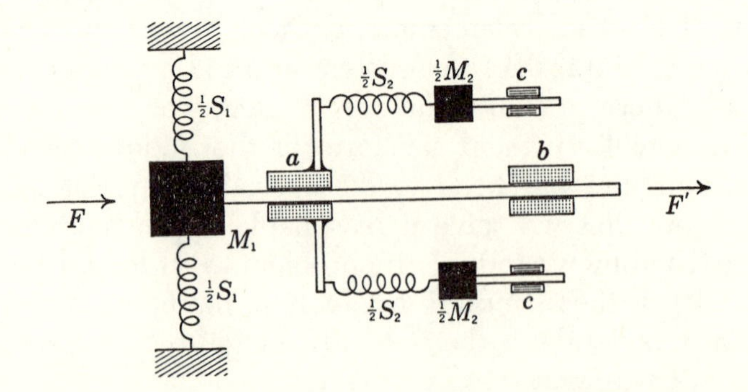

Fig. 144. The mechanical counterpart of the preceding circuit. S_1, S_2 are stiffnesses, M_1, M_2 are masses, and the bearings *b* and *a* are considered as producing dissipations D_1 and D_2, respectively. Bearings *c*, *c* are considered to be frictionless and all links are massless. *F* is the applied force and *F′* is the force transferred to the stapes.

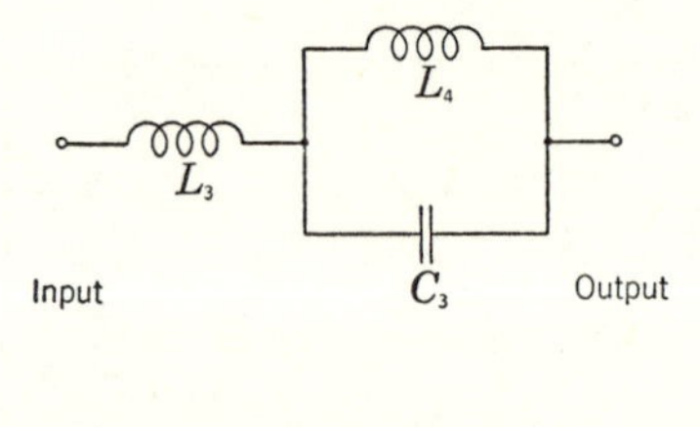

Fig. 145. An electric circuit analogy for the stapes and its related parts. C_3 is a capacitance and L_3, L_4 are inductances.

another inductance L_4 and a capacitance C_3. The mechanical counterpart of this system is shown in Fig. 146 and contains two masses M_3 and M_4 and one stiffness S_3. As in the foregoing mechanical analogy, S_3 and M_4 have been divided into two parts for purposes of symmetry and balance.

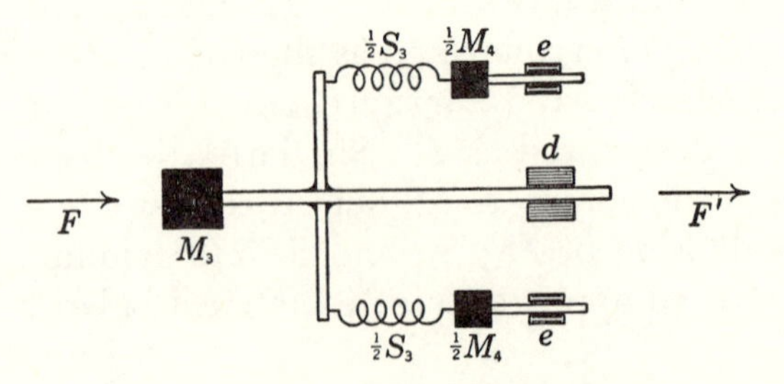

Fig. 146. The mechanical counterpart of Fig. 145. S_3 is a stiffness and M_3, M_4 are masses. Bearings *d* and *e* are considered frictionless and all links are massless. *F* is the applied force and *F′* is the force transferred to the inner ear.

To obtain an analogy of the middle ear as a whole we combine the systems just described and add a pair of transformers to represent the ear's impedance-matching facilities. The combined electrical system is shown in Fig. 147, with T_1 and T_2 representing the

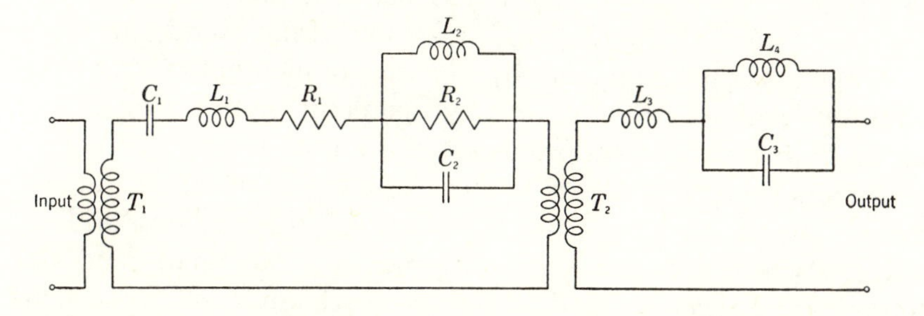

Fig. 147. An electric circuit analogy for the middle ear as a whole. T_1, T_2 represent impedance transformers, and the other parts are as designated in Figs. 143 and 145.

two transformers and other symbols having the same meanings as in preceding figures. The first transformer T_1 represents the pressure amplification afforded by the areal ratio of drum membrane and stapedial footplate and the second transformer T_2 represents the amplification produced by the ossicular lever system. Figure 148 presents the corresponding mechanical analogy,

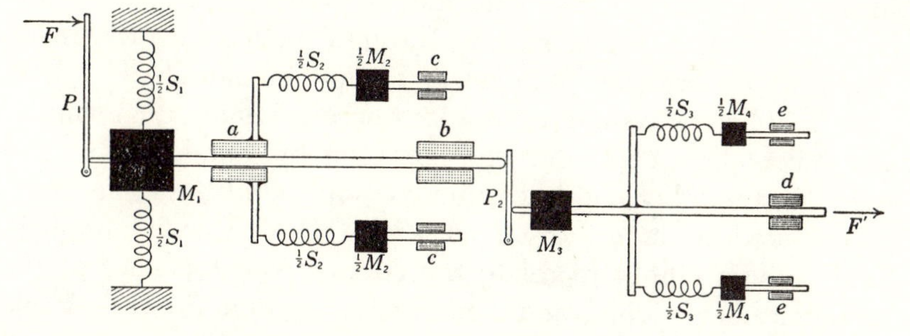

Fig. 148. The mechanical counterpart of the preceding figure. P_1, P_2 are mechanical levers of the amplitude-reducing type, and other parts are as designated in Figs. 144 and 146. F is the force produced by the sound waves and F' is the force delivered to the inner ear.

with levers shown in the positions of the transformers to serve the purpose of impedance matching. In an analytical treatment P_1 and P_2 can be combined and M_1 and M_3 can be combined also if the action of T_2 is taken into account in doing so. The treatment then will be concerned with three masses, three stiffnesses, two frictional elements, and one lever ratio.

The identification of these analogical elements in terms of anatomical structures in the ear is of course a matter of conjecture. We regard the following as a reasonable hypothesis. M_1 is the combined mass of the malleus, incus, and drum membrane. S_1 is a stiffness mainly supplied by the drum membrane, but contributed to also by the suspensory ligaments of the malleus and incus. D_1 is a dissipation produced in the movements of M_1. S_2 is a stiffness produced wholly in the suspensory system: in the anterior, lateral, and superior ligaments of the malleus and its anterior process, the posterior ligament of the incus, and the tensor tympani tendon and muscle. M_2 is the mass of the suspensory system and of other parts set in motion by way of S_2. These other parts contribute to the mass because they are not altogether rigid but are somewhat involved in the motions. D_2 is the friction generated in the movements of M_2, and thus arising in the more distal portions of the ligaments as their fibers move on one another or over fixed structures. M_3 is primarily the mass of the stapes, but to this mass we must add that of other parts—its annular ligament, the stapedial tendon and muscle, and some of the cochlear fluid—insofar as these parts move strictly with the stapes. S_3 is the stiffness of the annular ligament and stapedial tendon. M_4 is the mass of structures set in motion through S_3, namely, the distal portions of the annular ligament and stapedius muscle.

The middle ear system as analyzed here is a good deal simpler than the models set up for consideration by Frank, Esser, and Onchi. On the other hand, it is more complicated than one sometimes offered, as in the discussions by Johansen and Campbell. These authors sought to relate the conductive properties of the ear under both normal and abnormal conditions to the simple impedance formula, already given on page 20. Their treatment is based upon the assumption that the characteristics of the ear may be lumped into single values of mass, stiffness, and friction, and that the total impedance is a vector sum of the separate impedances produced by these three elements.

Under certain conditions this simplified treatment gives useful results, and often throughout this book we have dealt with the ear in such terms. The simplification works, we think, largely because M_1, the mass of the malleus, incus, and drum membrane, greatly exceeds the other moving masses of the middle ear. Because this mass is paramount the elasticity S_1 and the friction D_1 associated with it also take a predominant role. The other elements therefore may be neglected provided that we are content with only a rough and approximate solution of the vibratory problem. A more exact solution requires that we considerably complicate the analysis by bringing in a number of additional elements as indicated.

In this analysis of the middle ear we have the beginnings of an understanding of its performance as a dynamic system. This system fulfills its duties as a mechanical transformer with remarkably little disturbance of the transmitted sounds, and it does so over an extensive frequency range. It operates in this way because it is a multiply-tuned system, with three regions of resonance, and at the same time with sufficient frictional damping to make these resonances broad and prevent excessive emphasis on any one frequency region.

Our study has been carried out mainly on the ear of the cat. No doubt the human ear is about as well adapted to its task, yet we must expect some differences. A consideration of the form of the sensitivity function in man suggests that in this ear the advantages of uniformity have been moderated somewhat in favor of high sensitivity in the middle range.

18

The Present Status of Auditory Theory

In the foregoing chapters we have examined the mechanical processes of the ear from the initial application of sounds to the action upon the final receptor cells of the cochlea. We have considered in detail the operation of the middle ear as a sound conductor, and have found that this mechanism carries out its task with remarkable efficiency and fidelity. It acts as a mechanical transformer to match the acoustic properties of the external aerial medium with those of the cochlea, and at the same time it imposes only a minimum of its own peculiarities upon the transmitted vibrations. It is now well established that the distortions to which the ear is subject, which are prominently shown in the appearance of overtones and combination tones, are not produced in the peripheral conductive processes as has been generally believed, but arise in the further actions of the auditory receptive system: in the electromechanical and electrophysiological processes of the sensory and neural elements.

Our examination of the mechanical actions occurring within the cochlea has brought us face to face with some of the most difficult and controversial questions having to do with the ear: how sounds pass through the cochlea and distribute their effects over its extent, and what their particular actions are in exciting its sensory cells. The discussions of these questions over many years have centered upon the place theory and its claims of adequacy in explaining how sounds act upon the ear and give us our auditory experiences. It is well therefore to review here some of the major arguments and show the present status of this theory.

AN EVALUATION OF THE PLACE THEORY

There are many forms of place theory, and an exhaustive treatment of them has not been attempted in this book. We have discussed a number of its forms, however, and enough to show its singular dependence upon the principle of spatial representation in the determination of pitch. Every discriminable tone is considered as having its own distinctive place of action in the cochlea and hence in the fibers of the auditory nervous system. The nature and delicacy of our discriminations of pitch are considered to be a function simply of the mechanical selectivity and continuing spatial discreteness of this representation.

This conception of a spatial representation of the frequency of sounds is the place principle. It is important to note at this point that the use of this principle is not an exclusive prerogative of the place theories. Other theories accept this principle also, but they apply it more conservatively and do not make it the sole basis of explanation of pitch.

THE PRESENT FORM OF THE PLACE PRINCIPLE

In the original formulation of this principle and, we are sure, in many of the less critical considerations of it in more recent times, the selectivity of the spatial representation of tones was taken as absolute: every discriminable tone had its own exclusive place in the cochlea and in the auditory nerve fibers. A more serious contemplation of the anatomy of the ear and of the principles of resonance soon raised a doubt that this ideal form of tonal representation could exist in fact, and indicated that for every tone there must be a spreading of action over the responsive structures.

This doubt has continually grown stronger, and from a consideration of evidence such as that presented in Chapters 13 and 14 it is now rather generally appreciated that the simple conception of the place principle must be given up. There is no precise specificity of tonal action, but every tone spreads its effects throughout the cochlea. The tone produces a pattern consisting of local variations of magnitude along the basilar membrane, and of a form varying characteristically as a function of frequency. The high tones have patterns in which the major action is in the basal region and the apical regions take only a feeble part. The low tones have broader patterns in which all regions of the cochlea play a considerable role. In general, the region of most vigorous activity shifts progressively toward the basal end of the cochlea and this region becomes more accentuated as we go up the frequency scale.

With the form of place representation just described there is now a very large measure of agreement. At this point, however, opinion takes two diverging courses. There are some persons, and the present authors are among them, who accept the evidence as it stands and regard the place principle as providing only a limited basis for pitch perception. There are others who continue to rely upon this principle as the sole basis for an explanation of pitch. Although accepting the evidence as showing only a limited specificity in the initial action of sounds they look for secondary means by which a suitable degree of specificity may be developed. They postulate further processes in which the primary pattern of action is "sharpened up."

Two kinds of special assumptions are made in this situation accordingly as the secondary restriction of activity is considered as arising in mechanical or neural processes.

Restriction in the Mechanical Domain. The various proposals for a mechanical limitation of action beyond the primary response have already received detailed consideration in Chapter 14. Our discussion there showed that such mechanical processes, though conceivable, are generally of a complex nature and are subject to serious limitations of operation. An action involving higher orders of change, such as the derivatives of temporal and spatial patterns or secondary effects like eddies, are of doubtful reliability in the face of individual differences and accidental changes. And yet the pitch discrimination of the ear is one of its most stable functions, persisting with little alteration despite all sorts of structural and physiological disturbances, even including grave impairments of sensitivity.

Restriction in the Neural Domain. A number of suggestions have been offered for a restriction of action in the neural representation of the primary cochlear pattern. The first proposal of this kind was made by Hostinsky, who said that by a competitive interaction of tonal "sensations" a final unification of pitch is achieved. Gray, in what he called a principle of maximum stimulation, proposed a similar neural function, but he failed to indicate what the process might be. He merely asserted that the pitch of a tone is determined by the particular nerve fiber representing the peak of the action on the basilar membrane, and impulses from other nerve fibers excited at the same time are somehow suppressed or ignored. Gray's principle met with general acceptance and has figured prominently in most subsequent formulations of the place theory, although still without much consideration of what neural process might be the basis of appreciation of the maximum.

Békésy (*1*) later formulated a "law of contrast" that amounted to much the same thing as Gray's principle. He suggested that neural excitation ought to be particularly great in a region of transition from one degree of stimulation to another. This contrast effect is supposed to be most prominent at a maximum because here are two transitions, from weak to strong and from strong to weak.

Huggins and Licklider offered suggestions of a somewhat more palpable form. They indicated that a sharpening up of action might be achieved by a process of neural differentiation, or even better by a double differentiation, or by some process approximating these transformations of the primary action pattern. When a double neural differentiation is added to the beam hypothesis the final action represents the sixth space derivative of the basilar membrane displacement. What has been said about the uncertainties of mechanical limiting processes holds similarly for the neural processes and for combinations of the two.

Further Consideration of the Sharpening Processes. The mechanical and neural sharpening processes, if we admit the claims made for them, will bring about a certain transformation of the primary action pattern. How much sharpening up of the primary pattern might be achieved by any of these processes or by some combination of them is difficult to determine with any exactness. We should need to know the precise form of the primary pattern and also the rigor with which the secondary processes are carried out.

Huggins and Licklider in their treatment of this problem accepted the pattern of primary action as represented by Békésy and then considered various transformations of this pattern. According to their formulations, a single or double differentiation, carried out either by mechanical or neural means, achieves only a moderate restriction of the pattern of activity. It must be noted further that their mechanical transformations are based upon the assumption of traveling waves in the cochlea, an assumption that our evidence fails to support. The differentiation of a simple displacement function, one not involving a traveling wave, does not lead to any appreciable restriction of activity. Rather, it generates a function of greater complexity than the primary one, with a peak to one side of the original crest, zero activity at the crest, and a negative peak or valley on the other side of the crest. Further differentiations add to the number of peaks and valleys that are distributed over the extent of the original displacement wave. Interpreted on a place hypothesis, transformations like these would give several discrete pitches for every stimulating tone.

Only by means of the beam hypothesis, and by making special assumptions regarding the mode of stimulation of the sensory cells in the "beam" action, were Huggins and Licklider able to indicate any significant degree of sharpening effect. The particular amount that they expected to be obtained by this action was not explicitly stated, but their illustrative example suggests a reduction to about one-third of the original pattern. It was intimated that further neural processes might restrict this action still more.

If we accept all these suggestions, and thereby represent the place theory in its most favorable light, we still must ask whether the amount of restriction that conceivably may be gained by these processes is anything near what the facts of hearing require. For an answer to this question we must look a little more closely at the subjective phenomena.

PHENOMENOLOGICAL CONSIDERATIONS

In this book it has not been possible to present a full account of the subjective aspects of hearing or even to describe in detail the characteristics of pitch that need to be considered in a complete theory of hearing. The subject is a broad one; the functional selectivity of the ear is manifested in the phenomena of pitch, masking, auditory fatigue, and the analysis of complex sounds, and somewhat less obviously in certain aspects of the perception of beats, combination tones, and phase relations. A discussion of all of these topics would carry us much too far afield. Here we can only take a cursory view of some of the facts of pitch perception, and note especially the keenness of pitch discrimination and its variations with intensity.

Pitch Discrimination. For tones of comfortable loudness (40 db above threshold) we are able to discriminate changes of 2 to 4~ over the range up to 2000~, and thereafter the necessary number of cycles increases rapidly, reaching 187~ for a tone of 15,000~. If we begin at one end of the frequency range and lay out one after the other the values of these just noticeable differences until we reach the other end of the range, the total number comes to 1410.

This number of just noticeable differences has sometimes been taken as indicating the number of discriminable pitches and hence the number of discrete resonators or other processes involved in the auditory system, but the assumption underlying this conclusion is often overlooked. This conclusion can be drawn only if the pitch dimension is made up of quanta: if a continuous variation of frequency gives rise to stepwise alterations of pitch. Careful investigation shows that such is not the case, but pitch is a continuum. The number of elementary processes therefore must be many times greater than 1410.

The Intensity Relation. Pitch discrimination is poor for a tone that is barely audible and then improves rapidly with intensity for about 20 db and more slowly thereafter. The improvement of discrimination is about twofold as the loudness level rises from 10 to 50 db and from there on the improvement is only slight. It is significant that there is no evidence of a deterioration even at extreme levels.

These facts impose a severe burden upon a place theory. It is obvious that, regardless of the particular formulation of such a theory, the specificity of action must be the greatest at threshold intensity when only the peak elements are effective, and must grow progressively less as the intensity is raised and other elements come into action.

The facts just mentioned can be explained on a place theory only by saying that the selectivity is high enough to give an adequate separation of the action of two discriminable tones at their greatest intensities. This is a very high degree of selectivity indeed. A 1000~ tone at a level 90 db above threshold is distinguished from a tone of 997.4~ or from one of 1002.6~. Hence the elementary processes immediately adjacent to the process responsible for the perception of 1000~ must be operating at a magnitude less than 0.00003 of the magnitude of the 1000~ process. The response function then would have the incredible slope of 69 db or more per cycle. How severe are the requirements made upon the secondary processes can be appreciated on consideration of the slopes of the primary function as shown for example in Békésy's curve of Fig. 99. Here the slopes adjacent to the peak are of the order of 0.01 db per cycle. A sharpening up of something like 2800-fold is therefore needed.

The Maximum Hypothesis. It is obvious that even with the aid of various kinds of sharpening processes the ear's activity is never reduced to the absolute form, in which only a single mechanical element and a single nerve fiber would be representative of a tone. There is always a certain degree of spread of activity, so that a pattern is formed involving many elements, some in greater and others in lesser degree. Hence the modern versions of the place theory, like the older ones, rest finally upon Gray's principle of maximum stimulation. They assume that it is the peak of the wave of activity as represented to the higher centers that determines the pitch of a tone.

These considerations make it possible to arrive at another estimation of the amount of sharpening of the primary response that would be necessary on a place theory. This estimation is made by applying our knowledge about loudness discrimination.

The appreciation of the maximum point of a response curve obviously requires that the degree of activity at this point be discriminated from the smaller degrees of activity elsewhere on the curve. To make our ideas definite, let us say that the degree of activity at any point is finally represented in terms of the rapidity of firing of a certain group of nerve fibers, those supplying the point in question. Our problem then is to discriminate this neural discharge in a group of fibers at the peak from that in other groups on either side of the peak.

Now let us make the assumption that this discrimination of rates of firings taking place simultaneously in two different groups of nerve fibers is on a par with the discrimination of different rates that may be produced successively in any one group. Loudness discrimination data then may be used as a conservative indication of this ability. Such discrimination no doubt is assisted by a spreading of the activity to additional elements as the intensity is raised, but no harm is done if for our present purposes we attribute the ability altogether to the appreciation of rates of nerve discharge.

Now if we know the form of the curve of response in the cochlea we can proceed from this assumption to determine how far down from the maximum we must go to reach a just discriminable point, and also how far we have moved to one side or the other away from the peak in doing so. If finally we can locate these points on the frequency scale we have two tonal distances from the peak that ought to be just discriminable.

This procedure might be carried out with any of the response curves suggested by particular theories. It was first carried out (Wever, 7) with a curve for 1000~ whose form was indicated by two sets of data on guinea pig ears, data obtained in our studies on overstimulation and on maximum cochlear potentials. It was found with this curve that dropping below the peak by 3 per cent (which is the value of the just noticeable loudness difference at 1000~) located a point on one side of the maximum 48 per cent of the distance from the basal end of the basilar membrane and a point on the other side of the maximum 56.4 per cent of the distance from the basal end. As this 1000~ curve has its peak at a position 52 per cent of the distance from the basal end, we now have marked off two steps of 4 per cent and 4.4 per cent of the length of the basilar membrane. By reference to data on the probable locations of the peaks produced by different tones these steps are evaluated in terms of tonal frequencies. They are found to represent the intervals 1200 to 1000~ and 1000 to 860~. Now by our original assumption each of these intervals ought to be just discriminable in terms of pitch. In fact, however, pitch discrimination data show that the interval 1200 to 1000~ contains about 70 just noticeable differences and the interval 1000 to 860~ contains about 55.

When this same procedure is carried out by using the response curve for 1000~ as determined theoretically by Ranke the interval on one side of the maximum is found to be 1120 to 1000~ and that on the other side is 1000 to 960~, and these intervals contain 46 and 15 just discriminable differences, respectively.

It is plain from this approach to the problem that our appreciation of the point of maximum stimulation is crude as based on the primary response and that this response must be sharpened up to an enormous extent—to something like 1/46 to 1/70 of the primary width—in order to account for our ability of pitch discrimination. This amount of sharpening is less than that just indicated in our consideration of pitch discrimination itself, but still is far beyond anything that reasonably may be expected from the sorts of processes so far considered.

NEUROLOGICAL EVIDENCE

We can gain further understanding of the actual spread of activity beyond the primary representation on the basilar membrane by a consideration of response patterns at different levels of the auditory nervous system. Galambos and Davis observed that elements of the cochlear nucleus are responsive to a wide range of tonal stimulation when the intensity is strong. One such element, for example, was set in action by tones extending over 3½ octaves when the tones were presented at intensities 90 db above threshold.

In the temporal area of the cerebral cortex the action is still widespread, as shown in experiments by Woolsey and Walzl, Tunturi, and others. A small group of nerve fibers stimulated at one point in the cochlea may give measurable potentials over as much as a third of the exposed auditory area of the temporal lobe. A single tone likewise excites activity over an extensive region. Thus it seems that a truly specific response has not appeared in the acoustic system as far along as the primary projections of the cortex. What happens further, in cortical regions beyond this projection area, is yet unknown.

From all these considerations it seems that the place principle, with or without conceivable sharpening processes, is seriously inadequate by itself for the explanation of pitch perception. It must further be pointed out, and strongly emphasized, that the place theories throughout their history have so concentrated their attention upon the problem of pitch as almost to ignore the many other aspects of hearing. Yet the other phenomena have to be considered. Some of these phenomena have never been explained on a place theory, and we do not believe that they can be. Such a theory then is not a theory of hearing in a complete sense but only a theory of pitch, and with serious inadequacies for this restricted service as we have indicated.

THE VOLLEY THEORY

As already mentioned, the place principle can be accepted simply at its face value and applied for what it is worth as an explanatory concept. This principle then is supplemented by another, the frequency principle.

The Frequency Principle. Experimental observations show that the periodicity of a sound is maintained in the responses of the sensory cells and, within limits, in the further actions of the auditory nerve fibers. The frequency of impulses in the nerve reproduces the frequency of the sound waves from the lowest perceptible tones to the region of 4000 to 5000~. These synchronous relations are still preserved in the actions in the acoustic nuclei of the medulla oblongata. As the frequency is raised further, however, the synchronous relation becomes disturbed and gradually passes over into asynchronism. That the synchronous relation exists over the principal range of tones is the frequency principle, and that we make use of this relationship in our perceptions of sound is one of the basic postulates of the volley theory.

Place and Frequency Principles Combined. The volley theory is developed further on the assumption that the perceptive centers are able to utilize both spatial and temporal dimensions of the nerve action insofar as these are representative of properties of the stimuli. The particular nature of our tonal sensitivity reflects the combined adequacies of these two dimensional relations.

The representation in terms of nerve impulses gives a faithful indication of the stimulus frequency for all the low and intermediate tones, but it fails for the high tones. On the other hand, the representation in terms of place is only vague for the low tones and becomes progressively more specific as the frequency is raised. Thus for all but the highest tones both frequency and place representations are available, though these two vary in different ways along the frequency scale as to the specificity of the information that they give. The evidence indicates that in pitch discrimination the frequency pattern is the main determiner for the low and middle tones whereas the spatial pattern serves alone for the high tones.

Our appreciation of temporal phenomena, such as phase shifts, transients of frequency, modulation, and binaural differences of time and phase depend upon a precise rendition of the wave relations of the stimuli, and thus are mainly accounted for on the frequency principle. Most other auditory phenomena depend, as pitch does, upon a joint operation of place and frequency principles. It is the utilization in harmonious combination of both spatial and temporal representations of the nature of sounds that provides us with all the rich variety of our auditory experiences.

Appendixes, References and Index

GLOSSARY OF ABBREVIATIONS AND SYMBOLS

(Given in parentheses after each quantity is the most common unit in which it is measured.)

A Area (square centimeters)
cm Centimeters
d Displacement (centimeters)
db Decibels
e Electrical potential (volts)
E Energy (ergs)
f Frequency (cycles per second)
F Force (dynes)
ft Feet
i Electric current (amperes)
J Power (ergs per second per square centimeter)
M Mass (grams)
mg Milligrams
mm Millimeters
N Number of decibels
P Pressure (dynes per square centimeter)
r Ratio
R Specific acoustic resistance (mechanical ohms per square centimeter)
R_2 Apparent impedance; the impedance of the ear as presented at the external auditory meatus (mechanical ohms per square centimeter)
R_e Electrical resistance (ohms)
R_m Frictional resistance (mechanical ohms)
S Stiffness or elasticity (dynes per centimeter)
sq Square
t Time (seconds)
T Transmission
u Particle velocity (centimeters per second)
X Displacement (centimeters)
X_m Mass reactance (mechanical ohms)
X_s Elastic reactance (mechanical ohms)
Z Mechanical impedance (mechanical ohms)
θ Any angle (degrees or radians)
π 3.1416
ρ Density (grams per cubic centimeter)
ω Angular frequency or angular velocity, $2\pi f$ (radians per second)
°C Temperature, degrees Centigrade
$\sim$ Cycles per second
$\propto$ Is proportional to

APPENDIX A

Reference Levels for Sound Intensity Measurements

Designation	Numerical value	Decibels above ASA standard for power
Phonic level	10^{-6} watts per sq cm	100
Zero level	1 dyne per sq cm	73.8[a]
Mean threshold level[b]	10^{-3} dynes per sq cm	13.8[a]
1000~ threshold level[c]	2.4×10^{-16} watts per sq cm	3.8
ASA reference (power)[d]	10^{-16} watts per sq cm	0
ASA reference (pressure)	2×10^{-4} dynes per sq cm	—0.2[a]

[a] For sounds in air under standard conditions.
[b] For the range 500 to 8000~.
[c] For uniaural listening, after Sivian and White.
[d] ASA refers to the American Standards Association.

APPENDIX B

Formulas Pertaining to Plane Progressive Sinusoidal Waves

In any homogeneous medium	In air under standard conditions
$u = \omega d$	$u = \omega d$
$P = uR$	$P = 41.5u$
$J = P^2/R$	$J = P^2/41.5$
$J = \omega^2 d^2 R$	$J = 41.5\omega^2 d^2$

d = amplitude in cm
u = particle velocity in cm per second
P = pressure in dynes per sq cm
J = power in ergs per second per sq cm
ω = angular velocity = $2\pi f$
f = frequency in cycles per second
R = specific acoustic resistance (41.5 for air under standard conditions).

All values are root-mean-square values. To obtain maximum values, multiply by $\sqrt{2}$.

APPENDIX C

Relations between Pressure and Displacement in Aerial Sounds

Frequency	Pressure at threshold[a] (in 10^{-3} dynes per sq cm)	Displacement at threshold (in 10^{-9} cm)	Displacement for pressure of 1 dyne per sq cm (in 10^{-7} cm)
100	8.6	330	384
200	2.5	48.0	192
300	1.1	14.1	126
500	0.50	3.83	76.8
700	0.40	2.19	54.7
1,000	0.32	1.23	38.4
2,000	0.079	0.15	19.2
3,000	0.063	0.08	12.6
5,000	0.158	0.12	7.68
7,000	0.45	0.25	5.47
10,000	0.71	0.27	3.84
15,000	3.2	0.82	2.56

[a] Threshold values were interpolated from curve *a* of Fig. 9.

APPENDIX D

Measurements of Auditory Structures

(The data are for the adult human ear except as stated otherwise. The number of ears is indicated when it is known. When no source is given the observations are our own.)

OUTER AND MIDDLE EAR

External Auditory Meatus

Length, 2.3-2.97 cm, mean of 20 ears 2.57 cm (Bezold, *3*); 2.7 cm Békésy, *4*); 2.3 cm (Wiener and Ross).

Size of lumen at entrance, 0.9 cm vertically by 0.65 cm horizontally, mean of 21 ears (Bezold, *3*); mean diameter, 0.7 cm (Békésy, *4*).

Area of external opening, 0.3-0.5 sq cm (Békésy, *4*).

Volume, 1.04 cubic cm (Békésy, *4*).

Resonance frequency, 3400-3900~ (Wiener and Ross); 3000-4000~ (Fleming).

Tympanic Membrane

Diameter along the manubrium, 8.5-10 mm, mean of 9 ears, 9.2 mm (Bezold, *3*); 9-10 mm (Helmholtz, *3*); 8-10 mm (H. Gray).

Diameter perpendicular to the manubrium, 8-9 mm, mean of 8 ears 8.5 mm (Bezold, *3*); 7.5-9 mm (Helmholtz, *3*); 8-9 mm (H. Gray).

Height of cone (inward displacement of umbo), 2 mm (Siebenmann, *2*).

Area, 69.5 sq mm (Schwalbe); 65.0 sq mm (Keith); 85 sq mm (Békésy, *11*); 66 sq mm (Stuhlman, *1*); 55.8, 59.8, and 63.0 sq mm in 3 specimens. For the cat: 36.0-46.5 sq mm, mean of 4 ears 41.8 sq mm (Wever, Lawrence, and Smith, *1*); 32.3-47.6 sq mm, mean of 12 ears 39.8 sq mm (Payne and Githler). For the guinea pig: 23.5-28.0 sq mm, mean of 8 ears 24.8 sq mm (Whittle).

Effective area, 55 sq mm (Békésy, *11*); 42.9 sq mm (our calculation, see page 63).

Thickness of whole membrane, 0.1 mm (Helmholtz, *3*); of its fibrous layer, 0.05 mm (Helmholtz, *3*).

Weight, 14 mg (1 specimen).

Breaking strength, for positive air pressure, in cadaver specimens: for normal-appearing membranes 0.4-3.0 $\times 10^6$ dynes per sq cm, mean of 111 ears 1.61 $\times 10^6$ dynes per sq cm (Zalewski); for abnormally thin membranes, mean of 12 specimens 0.52 $\times 10^6$ dynes per sq cm; for scarred membranes, mean of 12 specimens 0.3 $\times 10^6$ dynes per sq cm (Zalewski). For the dog: 0.6-1.6 $\times$ 10^6 dynes per sq cm, mean 1.0 $\times 10^6$ dynes per sq cm (Zalewski); for young dogs 0.96×10^6 dynes per sq cm, for old dogs 0.90×10^6 dynes per sq cm (Zickero). For the cat: about 0.11×10^6 dynes per sq cm (Wever, Bray, and Lawrence, *7*).

Middle Ear Cavity

Total volume, 2.0 cubic cm (Békésy, *8*).

Volume of ossicles, 0.5-0.8 cubic cm (Békésy, *8*).

Malleus

Weight, 23 mg (Stuhlmann, *1*); 27 mg (1 specimen).

Length, from end of manubrium to end of lateral process, 5.8 mm (Stuhlmann, *1*); total length, 7.6-9.1 mm (Bast and Anson).

Incus

Weight, 25.0-30.0 mg (Stuhlman, *1*); 32 mg (1 specimen).

Length along long process, 7.0 mm (Stuhlman, *1*).

Length along short process, 5.0 mm (Stuhlman, *1*).

Stapes

Weight, 2.5 mg (Stuhlman, *1*); 2.05-4.35 mg, mean 2.86 mg (Bast and Anson); 3.9 mg (1 specimen).

Height, 4 mm (Stuhlman, *1*); 2.50-3.78 mm, mean 3.26 mm (Bast and Anson).

Length of footplate, 2.64-3.36 mm, mean of 75 ears 2.99 mm (Bast and Anson).

Width of footplate, 0.7 mm (Helmholtz, *3*); 1.4 mm (Stuhlman, *1*); 1.08-1.66 mm, mean of 75 ears 1.41 mm (Bast and Anson).

Area of footplate, 2.65-3.75 sq mm (Keith); 3.2 sq mm (Békésy, *11*); 3.2 sq mm (Stuhlman, *1*); 2.93 sq mm (1 specimen). For the cat: 1.07-1.33 sq mm, mean of 4 ears 1.15 sq mm (Wever, Lawrence, and Smith, *1*). For the guinea pig: 0.79-0.95 sq mm, mean of 8 ears 0.88 sq mm (Whittle).

Oval Window

Dimensions, 1.2 by 3 mm (Helmholtz, *3*); 2.0 by 3.7 mm (1 specimen).

Area, for the cat: 1.12-1.27 sq mm, mean of 4 ears 1.20 sq mm (Wever, Lawrence, and Smith, *1*). For the guinea pig: mean of 6 ears 1.41 sq mm (Fernández).

Round Window

Dimensions, 2.25 by 1.0-1.25 mm (Weber-Liel, *2*).

Area, 2 sq mm (Keith). For the cat: 2.78-3.29 sq mm, mean of 4 ears 3.01 sq mm (Wever, Lawrence, and Smith, *1*). For the guinea pig: mean of 6 ears 1.02 sq mm (Fernández).

Tensor Tympani Muscle

Length, 23-26 mm, mean of 4 specimens about 25 mm.

Cross-sectional area, 4.8 and 6.9 sq mm in 2 specimens.

Stapedius Muscle

Length, mean of 3 specimens 6.3 mm.

Cross-sectional area, mean of 3 specimens 4.9 sq mm.

APPENDIX D

INNER EAR

Cochlea

Number of turns, 2⅙-2⅞, for most ears 2⅝ (Hardy).

Volume, mean of 3 ears 98.1 cubic mm (including the vestibule proper).

Scala Vestibuli

Volume, including the vestibule proper, mean of 3 ears 54.0 cubic mm.

Scala Tympani

Volume, mean of 3 ears 37.4 cubic mm.

Cochlear Duct

Volume, mean of 3 ears 6.7 cubic mm.

Length, 35 mm (Retzius, *1*).

Shortest fluid pathway, estimated as 20 mm (Wilkinson and Gray). For the cat: 16.3 mm in a cochlea with a basilar membrane 21.1 mm long.

Helicotrema

Area, 0.08-0.2 sq mm, mean of 6 ears 0.15 sq mm (Keith).

Basilar Membrane

Length, 25.3-35.5 mm, mean of 68 ears 31.52 mm (Hardy). For the cat: 19.4-25.4 mm, mean of 6 ears 22.5 mm (Freedman). For the guinea pig: mean of 3 ears 16.4 mm (Suib); mean of 6 ears 18.8 mm (Fernández).

Width, varies on the average 6.25-fold from the basal end to a position near the apex, from a minimum of 0.08 mm to a maximum of 0.498 mm; range of maximums 0.423-0.651 mm in 25 ears (Wever, *1*). For the guinea pig: varies 3.35-fold, from a minimum of 0.062 mm to a maximum of 0.209 mm; range of maximums 0.194-0.228 mm in 35 ears (Guild, *1*); maximums 0.213-0.240 mm, mean of 10 ears 0.228 mm (Perlin).

Number of transverse fibers about 24,000 (Retzius, *1*). For the cat: 15,700 (Retzius, *1*).

Organ of Corti

Cross-sectional area varies about 4.6-fold (mean of 3 ears), from a minimum of 0.0053 sq mm to a maximum of 0.0223 sq mm (Wever, *5*). For the cat: varies about 3.6 fold, from a minimum of 0.0055 sq mm to a maximum of 0.0201 sq mm (Freedman).

Number of inner hair cells, 3500 (Retzius, *1*). For the cat: 2600 (Retzius, *1*).

Number of outer hair cells, 12,000 in 4 rows (Retzius, *1*). For the cat: 9900 in 3 rows (Retzius, *1*).

Spiral Ligament

Cross-sectional area varies about 13-fold (mean of 3 ears), from 0.543 sq mm near the basal end to 0.042 sq mm at the apex (Wever, *5*).

Ganglion Cells

Total number, mean of 3 children 29,019, mean of 4 adults with good hearing 25,614 (Guild, Crowe, Bunch, and Polvogt); mean of 23 ears 30,500 (Wever, 5).

Density of innervation varies from 583 cells per mm in the upper middle and apical region to 1274 cells per mm in the upper basal region (Guild, Crowe, Bunch, and Polvogt); in 23 ears, maximum density 1250 cells per mm, mean density 970 cells per mm (Wever, 5).

APPENDIX E

Instructions for Use of Decibel Tables

Included here are tables to facilitate the conversion of decibels to ratios and the reverse. One table goes from decibels to energy and power ratios and the other goes from decibels to voltage, pressure, current, and velocity ratios. In each section of the tables are three columns, with decibel values in the middle column and ratios on either side. To the left are ratios smaller than unity, for which the decibel values are negative, and to the right are ratios greater than unity, for which the decibel values are positive. The tables extend to 20 db by tenths. To obtain values outside this range, proceed as follows:

Given decibels, to find ratios, break the number of decibels into parts of 20 db and whatever remainder is necessary, look up the ratios of all these parts individually, and multiply these numbers to get the final ratio.

For example, to find the power ratio corresponding to +37 db, write 37 db = 20 + 17 db. The ratio $J/J_0 = 100 \times 50.12 = 5012$.

To find the voltage ratio corresponding to +67 db write 67 db = 20 + 20 + 20 + 7 db. The voltage ratio $e/e_0 = 10 \times 10 \times 10 \times 2.239 = 2239$. For —67 db write —67 db = —20 — 20 — 20 — 7 db. The voltage ratio $e/e_0 = 0.1 \times 0.1 \times 0.1 \times 0.4467 = 0.0004467$.

Given power ratios, to find decibels outside the table, move the decimal point to the left or to the right until a number is obtained that is included in the table. Find the decibel value for this number, and then add 10 db for every time you moved the decimal point to the left or subtract 10 db for every time you moved it to the right. Given voltage ratios, you do the same thing except that you add or subtract 20 db for each displacement of the decimal point.

For example, given the power ratio of 0.0031, write $0.0031 = 0.031 \times 0.1$; $N = -15.1 - 10 = -25.1$ db.

Given the voltage ratio of 507.0, write $507.0 = 5.070 \times 10 \times 10$; $N = 14.1 + 20 + 20 = 54.1$ db. For the voltage ratio 0.066, write $0.066 = 0.66 \times 0.1$; $N = -3.6 - 20 = -23.6$ db.

Conversion of Decibels to Energy or Power Ratios

Ratio (for −db)	Decibels	Ratio (for +db)	Ratio (for −db)	Decibels	Ratio (for +db)	Ratio (for −db)	Decibels	Ratio (for +db)	Ratio (for −db)	Decibels	Ratio (for +db)
.9772	0.1	1.023	.3090	5.1	3.236	.09772	10.1	10.23	.03090	15.1	32.3
.9550	0.2	1.047	.3020	5.2	3.311	.09550	10.2	10.47	.03020	15.2	33.1
.9333	0.3	1.072	.2951	5.3	3.388	.09333	10.3	10.72	.02951	15.3	33.8
.9120	0.4	1.096	.2884	5.4	3.467	.09120	10.4	10.96	.02884	15.4	34.6
.8913	0.5	1.122	.2818	5.5	3.548	.08913	10.5	11.22	.02818	15.5	35.4
.8710	0.6	1.148	.2754	5.6	3.631	.08710	10.6	11.48	.02754	15.6	36.3
.8511	0.7	1.175	.2692	5.7	3.715	.08511	10.7	11.75	.02692	15.7	37.1
.8318	0.8	1.202	.2630	5.8	3.802	.08318	10.8	12.02	.02630	15.8	38.0
.8128	0.9	1.230	.2570	5.9	3.890	.08128	10.9	12.30	.02570	15.9	38.9
.7943	1.0	1.259	.2512	6.0	3.981	.07943	11.0	12.59	.02512	16.0	39.8
.7762	1.1	1.288	.2455	6.1	4.074	.07762	11.1	12.88	.02455	16.1	40.7
.7586	1.2	1.318	.2399	6.2	4.169	.07586	11.2	13.18	.02399	16.2	41.6
.7413	1.3	1.349	.2344	6.3	4.266	.07413	11.3	13.49	.02344	16.3	42.6
.7244	1.4	1.380	.2291	6.4	4.365	.07244	11.4	13.80	.02291	16.4	43.6
.7079	1.5	1.413	.2239	6.5	4.467	.07079	11.5	14.13	.02239	16.5	44.6
.6918	1.6	1.445	.2188	6.6	4.571	.06918	11.6	14.45	.02188	16.6	45.7
.6761	1.7	1.479	.2138	6.7	4.677	.06761	11.7	14.79	.02138	16.7	46.7
.6607	1.8	1.514	.2089	6.8	4.786	.06607	11.8	15.14	.02089	16.8	47.8
.6457	1.9	1.549	.2042	6.9	4.898	.06457	11.9	15.49	.02042	16.9	48.9
.6310	2.0	1.585	.1995	7.0	5.012	.06310	12.0	15.85	.01995	17.0	50.1
.6166	2.1	1.622	.1950	7.1	5.129	.06166	12.1	16.22	.01950	17.1	51.2
.6026	2.2	1.660	.1905	7.2	5.248	.06026	12.2	16.60	.01905	17.2	52.4
.5888	2.3	1.698	.1862	7.3	5.370	.05888	12.3	16.98	.01862	17.3	53.7
.5754	2.4	1.738	.1820	7.4	5.495	.05754	12.4	17.38	.01820	17.4	54.9
.5623	2.5	1.778	.1778	7.5	5.623	.05623	12.5	17.78	.01778	17.5	56.2
.5495	2.6	1.820	.1738	7.6	5.754	.05495	12.6	18.20	.01738	17.6	57.5
.5370	2.7	1.862	.1698	7.7	5.888	.05370	12.7	18.62	.01698	17.7	58.8
.5248	2.8	1.905	.1660	7.8	6.026	.05248	12.8	19.05	.01660	17.8	60.2
.5129	2.9	1.950	.1622	7.9	6.166	.05129	12.9	19.50	.01622	17.9	61.6
.5012	3.0	1.995	.1585	8.0	6.310	.05012	13.0	19.95	.01585	18.0	63.1
.4898	3.1	2.042	.1549	8.1	6.457	.04898	13.1	20.42	.01549	18.1	64.5
.4786	3.2	2.089	.1514	8.2	6.607	.04786	13.2	20.89	.01514	18.2	66.0
.4677	3.3	2.138	.1479	8.3	6.761	.04677	13.3	21.38	.01479	18.3	67.6
.4571	3.4	2.188	.1445	8.4	6.918	.04571	13.4	21.88	.01445	18.4	69.1
.4467	3.5	2.239	.1413	8.5	7.079	.04467	13.5	22.39	.01413	18.5	70.7
.4365	3.6	2.291	.1380	8.6	7.244	.04365	13.6	22.91	.01380	18.6	72.4
.4266	3.7	2.344	.1349	8.7	7.413	.04266	13.7	23.44	.01349	18.7	74.1
.4169	3.8	2.399	.1318	8.8	7.586	.04169	13.8	23.99	.01318	18.8	75.8
.4074	3.9	2.455	.1288	8.9	7.762	.04074	13.9	24.55	.01288	18.9	77.6
.3981	4.0	2.512	.1259	9.0	7.943	.03981	14.0	25.12	.01259	19.0	79.4
.3890	4.1	2.570	.1230	9.1	8.128	.03890	14.1	25.70	.01230	19.1	81.2
.3802	4.2	2.630	.1202	9.2	8.318	.03802	14.2	26.30	.01202	19.2	83.1
.3715	4.3	2.692	.1175	9.3	8.511	.03715	14.3	26.92	.01175	19.3	85.1
.3631	4.4	2.754	.1148	9.4	8.710	.03631	14.4	27.54	.01148	19.4	87.1
.3548	4.5	2.818	.1122	9.5	8.913	.03548	14.5	28.18	.01122	19.5	89.1
.3467	4.6	2.884	.1096	9.6	9.120	.03467	14.6	28.84	.01096	19.6	91.2
.3388	4.7	2.951	.1072	9.7	9.333	.03388	14.7	29.51	.01072	19.7	93.3
.3311	4.8	3.020	.1047	9.8	9.550	.03311	14.8	30.20	.01047	19.8	95.5
.3236	4.9	3.090	.1023	9.9	9.772	.03236	14.9	30.90	.01023	19.9	97.7
.3162	5.0	3.162	.1000	10.0	10.000	.03162	15.0	31.62	.01000	20.0	100.0

Conversion of Decibels to Voltage or Pressure Ratios

Deci-bels	Ratio (for +db)	Ratio (for -db)	Deci-bels	Ratio (for +db)	Ratio (for -db)	Deci-bels	Ratio (for +db)	Ratio (for -db)	Deci-bels	Ratio (for +db)
0.1	1.012	.5559	5.1	1.799	.3126	10.1	3.199	.1758	15.1	5.689
0.2	1.023	.5495	5.2	1.820	.3090	10.2	3.236	.1738	15.2	5.754
0.3	1.035	.5433	5.3	1.841	.3055	10.3	3.273	.1718	15.3	5.821
0.4	1.047	.5370	5.4	1.862	.3020	10.4	3.311	.1698	15.4	5.888
0.5	1.059	.5309	5.5	1.884	.2985	10.5	3.350	.1679	15.5	5.957
0.6	1.072	.5248	5.6	1.905	.2951	10.6	3.388	.1660	15.6	6.026
0.7	1.084	.5188	5.7	1.928	.2917	10.7	3.428	.1641	15.7	6.095
0.8	1.096	.5129	5.8	1.950	.2884	10.8	3.467	.1622	15.8	6.166
0.9	1.109	.5070	5.9	1.972	.2851	10.9	3.508	.1603	15.9	6.237
1.0	1.122	.5012	6.0	1.995	.2818	11.0	3.548	.1585	16.0	6.310
1.1	1.135	.4955	6.1	2.018	.2786	11.1	3.589	.1567	16.1	6.383
1.2	1.148	.4898	6.2	2.042	.2754	11.2	3.631	.1549	16.2	6.457
1.3	1.161	.4842	6.3	2.065	.2723	11.3	3.673	.1531	16.3	6.531
1.4	1.175	.4786	6.4	2.089	.2692	11.4	3.715	.1514	16.4	6.607
1.5	1.189	.4732	6.5	2.113	.2661	11.5	3.758	.1496	16.5	6.683
1.6	1.202	.4677	6.6	2.138	.2630	11.6	3.802	.1479	16.6	6.761
1.7	1.216	.4624	6.7	2.163	.2600	11.7	3.846	.1462	16.7	6.839
1.8	1.230	.4571	6.8	2.188	.2570	11.8	3.890	.1445	16.8	6.918
1.9	1.245	.4519	6.9	2.213	.2541	11.9	3.936	.1429	16.9	6.998
2.0	1.259	.4467	7.0	2.239	.2512	12.0	3.981	.1413	17.0	7.079
2.1	1.274	.4416	7.1	2.265	.2483	12.1	4.027	.1396	17.1	7.161
2.2	1.288	.4365	7.2	2.291	.2455	12.2	4.074	.1380	17.2	7.244
2.3	1.303	.4315	7.3	2.317	.2427	12.3	4.121	.1365	17.3	7.328
2.4	1.318	.4266	7.4	2.344	.2399	12.4	4.169	.1349	17.4	7.413
2.5	1.334	.4217	7.5	2.371	.2371	12.5	4.217	.1334	17.5	7.499
2.6	1.349	.4169	7.6	2.399	.2344	12.6	4.266	.1318	17.6	7.586
2.7	1.365	.4121	7.7	2.427	.2317	12.7	4.315	.1303	17.7	7.674
2.8	1.380	.4074	7.8	2.455	.2291	12.8	4.365	.1288	17.8	7.762
2.9	1.396	.4027	7.9	2.483	.2265	12.9	4.416	.1274	17.9	7.852
3.0	1.413	.3981	8.0	2.512	.2239	13.0	4.467	.1259	18.0	7.943
3.1	1.429	.3936	8.1	2.541	.2213	13.1	4.519	.1245	18.1	8.035
3.2	1.445	.3890	8.2	2.570	.2188	13.2	4.571	.1230	18.2	8.128
3.3	1.462	.3846	8.3	2.600	.2163	13.3	4.624	.1216	18.3	8.222
3.4	1.479	.3802	8.4	2.630	.2138	13.4	4.677	.1202	18.4	8.318
3.5	1.496	.3758	8.5	2.661	.2113	13.5	4.732	.1189	18.5	8.414
3.6	1.514	.3715	8.6	2.692	.2089	13.6	4.786	.1175	18.6	8.511
3.7	1.531	.3673	8.7	2.723	.2065	13.7	4.842	.1161	18.7	8.610
3.8	1.549	.3631	8.8	2.754	.2042	13.8	4.898	.1148	18.8	8.710
3.9	1.567	.3589	8.9	2.786	.2018	13.9	4.955	.1135	18.9	8.811
4.0	1.585	.3548	9.0	2.818	.1995	14.0	5.012	.1122	19.0	8.913
4.1	1.603	.3508	9.1	2.851	.1972	14.1	5.070	.1109	19.1	9.016
4.2	1.622	.3467	9.2	2.884	.1950	14.2	5.129	.1096	19.2	9.120
4.3	1.641	.3428	9.3	2.917	.1928	14.3	5.188	.1084	19.3	9.226
4.4	1.660	.3388	9.4	2.951	.1905	14.4	5.248	.1072	19.4	9.333
4.5	1.679	.3350	9.5	2.985	.1884	14.5	5.309	.1059	19.5	9.441
4.6	1.698	.3311	9.6	3.020	.1862	14.6	5.370	.1047	19.6	9.550
4.7	1.718	.3273	9.7	3.055	.1841	14.7	5.433	.1035	19.7	9.661
4.8	1.738	.3236	9.8	3.090	.1820	14.8	5.495	.1023	19.8	9.772
4.9	1.758	.3199	9.9	3.126	.1799	14.9	5.559	.1012	19.9	9.886
5.0	1.778	.3162	10.0	3.162	.1778	15.0	5.623	.1000	20.0	10.000

References

Adrian, E. D. The microphonic action of the cochlea in relation to theories of hearing, *Report of a discussion on audition*, Physical Society of London, June 1931, 5-9.

Alexander, G. Die Theorie der Luftleitung mit besonderer Berücksichtigung der Anatomie und Klinik, *Monatsschr. f. Ohrenheilk.*, 1931, 65, 1-20, 173-200.

Alexander, I. E., and Githler, F. J. (*1*) Histological examination of cochlear structure following exposure to jet engine noise, *J. Comp. Physiol. Psychol.*, 1951, 44, 513-524. (*2*) Chronic effects of jet engine noise on the structure and function of the cochlear apparatus, *J. Comp. Physiol. Psychol.*, 1952, 45, 381-391.

Ayers, H. (*1*) Die Membrana tectoria—was sie ist, und die Membrana basilaris—was sie verrichtet, *Anat. Anzeiger*, 1891, 6, 219-220. (*2*) Vertebrate cephalogenesis; II, A contribution to the morphology of the vertebrate ear, with a reconsideration of its functions, *J. Morphol.*, 1892, 6, 1-360.

Bárány, E. A contribution to the physiology of bone conduction, *Acta oto-laryngol.*, 1938, Suppl. 26, 223 pp.

Bast, Theodore H., and Anson, Barry J. *The temporal bone and the ear*, 1949.

Békésy, G. von. (*1*) Zur Theorie des Hörens; Die Schwingungsform der Basilarmembran, *Phys. Zeits.*, 1928, 29, 793-810. (*2*) Zur Theorie des Hörens; Über die Bestimmung des einem reinen Tonempfinden entsprechenden Erregungsgebietes der Basilarmembran vermittelst Ermüdungserscheinungen, *Phys. Zeits.*, 1929, 30, 115-125. (*3*) Zur Theorie des Hörens bei der Schallaufnahme durch Knochenleitung, *Ann. d. Phys.*, 1932, 13, 111-136. (*4*) Über den Einfluss der durch den Kopf und den Gehörgang bewirkten Schallfeldverzerrungen auf die Hörschwelle, *Ann. d. Phys.*, 1932, 14, 51-56. (*5*) Über den Knall und die Theorie des Hörens, *Phys. Zeits.*, 1933, 34, 577-582. (*6*) Über die nichtlinearen Verzerrungen des Ohres, *Ann. d. Phys.*, 1934, 20, 809-827. (*7*) Physikalische Probleme der Hörphysiologie, *Elektr. Nachr. Techn.*, 1935, 12, 71-83. (*8*) Zur Physik des Mittelohres und über das Hören bei fehlhaftem Trommelfell, *Akust. Zeits.*, 1936, 1, 13-23. (*9*) Fortschritte der Hörphysiologie, *Zeits. f. techn. Phys.*, 1936, 17, 522-528. (*10*) Über die mechanisch-akustischen Vorgänge beim Hören, *Acta oto-laryngol.*, 1939, 27, 281-296, 388-396. (*11*) Über die Messung der Schwingungsamplitude der Gehörknochelchen mittels einer kapazitiven Sonde, *Akust. Zeits.*, 1941, 6, 1-16. (*12*) Über die Schwingungen der Schneckentrennwand beim Präparat und Ohrenmodell, *Akust. Zeits.*, 1942, 7, 173-186; trans., The vibration of the cochlear partition in anatomical preparations and in models of the inner ear, *J. Acoust. Soc. Amer.*, 1949, 21, 233-245. (*13*) Über die Resonanzkurve und die Abklingzeit der verschiedenen Stellen der Schneckentrennwand, *Akust. Zeits.*, 1943, 8, 66-76; *trans.*, On the resonance curve and the decay period at various points on the cochlear partition, *J. Acoust. Soc. Amer.*, 1949, 21, 245-254. (*14*) Über die Frequenzauflösung in der menschlichen Schnecke, *Acta oto-laryngol.*, 1944, 32, 60-84. (*15*) The sound pressure difference between the round and the oval windows and the artificial window of labyrinthine fenestration, *Acta oto-laryngol.*, 1947, 35, 301-315. (*16*) The variation of phase along the basilar membrane with sinusoidal vibrations, *J. Acoust. Soc. Amer.*, 1947, 19, 452-460. (*17*) Vibration of the head in a sound field and its role in hearing by bone conduction, *J. Acoust. Soc. Amer.*, 1948, 20, 749-760. (*18*) The structure of the middle ear and the hearing of one's own voice by bone conduction, *J. Acoust. Soc. Amer.*, 1949, 21, 217-232. (*19*) The coarse pattern of the electrical resistance in the cochlea of the guinea pig (electroanatomy of the cochlea), *J. Acoust. Soc. Amer.*, 1951, 23, 18-28. (*20*) Microphonics produced by touching the cochlear partition with a vibrating electrode, *J. Acoust. Soc. Amer.*, 1951, 23, 29-35. (*21*) DC potentials and energy balance of the cochlear partition, *J. Acoust. Soc. Amer.*, 1951, 23, 576-582. (*22*) DC resting potentials inside the cochlear partition, *J. Acoust. Soc. Amer.*, 1952, 24, 72-76. (*23*) Gross localization of the place of origin of the cochlear microphonics, *J. Acoust. Soc. Amer.*, 1952, 24, 399-409. (*24*) Direct observation of the vibrations of the cochlear partition under a microscope, *Acta oto-laryngol.*, 1952, 42, 197-201.

Békésy, G. von, and Rosenblith, W. A. The mechanical properties of the ear, in S. S. Stevens' *Handbook of experimental psychology*, 1951, 1075-1115.

Bell, Charles. *The anatomy of the human body*, 1803, III, 373-453.

Beranek, L. L., and Sleeper, H. P., Jr. The design and construction of anechoic sound chambers, *J. Acoust. Soc. Amer.*, 1946, 18, 140-150.

Berengario da Carpi, Jacopo. [Not seen; quoted by Fallopius and Schelhammer.]

Bezold, Friedrich. (*1*) Statistische Ergebnisse über die diagnostische Verwendbarkeit des Rinneschen Versuches, *Zeits. f. Ohrenheilk.*, 1887, 17, 153-237. (*2*) Weitere Untersuchungen über "Knochenleitung" und Schallleitungsapparat im Ohr, *Zeits. f. Ohrenheilk.*, 1904, 48, 107-171. (*3*) *Die Corrosions-Anatomie des Ohres*, 1882.

Bingham, W. V. D. The rôle of the tympanic membrane in audition, *Psychol. Rev.*, 1907, 14, 229-243.

Blake, C. J. Operation for removal of the stapes, *Boston Med. and Surg. J.*, 1892, 127, 469-470.

Bockendahl, A. Ueber die Bewegungen des M. tensor tympani nach Beobachtungen am Hunde, *Arch. f. Ohrenheilk*, 1880, 16, 241-259.

Bogert, B. P. Determination of the effects of dissipation in the cochlear partition by means of a network representing the basilar membrane, *J. Acoust. Soc. Amer.*, 1951, 23, 151-154.

Bonnier, Pierre. (*1*) Le limaçon membraneux considéré comme appareil enregistreur, *Compt. r. soc. biol.*, 1895, 47, 127-129. (*2*) Fonctions de la membrane de Corti, *Compt. r. soc. biol.*, 1895, 47, 130-131. (*3*) De la nature des phénomènes auditifs, *Bull. sci. de France et Belg.*, 1895, 25, 367-397. (*4*) *L'audition*, 1901.

Bordley, J. E., and Hardy, M. Effect of lesions of the tympanic membrane on the hearing acuity, *Arch. of Otolaryngol.*, 1937, 26, 649-657.

Borghesan, E. Sulla modalità di eccitamento delle cellule acustiche e limite fra i sistemi di conduzione e di percezione, *Atti d. clin. oto-rino-laring. Univ. Palermo* (Anno 1949-50), 1951, 4.

Bornschein, H., and Krejci, F. Über die Frequenzabhängigkeit reversibler Änderungen der Cochlearpotentiale bei temporärer Anoxie, *Experientia*, 1949, 5, 359-360.

Bouman, H. D. Experiments on the Weber-test as used in otology, *Acta brevia neérl. physiol.*, 1934, 4, 44-46.

Bryant, W. S. Die Lehre von den schallempfindlichen Haarzellen, *Arch. f. Ohrenheilk.*, 1909, 79, 93-102.

Buck, A. H. (*1*) On the mechanism of the ossicles of the ear, *Arch. of Ophthalmol. Otol.*, 1869, 1, 603-620. (*2*) Untersuchungen über den Mechanismus der Gehörknöchelchen, *Arch. f. Augen- Ohrenheilk.*, 1870, 1, Abt. 2, 121-136.

Budde, E. (*1*) Über die Resonanztheorie des Hörens, *Phys. Zeits.*, 1917, 18, 225-236, 249-260. (*2*) Mathematische Theorie der Gehörsempfindung, in E. Abderhalden's *Handbuch der biologischen Arbeitsmethoden*, Abt. 5, Teil 7, Lief. 12, 1-194.

Bütschli, Otto. *Vorlesungen über vergleichende Anatomie*, 1921, I, 749.

Bunch, Cordia C. *Clinical audiometry*, 1943.

Burnett, C. H. Untersuchungen über den Mechanismus der Gehörknöchelchen und der Membran des runden Fensters, *Arch. f. Augen- Ohrenheilk.*, 1872, 2, Abt. 2, 64-74.

Campbell, P. A. The importance of the impedance formula in the interpretation of audiograms, *Trans. Amer. Acad. Ophthalmol. Otolaryngol.*, 1950, 54, 245-252.

Coiter, Volcher. *De auditus instrumento*, in *Externarum et internarum principalium humani corporis*, partium tabulae, Noribergae, 1573, 88-105. [First appeared separately, 1566.]

Corti, A. Recherches sur l'organe de l'ouïe des mammifères, *Zeits. f. wiss. Zool.*, 1851, 3, 109-169.

Cotton, J. C. Beats and combination tones at intervals between the unison and the octave, *J. Acoust. Soc. Amer.*, 1935, 7, 44-50.

Cotugno, Dominico. *De aquaeductibus auris humanae internae*, Viennae, 1774. [First ed., 1760.]

Crowe, S. J., Guild, S. R., and Polvogt, L. M. Observations on the pathology of high-tone deafness, *Bull. Johns Hopkins Hosp.*, 1934, 54, 315-379.

Crowe, S. J., Hughson, W., and Witting, E. G. Function of the tensor tympani muscle, *Trans. Amer. Otol. Soc.*, 1931, 21, 274-287.

Culler, E., Finch, G., and Girden, E. S. (*1*) Function of the round window, *Science*, 1933, 78, 269-270. (*2*) Correlation of auditory acuity with peripheral electrical response of the acoustic mechanism, *J. of Psychol.*, 1936, 2, 409-419.

Dahmann, H. Zur Physiologie des Hörens; experimentelle Untersuchungen über die Mechanik der Gehörknöchelchenkette, sowie über deren Verhalten auf Ton und Luftdruck, *Zeits. f. Hals-Nasen- Ohrenheilk.*, 1929, 24, 462-497, and 1930, 27, 329-368.

Davis, H., Fernández, C., and McAuliffe, D. R. The excitatory process in the cochlea, *Proc. Nat. Acad. Sci.*, Washington, 1950, 36, 580-587.

Davis, H., and Walsh, T. E. The limits of improvement of hearing following the fenestration operation, *Laryngoscope*, 1950, 60, 273-295.

Dennert, H. Akustisch-physiologische Untersuchungen, *Arch. f. Ohrenheilk*, 1887, 24, 171-184.

Derbyshire, A. J., and Davis, H. The action potentials of the auditory nerve, *Amer. J. Physiol.*, 1935, 113, 476-504.

DeRosa, L. A. A theory as to the function of the scala tympani in hearing, *J. Acoust. Soc. Amer.*, 1947, 19, 623-628.

DeVries, H. (*1*) The minimum audible energy, *Acta oto-laryngol.*, 1948, 36, 230-235. (*2*) Struktur und Lage der Tektorialmembran in der Schnecke, untersucht mit neueren Hilfsmitteln, *Acta oto-laryngol.*, 1949, 37, 334-338. (*3*) Brownian motion and the transmission of energy in the cochlea, *J. Acoust. Soc. Amer.*, 1952, 24, 527-533.

Dishoeck, H. A. E. van. Theorie und Praxis des Pneumophons, *Arch. f. Ohren- Nasen- Kehlkopfheilk.*, 1939, 146, 5-11.

Dishoeck, H. A. E. van, and DeWit, G. Loading and covering of the tympanic membrane and obstruction of the external auditory canal, *Acta oto-laryngol.*, 1944, 32, 99-111.

DuVerney, Joseph Guichard. *Traité de l'organe de l'ouie*, Paris, 1683.

Dworkin, S., Katzman, J., Hutchison, G. A., and McCabe, J. R. Hearing acuity of animals as measured by conditioning methods, *J. Exper. Psychol.*, 1940, 26, 281-298.

Esser, M. H. M. (*1*) The mechanism of the middle ear; I, The two piston problem, *Bull. of Math. Biophys.*, 1947, 9, 29-40. (*2*) The mechanism of the middle ear; Part II, The drum, *Bull. of Math. Biophys.*, 1947, 9, 75-91.

Eustachius, Bartholomaeus. *Opuscula anatomica*, Venetiis, 1564, 148-164.

Ewald, J. R. (*1*) Ueber eine neue Hörtheorie, *Wien. klin. Wochenschr.*,

1898, 11, 721. (*2*) Zur Physiologie des Labyrinths; VI, Eine neue Hörtheorie, *Pflüg. Arch. ges. Physiol.*, 1899, 76, 147-188. (*3*) Zur Physiologie des Labyrinths; VII Mitt., Die Erzeugung von Schallbildern in der Camera acustica, *Pflüg. Arch. ges. Physiol.*, 1903, 93, 485-500.

Eysell, A. Beiträge zur Anatomie des Steigbügels und seiner Verbindungen, *Arch. f. Ohrenheilk*, 1870, 5, 237-249.

Fabricius ab Aquapendente, Hieronymus. *De visione, voce, auditu*, Venitiis, 1600.

Fallopius, Gabriel. *Observationes anatomicae*, ad Petrum Mannam, Venetiis, 1561.

Fernández, C. Dimensions of the cochlea (guinea pig), *J. Acoust. Soc. Amer.*, 1952, 24, 519-523.

Fischer, O. Über ein von Max Wien geäussertes Bedenken gegen die Helmholtzsche Resonanztheorie des Hörens, *Ann. d. Phys.*, 1908, 25, 118-134.

Fleming, N. Resonance in the external auditory meatus, *Nature*, London, 1939, 143, 642-643.

Fletcher, Harvey. (*1*) *Speech and hearing*, 1929. (*2*) The mechanism of hearing as revealed through experiment on the masking effect of thermal noise, *Proc. Nat. Acad. Sci.*, Washington, 1938, 24, 265-274. (*3*) On the dynamics of the cochlea, *J. Acoust. Soc. Amer.*, 1951, 23, 637-645. (*4*) The dynamics of the middle ear and its relation to the acuity of hearing, *J. Acoust. Soc. Amer.*, 1952, 24, 129-131.

Fowler, E. P. Drum tension and middle ear air pressures, their determination, significance and effect upon the hearing, *Ann. of Otol. Rhinol. Laryngol.*, 1920, 29, 688-694.

Frank, O. (*1*) Anwendung des Prinzips der gekoppelten Schwingungen auf einige physiologische Probleme, *Sitzungsber. Akad. Wiss. München*, 1915, 289-315, and 1918, 107-158. (*2*) Die Leitung des Schalles im Ohr, *Sitzungsber. Akad. Wiss. München*, 1923, 11-77. (*3*) Die Theorie der Pulswellen, *Zeits. f. Biol.*, 1926, 85, 91-130.

Freedman, Richard. *The area of the auditory sense organ in the cat*, Thesis, Princeton Univ., 1947.

Frey, H. Zur Mechanik der Gehörknöchelchenkette, *Verhandl. deutsch. otol. Gesellsch.*, 1910, 19, 113-120.

Fumagalli, Z. (*1*) Ricerche morfologische sull' apparato di trasmissione del suono, *Arch. ital. di otol. rinol. laringol.*, 1949, 60, Suppl. 1, 323 pp. (*2*) Mécanique de l'appareil de transmission du son, *Revue de laryngol. otol. rhinol.*, 1951, Suppl. 72, 320-331.

Galambos, R., and Davis, H. (*1*) The response of single auditory-nerve fibers to acoustic stimulation, *J. Neurophysiol.*, 1943, 6, 39-57. (*2*) Inhibition of activity in single auditory nerve fibers by acoustic stimulation, *J. Neurophysiol.*, 1944, 7, 287-303. (*3*) Action potentials from single auditory-nerve fibers? *Science*, 1948, 108, 513.

Galen, Claudius [sic]. *Opera omnia*, medicorum Graecorum opera quae exstant, ed. by D. C. G. Kühn, Lipsiae, 1822, II, 837*ff*, III, 644*ff*.

Garner, W. R. The effect of frequency spectrum on temporal integration of energy in the ear, *J. Acoust. Soc. Amer.*, 1947, 19, 808-815.

Geffcken, W. Untersuchungen über Schwellenwerte; III, Über die Bestimmung der Reizschwelle der Hörempfindung aus Schwellendruck und Trommelfellimpedanz, *Ann. d. Phys.*, 1934, 19, 829-848.

Gildemeister, M. Probleme und Ergebnisse der neueren Akustik, *Zeits. f. Hals- Nasen- Ohrenheilk.*, 1930, 27, 299-328.

Gill, N. W. Some observations on the conduction mechanism of the ear, *J. of Laryngol. Otol.*, 1951, 65, 404-413.

Gisselsson, L. Experimental investigation into the problem of humoral transmission in the cochlea, *Acta oto-laryngol.*, 1950, Suppl. 82, 78 pp.

Gray, A. A. (*1*) On a modification of the Helmholtz theory of hearing, *J. of Anat. Physiol.*, 1900, 34, 324-350. (*2*) The application of the principle of maximum stimulation to clinical otology, *J. of Laryngol. Otol.*, 1929, 44, 817-826.

Gray, Henry. *Anatomy of the human body*, 20th ed., rev. by W. H. Lewis, 1918.

Griessmann, B. Neue Methoden zur Hörprüfung, *Passow-Schaefer's Beitr. z. Anat. Physiol. des Ohres*, 1921, 16, 47-55.

Guild, S. R. (*1*) The width of the basilar membrane, *Science*, 1927, 65, 67-69. (*2*) Early stages of otosclerosis, *Arch. of Otolaryngol.*, 1930, 12, 457-483. (*3*) Correlations of histologic observations and the acuity of hearing, *Acta oto-laryngol.*, 1932, 17, 207-249. (*4*) Hearing by bone conduction: the pathways of transmission of sound, *Ann. of Otol. Rhinol. Laryngol.*, 1936, 45, 736-754. (*5*) Symposium: The neural mechanism of hearing; Comments on the physiology of hearing and the anatomy of the inner ear, *Laryngoscope*, 1937, 47, 365-372. (*6*) Histologic otosclerosis, *Ann. of Otol. Rhinol. Laryngol.*, 1944, 53, 246-266. (*7*) Interpretation of hearing tests, *J. Amer. Med. Assoc.*, 1950, 142, 466-469.

Guild, S. R., Crowe, S. J., Bunch, C. C., and Polvogt, L. M. Correlations of differences in the density of innervation of the organ of Corti with differences in the acuity of hearing, *Acta oto-laryngol.*, 1931, 15, 269-308.

Habermann, J. Zur Pathologie der Taubstummheit und der Fensternischen, *Arch. f. Ohrenheilk.*, 1901, 53, 52-67.

Hällström, G. G. Von den Combinationstönen, *Ann. d. Phys.*, 1832, 24, 438-466.

Haller, Albertus von. (*1*) *Primae lineae physiologiae*, Gottingae, 1751. (*2*) *Elementa physiologiae corporis humani*, Lausannae, 1763.

Hallpike, C. S., and Rawdon-Smith, A. F. The "Wever and Bray phenomenon." A study of the electrical response in the cochlea with especial reference to its origin, *J. of Physiol.*, 1934, 81, 395-408.

Hallpike, C. S., and Scott, P. Observations on the function of the round window, *J. of Physiol.*, 1940, 99, 76-82.

Hamberger, C. A., and Hydén, H. Cytochemical changes in the cochlear ganglion caused by acoustic stimulation and trauma, *Acta oto-laryngol.*, 1945, Suppl. 61, 89 pp.

Hamberger, C. A., Hydén, H., and Nilsson, G. The correlation between cytochemical changes in the cochlear ganglion and functional tests after acoustic stimulation and trauma, *Acta oto-laryngol.*, 1949, Suppl. 75, 124-133.

Hammerschlag, V. (*1*) Ueber den Tensorreflex, *Arch. f. Ohrenheilk.*, 1899, 46, 1-13. (*2*) Ueber die Reflexbewegung des Musculus tensor tympani und ihre centralen Bahnen, *Arch. f. Ohrenheilk.*, 1899, 47, 251-275.

Hardy, M. The length of the organ of Corti in man, *Amer. J. Anat.*, 1938, 62, 291-311.

Held, H. (*1*) Zur Kenntniss der peripheren Gehörleitung, *Arch. f. Anat. Physiol., Anat. Abt.*, 1897, 350-360. (*2*) Untersuchungen über den feineren Bau des Ohrlabyrinthes der Wirbeltiere, I, *Akad. Wiss. Leipzig, Abh. k. Saechs. Gesellsch.*, math.-phys. Kl., 1904, 28, 1-74. (*3*) Die Cochlea der Säuger und der Vögel, ihre Entwicklung und ihr Bau, in A. Bethe's *Handbuch der normalen und pathologischen Physiologie*, 11, *Receptionsorgane*, 1926, I, 467-534.

Helmholtz, Hermann L. F. (*1*) Ueber Combinationstöne, *Ann. d. Phys.*, 1856, 99, 497-540. (*2*) Über die Combinationstöne, *Monatsber. Akad. Wiss. Berlin*, 1856, 279-285. (*3*) *Die Lehre von den Tonempfindungen als physiologische Grundlage für die Theorie der Musik*, 1st ed., 1863; trans. by A. J. Ellis, *On the sensations of tone*, 2nd Eng. ed., 1885. (*4*) Die Mechanik der Gehörknöchelchen und des Trommelfells, *Pflüg. Arch. ges. Physiol.*, 1868, 1, 1-60; trans. by A. H. Buck and N. Smith, *The mechanism of the ossicles of the ear and membrana tympani*, 1873.

Hensen, V. (*1*) Zur Morphologie der Schnecke des Menschen und der Säugethiere, *Zeits. f. wiss. Zool.*, 1863, 13, 481-512. (*2*) Beobachtungen über die Thätigkeit des Trommelfellspanners bei Hund und Katze, *Arch. f. Physiol.*, 1878, 2, 312-319.

Herzog, H. (*1*) Das Knochenleitungsproblem; theoretische Erwägungen, *Zeits. f. Hals- Nasen- Ohrenheilk.*, 1926, 15, 300-306. (*2*) Die Mechanik der Knochenleitung im Modellversuch, *Zeits. f. Hals- Nasen- Ohrenheilk.*, 1930, 27, 402-408.

Holmgren, G. (*1*) Some experiences in surgery of otosclerosis, *Acta oto-laryngol.*, 1923, 5, 460-466. (*2*) Zur Chirurgie der Otosklerose, *Acta oto-laryngol.*, 1931, 15, 7-12. (*3*) The surgery of otosclerosis, *Ann. of Otol. Rhinol. Laryngol.*, 1937, 46, 3-12.

Hostinsky, Ottokar. *Die Lehre von den musikalischen Klängen*, 1879.

Howe, H. A. The relation of the organ of Corti to audioelectric phenomena in deaf albino cats, *Amer. J. Physiol.*, 1935, 111, 187-191.

Howe, H. A., and Guild, S. R. Absence of the organ of Corti and its possible relation to electric auditory nerve responses, *Anat. Rec.*, 1932-33, 55, Suppl., 20-21.

Huggins, W. H. Theory of cochlear frequency discrimination, *Quart. Prog. Report, Res. Lab. of Electronics, Mass. Inst. of Technol.*, Oct. 15, 1950, 54-59.

Huggins, W. H., and Licklider, J. C. R. Place mechanisms of auditory frequency analysis, *J. Acoust. Soc. Amer.*, 1951, 23, 290-299.

Hughson, W. What can be done for chronic progressive deafness? Rationale, technique, case reports and observations with grafts in the round window, *Laryngoscope*, 1938, 48, 533-551.

Hughson, W., and Crowe, S. J. (*1*) Function of the round window; an experimental study, *J. Amer. Med. Assoc.*, 1931, 96, 2027-2028. (*2*) Immobilization of the round window membrane: a further experimental study, *Ann. of Otol. Rhinol. Laryngol.*, 1932, 41, 332-348. (*3*) Experimental investigation of the physiology of the ear, *Acta oto-laryngol.*, 1933, 18, 291-339.

Hurst, C. H. A new theory of hearing, *Trans. Liverpool Biol. Soc.*, 1895, 9, 321-353.

Huschke, E. (*1*) Ueber die Gehörzähne, einen eigenthümlichen Apparat in der Schnecke des Vogelohrs, *Arch. f. Anat. Physiol.*, 1835, 335-346. (*2*) Lehre von den Eingeweiden und Sinnesorganen des menschlichen Körpers, in S. T. von Sömmerring's *Von Bau des menschlichen Körpers*, 1844, V.

Ingrassia, Giovan Filippo. [Not seen; quoted by Schelhammer.]

Jack, F. L. (*1*) Remarkable improvement of hearing by removal of the stapes, *Trans. Amer. Otol. Soc.*, 1892, 5, part 2, 284-305. (*2*) Further observations on removal of the stapes, *Trans. Amer. Otol. Soc.*, 1893, 5, part 3, 474-487. (*3*) Remarks on stapedectomy, *Trans. Amer. Otol. Soc.*, 1894, 6, part 1, 102-106.

Jenkins, G. J. Otosclerosis, certain clinical features and experimental operative procedures, *Trans. XVIIth Internat. Cong. Med. London: 1913*, Section XVI, Otology, part II, 1914, 609-617.

Jepsen, O. The threshold of the reflexes of the intratympanic muscles in a normal material examined by means of the impedance method, *Acta oto-laryngol.*, 1951, 39, 406-408.

Johansen, H. Relation of audiograms to the impedance formula, *Acta oto-laryngol.*, 1948, Suppl. 74, 65-75.

Jones, A. T. The discovery of difference tones, *Amer. Physics Teacher*, 1935, 3, 49-51.

Juers, A. L. Observations on bone conduction in fenestration cases, *Ann. of Otol. Rhinol. Laryngol.*, 1948, 57, 28-40.

Kahana, L., Rosenblith, W. A., and Galambos, R. Effect of temperature change on round window response in the hamster, *Amer. J. Physiol.*, 1950, 163, 213-223.

Kato, T. Zur Physiologie der Binnenmuskeln des Ohres, *Pflüg. Arch. ges. Physiol.*, 1913, 150, 569-625.

Keibs, L. Methode zur Messung von Schwellendrucken und Trommelfellimpedanzen in fortschreitenden Wellen, *Ann. d. Phys.*, 1936, 26, 585-608.

Keith, A. An appendix on the structures concerned in the mechanism of hearing, in Thomas Wrightson and Arthur Keith's *An enquiry into the analytical mechanism of the internal ear*, 1918, 161-254.

Kobrak, H. (*1*) Zur Physiologie der Binnenmuskeln des Ohres, *Passow-Schaefer's Beitr. zur Anat. Physiol. des Ohres*, 1930, 28, 138-160. (*2*) Influence of the middle ear on labyrinthine pressure, *Arch. of Otolaryngol.*, 1935, 21, 547-560. (*3*) Construction material of the sound conduction system of the human ear, *J. Acoust. Soc. Amer.*, 1948, 20, 125-130.

Koch, H. (*1*) Die Ewaldsche Hörtheorie, *Zeits. f. Sinnesphysiol.*, 1928, 59, 15-54. (*2*) Die Schwingungsformen der Basilarmembran, *Ber. ges. Physiol.*, 1931, 61, 362-363.

Köhler, W. Akustische Untersuchungen, I, *Zeits. f. Psychol.*, 1910, 54, 241-289.

Kucharski, P. Sur le loi d'excitation de l'oreille, *Compt. r. soc. biol.*, 1925, 92, 690-693.

Kuile, E. ter. (*1*) Die Uebertragung der Energie von der Grundmembran auf die Haarzellen, *Pflüg. Arch. ges. Physiol.*, 1900, 79, 146-157. (*2*) Die richtige Bewegungsform der Membrana basilaris, *Pflüg. Arch. ges. Physiol.*, 1900, 79, 484-509.

Kurtz, R. Zur Messung von Absorptions- und Empfindlichkeitskurven des menschlichen Ohres, *Akust. Zeits.*, 1938, 3, 74-79.

Lawrence, M. Recent investigations of sound conduction, I, The normal ear, *Ann. of Otol. Rhinol. Laryngol.*, 1950, 59, 1020-1036.

Lawrence, M., and Wever, E. G. Effects of oxygen deprivation upon the structure of the organ of Corti, *Arch. of Otolaryngol.*, 1952, 55, 31-37.

Lehmann, A. Ueber die Schwingungen der Basilarmembran und die Helmholtzsche Resonanztheorie, *Folia Neurobiol.*, 1910, 4, 116-132.

Lempert, J. (*1*) Improvement of hearing in cases of otosclerosis: a new, one stage surgical technic, *Arch. of Otolaryngol.*, 1938, 28, 42-97. (*2*) Endaural fenestration of external semicircular canal for restoration of hearing in cases of otosclerosis, *Arch. of Otolaryngol.*, 1940, 31, 711-779. (*3*) Endaural fenestration of the horizontal semicircular canal for otosclerosis, *Laryngoscope*, 1941, 51, 330-362. (*4*) Fenestra nov-ovalis; a new oval window for the improvement of hearing in cases of otosclerosis, *Arch. of Otolaryngol.*, 1941, 34, 880-912. (*5*) Lempert fenestra nov-ovalis with mobile stopple, *Arch. of Otolaryngol.*, 1945, 41, 1-41. (*6*) Lempert fenestra nov-ovalis for the restoration of practical unaided hearing in clinical otosclerosis: its present status, *Proc. Royal Soc. Med.*, 1948, 41, 617-630. (*7*) Physiology of hearing; what have we learned about it following fenestration surgery? *Arch. of Otolaryngol.*, 1952, 56, 101-113.

Lempert, J., Meltzer, P. E., Wever, E. G., and Lawrence, M. (*1*) The cochleogram and its clinical application; concluding observations, *Arch. of Otolaryngol.*, 1950, 51, 307-311. (*2*) Unpublished observations.

Lempert, J., Wever, E. G., and Lawrence, M. The cochleogram and its clinical applications; a preliminary report, *Arch. of Otolaryngol.*, 1947, 45, 61-67.

Lempert, J., Wever, E. G., Lawrence, M., and Meltzer, P. E. Perilymph: its relation to the improvement of hearing which follows fenestration of the vestibular labyrinth in clinical otosclerosis, *Arch. of Otolaryngol.*, 1949, 50, 377-387.

Lempert, J., and Wolff, D. Otosclerosis: theory of its origin and development, *Arch of Otolaryngol.*, 1949, 50, 115-155.

Lewis, D., and Reger, S. N. An experimental study of the rôle of the tympanic membrane and the ossicles in the hearing of certain subjective tones, *J. Acoust. Soc. Amer.*, 1933, 5, 153-158.

Lindsay, J. R., Kobrak, H., and Perlman, H. B. Relation of the stapedius reflex to hearing sensation in man, *Arch. of Otolaryngol.*, 1936, 23, 671-678.

Link, R. Zur Funktion der Membran des runden Fensters und des Schalleitungsapparates, *Der Hals- Nasen- Ohrenarzt*, 1943, 33, 241-267.

Loch, W. E. (*1*) The effect on hearing of experimental occlusion of the Eustachian tube in man, *Ann. of Otol. Rhinol. Laryngol.*, 1942, 51, 396-405. (*2*) Effect of experimentally altered air pressure in the middle ear on hearing acuity in man, *Ann. of Otol. Rhinol. Laryngol.*, 1942, 51, 995-1006.

Lorente de Nó, R. The reflex contractions of the muscles of the middle ear as a hearing test in experimental animals, *Trans. Amer. Laryngol. Rhinol. Otol. Soc.*, 1933, 26-42.

Lorente de Nó, R., and Harris, A. S. Experimental studies in hearing, I, The thresholds of the reflexes of the middle ear; II, The hearing-loss after extirpation of the tympanic membrane, *Laryngoscope*, 1933, 43, 315-326.

Lowy, K. Cancellation of the electrical cochlear response with air- and bone-conducted sound, *J. Acoust. Soc. Amer.*, 1942, 14, 156-158.

Lucae, A. Untersuchungen über die sogenannte "Knochenleitung" und deren Verhältniss zur Schallfortpflanzung durch die Luft, im gesunden und kranken Zustande, *Arch. f. Ohrenheilk.*, 1864, 1, 303-317.

Luescher, E. (*1*) Die Funktion des Musculus stapedius beim Menschen, *Zeits. f. Hals- Nasen- Ohrenheilk.*, 1929, 23, 105-132. (*2*) Untersuchungen über die Beeinflussung der Hörfähigkeit durch Trommelfellbelastung, *Acta oto-laryngol.*, 1939, 27, 250-266. (*3*) Experimentelle Trommelfellbelastungen und Luftleitungsaudiogramme mit allgemeinen Betrachtungen zur normalen und pathologischen Physiologie des Schalleitungsapparates, *Arch. f. Ohren- Nasen- Kehlkopfheilk.*, 1939, 146, 372-401.

Lux, F., see Budde, E.

Mach, E. (*1*) Zur Theorie des Gehörorgans, *Sitzungsber. Akad. Wiss. Wien*, math.-nat. Cl., Abt. 2, 1863, 48, 283-300. (*2*) Bemerkungen über die Accommodation des Ohres, *Sitzungsber. Akad. Wiss. Wien*, math.-nat. Cl., Abt. 2, 1865, 51, 343-346.

Mach, E., and Kessel, J. (*1*) Die Function der Trommelhöhle und der Tuba Eustachii, *Sitzungsber. Akad. Wiss. Wien*, math.-nat. Cl., Abt. 3, 1872, 66, 329-336. (*2*) Versuche über die Accommodation des Ohres, *Sitzungsber. Akad. Wiss. Wien*, math.-nat. Cl., Abt. 3, 1872, 66, 337-343. (*3*) Beiträge zur Topographie und Mechanik des Mittelohres, *Sitzungsber. Akad. Wiss. Wien*, math.-nat. Cl., Abt. 3, 1874, 69, 221-243.

MacNaughton-Jones, H. *Hearing and equilibrium*, 1940.

Magnus, A. Beiträge zur Anatomie des mittleren Ohres, *Arch. f. path. Anat. Physiol.*, 1861, 20, 79-132.

Mangold, E. (*1*) Willkürliche Kontraktionen des Tensor tympani und die graphische Registrierung von Druckschwankungen im äusseren Gehörgang, *Pflüg. Arch. ges. Physiol.*, 1913, 149, 539-587. (*2*) Das äussere und mittlere Ohr und ihre physiologischen Funktionen, in A. Bethe's *Handbuch der normalen und pathologischen Physiologie*, 11, *Receptionsorgane*, I, 1926, 421-431.

Mangold, E., and Ekstein, A. Reflektorische Kontraktionen des Tensor tympani beim Menschen, *Pflüg. Arch. ges. Physiol.*, 1913, 152, 589-615.

Martini, V. (*1*) Liberazione di sostanza acetilcolino-simile nell'orecchio interno durante la stimolazione sonora, *Archivio di scienze biol.*, 1941, 27, 94-100. (*2*) Sull'azione di farmaci del sistema nervoso autonomo sull'orecchio interno; azione dell'eserina, *Boll. soc. ital. biol. sper.*, 1942, 17, 12-13.

Mayer, O. Das anatomische Substrat der Altersschwerhörigkeit, *Arch. f. Ohren- Nasen- Kehlkopfheilk.*, 1919, 105, 1-13.

Meckel, Philippus Fridericus. *De labyrinthi auris contentis*, Argentorati, 1777.

Menzel, W. Messungen der Hörschwelle und der Trommelfellabsorption an gesunden und kranken Ohren, *Akust. Zeits.*, 1940, 5, 257-268.

Metz, O. (*1*) The acoustic impedance measured on normal and pathological ears, *Acta oto-laryngol.*, 1946, Suppl. 63, 254 pp. (*2*) Studies on the contraction of the tympanic muscles as indicated by changes in the impedance of the ear, *Acta oto-laryngol.*, 1951, 39, 397-405.

Milstein, T. N. Zur Technik der Einmauerung des runden Fensters und über die im Zusammenhang damit stehenden Fragen der Physiologie und Pathologie des Ohres, *Acta oto-laryngol.*, 1937, 25, 387-402.

Miot, C. De la mobilisation de l'étrier, *Rev. de laryngol. otol. rhinol.*, 1890, 10, 49-64, 83-99, 113-130, 145-162, 200-216.

Moe, C. R. An experimental study of subjective tones produced within the human ear, *J. Acoust. Soc. Amer.*, 1942, 13, 159-166.

Moos, S. Hyperostose des Schädels und der beide Felsenbeine, *Arch. f. Augen- Ohrenheilk.*, 1871, 2, 108-113.

Morgagni, J. Baptistae. *Epistolae anatomicae*, Patavii, 1764, *vi.*

Müller, Johannes. *Handbuch der Physiologie des Menschen*, 1837, II, Abt. 1, 393-483; trans. by W. Baly, *Elements of physiology*, 1843, 744-771.

Munson, W. A. (*1*) The growth of auditory sensation, *J. Acoust. Soc. Amer.*, 1947, 19, 584-591. (*2*) Sound and hearing, in Otto Glasser's *Medical physics*, 1950, II, 995-1006.

Munson, W. A., and Wiener, F. M. In search of the missing 6 db, *J. Acoust. Soc. Amer.*, 1952, 24, 498-501.

Nilsson, G. (*1*) The immediate improvement of hearing following fenestration operation, *Acta oto-laryngol.*, 1951, 39, 329-337. (*2*) The immediate improvement of hearing following the fenestration operation, *Acta oto-laryngol.*, 1952, Suppl. 98, 72 pp.

Nylén, B. Histopathological investigations on the localization, number, activity, and extent of otosclerotic foci, *Upsala Läkaref. förhandl.*, 1949, 54, 1-52.

Onchi, Y. A study of the mechanism of the middle ear, *J. Acoust. Soc. Amer.*, 1949, 21, 404-410.

Oppikofer, E. Über das Verkommen von Fett in der runden Fensternische, *Zeits. f. Ohrenheilk.*, 1917, 75, 50-65.

Payne, M. C., and Githler, F. J. Effects of perforations of the tympanic membrane on cochlear potentials, *Arch. of Otolaryngol.*, 1951, 54, 666-674.

Perlin, S. The width of the basilar membrane in the guinea pig, *J. Exper. Psychol.*, 1946, 36, 127-133.

Perlman, H. B., and Case, T. J. Latent period of the crossed stapedius reflex in man, *Ann. of Otol. Rhinol. Laryngol.*, 1939, 48, 663-675.

Perrault, Claude. *Du bruit*, in C. and P. Perrault's *Oeuvres diverses de physique et de mechanique*, Aleide, 1721, I. [*Du bruit* first appeared in 1680.]

Peterson, J. Origin of higher orders of combination tones, *Psychol. Rev.*, 1915, 22, 512-518.

Peterson, L. C., and Bogert, B. P. A dynamical theory of the cochlea, *J. Acoust. Soc. Amer.*, 1950, 22, 369-381.

Philip, R. La physiologie de l'oreille moyenne, *Rev. de laryngol. otol. rhinol.*, 1932, 53, 695-748.

Pierce, N. H., *et al. Otosclerosis, a resume of the literature to July, 1928*; 1929, II.

Pohlman, A. G. (*1*) The problem of middle ear mechanics, *Ann. of Otol. Rhinol. Laryngol.*, 1922, 31, 1-45. (*2*) The problem of middle ear mechanics, Chapter II, *Ann. of Otol. Rhinol. Laryngol.*, 1922, 31, 430-481. (*3*) Neue Betrachtungen über den Mechanismus des Gehörorgans, *Monatsschr. f. Ohrenheilk.*, 1932, 66, 1025-1057; and A reconsideration of the mechanics of the auditory apparatus, *J. of Laryngol. Otol.*, 1933, 48, 156-195.

Pohlman, A. G., and Kranz, F. W. The problem of middle ear mechanics, Chapter III, Binaural acuity for air and bone transmitted sound under varying conditions in the external auditory canal, *Ann. of Otol. Rhinol. Laryngol.*, 1925, 34, 1224-1248.

Politzer, Adam. (*1*) Beiträge zur Physiologie des Gehörorgans, *Sitzungsber. Akad. Wiss. Wien*, math.-nat. Cl., Abt. 2, 1861, 43, 427-438. (*2*) Untersuchungen über Schallfortpflanzung und Schallleitung im Gehörorgane im gesunden und kranken Zustande, *Arch. f. Ohrenheilk.*, 1864, 1, 59-73. (*3*) Untersuchungen über Schallfortpflanzung und Schallleitung im Gehörorgane im gesunden und kranken Zustande, II, Ueber Schallleitung durch die Kopfknochen, *Arch. f. Ohrenheilk.*, 1864, 1, 318-352. (*4*) Primary disease of the bony labyrinthine capsule, *Arch. of Otol.*, 1894, 23, 255-270. (*5*) *A text-book of the diseases of the ear*, trans. by M. J. Ballin and C. L. Heller, 5th ed., 1909.

Pollak, J. Ueber die Function des Musculus tensor tympani, *Medez. Jahrb.*, 1886, 82, 555-582.

Polvogt, L. M. Histologic variations in the middle and inner ear of patients with normal hearing, *Arch. of Otolaryngol.*, 1936, 23, 48-56.

Polvogt, L. M., and Bordley, J. E. Pathologic changes in the middle ear of patients with normal hearing and of patients with a conduction type of deafness, *Ann. of Otol. Rhinol. Laryngol.*, 1936, 45, 760-768.

Potter, A. B. Function of the stapedius muscle, *Ann. of Otol. Rhinol. Laryngol.*, 1936, 45, 638-643.

Preyer, William. *Akustische Untersuchungen*, Jena, 1879.

Ranke, Otto F. (*1*) *Die Gleichrichter-Resonanztheorie*, München, 1931. (*2*) Das Massenverhältnis zwischen Membran und Flüssigkeit im Innenohr, *Akust. Zeits.*, 1942, 7, 1-11. (*3*) Theory of operation of the cochlea: a contribution to the hydrodynamics of the cochlea, *J. Acoust. Soc. Amer.*, 1950, 22, 772-777. (*4*) Folgerungen aus der

Theorie der Flüssigkeitsschwingungen in der Schnecke, *Berichte ges. Physiol.*, 1950, 139, 183-184.

Rasmussen, H. (*1*) Studies on the effect upon the hearing through air conduction brought about by variations of the pressure in the auditory meatus, *Acta oto-laryngol.*, 1946, 34, 415-424. (*2*) Studies on the effect on the air conduction and bone conduction from changes in the meatal pressure in normal subjects and otosclerotic patients, *Acta oto-laryngol.*, 1948, Suppl. 74, 54-64.

Reboul, J. A. (*1*) *Le phénomène de Wever et Bray*, Montpellier, 1937. (*2*) Théorie des phénomènes mécaniques se passant dans l'oreille interne, *J. de phys. radium*, 1938, 9, 185-194. (*3*) Remarques sur les théories de l'audition, *Rev. de laryngol. otol. rhinol.*, 1939, 60, 144-158.

Reissner, Ernest. *De auris internae formatione*, Dorpat (Livonia), 1851.

Retzius, Gustav. (*1*) *Das Gehörorgan der Wirbelthiere*, 1884, II. (*2*) Die Endigungsweise des Gehörnerven, *Biol. Unters.*, 1892, 3, 29-36. (*3*) Weiteres über die Endigungsweise des Gehörnerven, *Biol. Unters.*, 1893, 5, 35-38.

Riemann, B. Mechanik des Ohres, *Zeits. f. rat. Med.*, 1867, 29, 129-143.

Rinne, A. Beiträge zur Physiologie des menschlichen Ohres, *Vierteljahrschr. f. prakt. Heilk.*, Prag, 1855, 45, No. 1, 71-123; 46, No. 2, 45-72.

Rudmose, H. W., Clark, K. C., Carlson, F. D., Eisenstein, J. C., and Walker, R. A. The effect of high altitude on the threshold of hearing, *J. Acoust. Soc. Amer.*, 1948, 20, 766-770.

Schaefer, K. L. Eine neue Erklärung der subjectiven Combinationstöne auf Grund der Helmholtz'schen Resonanzhypothese, *Pflüg. Arch. ges. Physiol.*, 1899, 78, 505-526.

Schapringer, A. Über die Contraction des Trommelfellspanners, *Sitzungsber. Akad. Wiss. Wien*, math.-nat. Kl., Abt. 2, 1870, 62, 571-574.

Schelhammer, Gunther C. *De auditu*, liber unus, 1684.

Schwalbe, G. *Lehrbuch der Anatomie der Sinnesorgane*, 1887.

Shaw, W. A., Newman, E. B., and Hirsh, I. J. The difference between monaural and binaural thresholds, *J. Exper. Psychol.*, 1947, 37, 229-242.

Siebenmann, Friedrich. (*1*) Beiträge zur funktionellen Prüfung des normalen Ohres, *Zeits. f. Ohrenheilk*, 1892, 22, 285-307. (*2*) Mittelohr und Labyrinth, in Karl von Bardeleben's *Handbuch der Anatomie des Menschen*, 1897, V, Abt. 2, Lief. 6, 195-324. (*3*) Traitement chirurgical de la sclérose otique, *XIII Internat. cong. med.*, Paris, 1900, sect. d'otol., 170-177; and *Ann. des mal. oreille*, 1900, 26, *ii*, 467-474.

Sivian, L. J., and White, S. D. On minimum audible sound fields, *J. Acoust. Soc. Amer.*, 1933, 4, 288-321.

Skoog, T. On the transmission of air borne sounds in the fenestrated ear in clinical otosclerosis, *Acta oto-laryngol.*, 1952, Suppl. 100, 134-143.

Smith, K. R. (*1*) Bone conduction during experimental fixation of the stapes, *J. Exper. Psychol.*, 1943, 33, 96-107. (*2*) The problem of stimulation deafness; II, Histological changes in the cochlea as a function of tonal frequency, *J. Exper. Psychol.*, 1947, 37, 304-317.

Smith, K. R., and Wever, E. G. The problem of stimulation deafness; III, The functional and histological effects of a high-frequency stimulus, *J. Exper. Psychol.*, 1949, 39, 238-241.

Sourdille, M. (*1*) Nouvelles techniques chirurgicales pour le traitement des surdités de conduction, *Ann. des mal. oreille*, 1930, 49, 10-21. (*2*) Nouvelles techniques operatoires pour le traitement des surdites chroniques progressives ou otosclerose, *Acta oto-laryngol.*, 1931, 15, 13-34. (*3*) New technique in the surgical treatment of severe and progressive deafness from otosclerosis, *Laryngoscope*, 1937, 47, 853-873.

Stuhlman, Otto, Jr. (*1*) The nonlinear transmission characteristics of the auditory ossicles, *J. Acoust. Soc. Amer.*, 1937, 9, 119-128. (*2*) The asymmetrical response of the human ear in relation to the problem of combination tones, *Bull. Amer. Musicol. Soc.*, 1941, 5, 19-20. (*3*) *An introduction to biophysics*, 1943.

Suib, Edwin. *The area of the cochlear duct, scala vestibuli, and scala tympani of the guinea pig*, Thesis, Princeton Univ., 1947.

Tasaki, I., Davis, H., and Legouix, J. P. The space-time pattern of the cochlear microphonics (guinea pig), as recorded by differential electrodes, *J. Acoust. Soc. Amer.*, 1952, 24, 502-519.

Tasaki, I., and Fernández, C. Modifications of cochlear microphonics and action potentials by KCl solution and by direct currents, *J. of Neurophysiol.*, 1952, 15, 497-512.

Thompson, E., Howe, H. A., and Hughson, W. Middle ear pressure and auditory acuity, *Amer. J. Physiol.*, 1934, 110, 312-319.

Toynbee, Joseph. *The diseases of the ear*, 1860.

Tröger, J. Die Schallaufnahme durch das äussere Ohr, *Phys. Zeits.*, 1930, 31, 26-47.

Tsukamoto, H. Zur Physiologie der Binnenohrmuskeln, *Zeits. f. Biol.*, 1934, 95, 146-154.

Tsukamoto, H., Shinomiya, M., and Toida, M. Binnenohrmuskeln und Trommelfell, *Arch. f. Ohrenheilk.*, 1936, 141, 185-194.

Tunturi, A. R. (*1*) Audiofrequency localization in the acoustic cortex of the dog, *Amer. J. Physiol.*, 1944, 141, 397-403. (*2*) Further afferent connections to the acoustic cortex of the dog, *Amer. J. Physiol.*, 1945, 144, 389-394. (*3*) A study on the pathway from the medial geniculate body to the acoustic cortex in the dog, *Amer. J. Physiol.*, 1946, 147, 311-319.

Turnbull, W. W. Pitch discrimination as a function of tonal duration, *J. Exper. Psychol.*, 1944, 34, 302-316.

Valsalva, Antonio Maria. *De aure humana tractatus*, Rhenum, 1707.

Van Eyck, M. (*1*) Du rôle de la fenêtre ronde dans la transmission des ondes sonores après la fenestration du labyrinth, *Ann d'oto-laryngol.*, 1950, 67, 777-781. (*2*) Le role de la fenêtre ronde dans la transmission des ondes sonores a la cochlée après fenestration du labyrinthe, *Rev. de laryngol. otol. rhinol.*, Suppl., 1951, 72, 222-226.

Varolius, Constantius. *Anatomiae sive de resolutione corporis humani*, libri IIII, Francofurti, 1591.

Vesalius, Andreas. *De humani corporis fabrica*, libri septum, Basileae, 1543.

Waar, A. C. H. Mikroskopische Wahrnehmungen der Funktion der Mittelohrmuskeln beim Menschen, *Acta oto-laryngol.*, 1923, 5, 335-358.

Waetzmann, E. Absorptionsmessungen am Trommelfell mit der Schusterschen Brücke, *Akust. Zeits.*, 1938, 3, 1-6.

Waetzmann, E., and Keibs, L. Hörschwellenbestimmungen mit dem Thermophon und Messungen am Trommelfell, *Ann. d. Phys.*, 1935, 26, 141-144.

Wagner, V. R. *A study of temperature effects on cochlear potentials*, Thesis, Princeton Univ., 1949.

Watt, Henry J. (*1*) Psychological analysis and theory of hearing, *Brit. J. Psychol.*, 1914, 7, 1-43. (*2*) *The psychology of sound*, 1917.

Weber, Eduard F. (*1*) Über den Zweck der fenestra rotunda und die Vorrichtung der Schnecke im Gehörorgane der Menschen und der Säugethiere, *Amtlicher Bericht Versamml. deutsch. Naturf. Aerzte*, 19te Versamml. Gesellsch. deutsch. Naturf. Aerzte, 1841 (Braunschweig, 1842), 83-84. (*2*) Über den Mechanismus des Gehörorgans, *Berichte der Gesellsch. Wiss. Leipzig*, math.-phys. Cl., 1851, 3, 29-31.

Weber, Ernst Heinrich. *De pulsu, resorptione, auditu et tactu*, Lipsiae, 1834, 25-44.

Weber-Liel, F. E. (*1*) Zur Function der Membran des runden Fensters (Membrana tympani secundaria), *Centralbl. f. med. Wiss.*, 1876, 14, No. 2, 17-20. (*2*) Weitere anatomische und physicalische Untersuchungen über die Membrana tympani secundaria, *Monatsschr. f. Ohrenheilk.*, 1876, 10, 53-57, 72-76.

Wegel, R. L. Physical data and physiology of excitation of the auditory nerve, *Ann. of Otol. Rhinol. Laryngol.*, 1932, 41, 740-779.

Wegel, R. L., and Lane, C. E. The auditory masking of one pure tone by another and its possible relation to the dynamics of the inner ear, *Phys. Rev.*, 1924, 23, 266-285.

Wever, E. G. (*1*) The width of the basilar membrane in man, *Ann. of Otol. Rhinol. Laryngol.*, 1938, 47, 37-47. (*2*) The electrical responses of the ear, *Psychol. Bull.*, 1939, 36, 143-187. (*3*) The designation of combination tones, *Psychol. Rev.*, 1941, 48, 93-104. (*4*) The acoustic characteristics of the ear, in P. L. Harriman's *Twentieth century psychology*, 1946, 371-386. (*5*) *Theory of hearing*, 1949. (*6*) Recent investigations of sound conduction; II, The ear with conductive impairment, *Ann. of Otol. Rhinol. Laryngol.*, 1950, 59, 1037-1061. (*7*) Some remarks on the modern status of auditory theory, *J. Acoust. Soc. Amer.*, 1951, 23, 287-289.

Wever, E. G., and Bray, C. W. (*1*) The nature of acoustic response: the relation between sound intensity and the magnitude of responses in the cochlea, *J. Exper. Psychol.*, 1936, 19, 129-143. (*2*) The nature of bone conduction as shown in the electrical response of the cochlea, *Ann. of Otol. Rhinol. Laryngol.*, 1936, 45, 822-830. (*3*) The tensor tympani muscle and its relation to sound conduction, *Ann. of Otol. Rhinol. Laryngol.*, 1937, 46, 947-961. (*4*) The nature of acoustic response: the relation between stimulus intensity and the magnitude of cochlear responses in the cat, *J. Exper. Psychol.*, 1938, 22, 1-16. (*5*) Distortion in the ear as shown by the electrical responses of the cochlea, *J. Acoust. Soc. Amer.*, 1938, 9, 227-233. (*6*) The stapedius muscle in relation to sound conduction, *J. Exper. Psychol.*, 1942, 31, 35-43.

Wever, E. G., Bray, C. W., and Lawrence, M. (*1*) The locus of distortion in the ear, *J. Acoust. Soc. Amer.*, 1940, 11, 427-433. (*2*) The origin of combination tones, *J. Exper. Psychol.*, 1940, 27, 217-226. (*3*) A quantitative study of combination tones, *J. Exper. Psychol.*, 1940, 27, 469-496. (*4*) The interference of tones in the cochlea, *J. Acoust. Soc. Amer.*, 1940, 12, 268-280. (*5*) The nature of cochlear activity after death, *Ann. of Otol. Rhinol. Laryngol.*, 1941, 50, 317-329. (*6*) The effect of middle ear pressure upon distortion, *J. Acoust. Soc. Amer.*, 1941, 13, 182-187. (*7*) The effects of pressure in the middle ear, *J. Exper. Psychol.*, 1942, 30, 40-52.

Wever, E. G., and Lawrence, M. (*1*) Tonal interference in relation to cochlear injury, *J. Exper. Psychol.*, 1941, 29, 283-295. (*2*) The functions of the round window, *Ann of Otol. Rhinol. Laryngol.*, 1948, 57, 579-589. (*3*) The patterns of response in the cochlea, *J. Acoust. Soc. Amer.*, 1949, 21, 127-134. (*4*) The transmission properties of the middle ear, *Ann. of Otol. Rhinol. Laryngol.*, 1950, 59, 5-18. (*5*) The transmission properties of the stapes, *Ann. of Otol. Rhinol. Laryngol.*, 1950, 59, 322-330. (*6*) The acoustic pathways to the cochlea, *J. Acoust. Soc. Amer.*, 1950, 22, 460-467. (*7*) The place principle in auditory theory, *Proc. Nat. Acad. Sci.*, Washington, 1952, 38, 133-138. (*8*) Sound conduction in the cochlea, *Ann. of Otol. Rhinol. Laryngol.*, 1952, 61, 824-835.

Wever, E. G., Lawrence, M., Hemphill, R. W., and Straut, C. B. The effects of oxygen deprivation upon the cochlear potentials, *Amer. J. Physiol.*, 1949, 159, 199-208.

Wever, E. G., Lawrence, M., and Smith, K. R. (*1*) The middle ear in sound conduction, *Arch. of Otolaryngol.*, 1948, 48, 19-35. (*2*) The effects of negative air pressure in the middle ear, *Ann. of Otol. Rhinol. Laryngol.*, 1948, 57, 418-428.

Wever, E. G., and Smith, K. R. The problem of stimulation deafness; I, Cochlear impairment as a function of tonal frequency, *J. Exper. Psychol.*, 1944, 34, 239-245.

Whittle, G. T. *The drum–oval window relationship*, Thesis, Princeton Univ., 1946.

Wien, M. Ein Bedenken gegen die Helmholtzsche Resonanztheorie des Hörens, *Festschrift Adolph Wüllner*, Leipzig, 1905, 28-35.

Wiener, F. M. (*1*) On the diffraction of a progressive sound wave by the human head, *J. Acoust. Soc. Amer.*, 1947, 19, 143-146. (*2*) Sound diffraction by rigid spheres and circular cylinders, *J. Acoust. Soc. Amer.*, 1947, 19, 444-451.

Wiener, F. M., and Ross, D. A. The pressure distribution in the auditory canal in a progressive sound field, *J. Acoust. Soc. Amer.*, 1946, 18, 401-408.

Wiggers, H. C. The functions of the intra-aural muscles, *Amer. J. Physiol.*, 1937, 120, 771-780.

Wilkinson, George, and Gray, A. A. *The mechanism of the cochlea*, 1924.

Wilska, A. Eine Methode zur Bestimmung der Hörschwellenamplituden des Trommelfells bei verschiedenen Frequenzen, *Skand. Arch. Physiol.*, 1935, 72, 161-165.

Wing, K., Harris, J. D., Stover, A. D., and Brouillette, J. H. Effects of changes in arterial oxygen and carbon dioxide upon cochlear microphonics, *Med. Res. Lab. Report, U.S. Submarine Base*, New London, 1952, 11, 37 pp.

Wishart, D. E. S. Discussion of W. Hughson's paper, What can be done for chronic progressive deafness? *Laryngoscope*, 1938, 48, 545-549.

Woodman, De G. Congenital atresia of the auditory canal, *Arch. of Otolaryngol.*, 1952, 55, 172-178.

Woods, R. R. Bone conduction in otosclerosis, *Arch. of Otolaryngol.*, 1950, 51, 485-499.

Woolsey, C. N., and Walzl, E. M. Topical projection of nerve fibers from local regions of the cochlea to the cerebral cortex of the cat, *Bull. Johns Hopkins Hosp.*, 1942, 71, 315-344.

Wrightson, Thomas, and Keith, Arthur. *An enquiry into the analytical mechanism of the internal ear*, 1918.

Young, T. Outlines of experiments and inquiries respecting sound and light, *Phil. Trans. Royal Soc. London*, 1800, 90, part 1, 106-150.

Zalewski, T. Experimentelle Untersuchungen über die Resistenzfähigkeit des Trommelfells, *Zeits. f. Ohrenheilk.*, 1906, 52, 109-128.

Zimmermann, G. (*1*) *Die Mechanik des Hörens und ihre Störungen*,

1900. (*2*) Die Akkommodation im Ohr, *Arch. f. Anat. Physiol.*, 1908, 23-42.

Zöllner, F. Versteckte Störungen des Druckausgleiches im Mittelohr, *Arch. f. Ohren- Nasen- Kehlkopfheilk.*, 1939, 146, 12-15.

Zwislocki, Józef. (*1*) Über die mechanische Klanganalyse des Ohrs, *Experientia*, 1946, 2, 415-417. (*2*) *Theorie der Schneckenmechanik*, Solothurn, 1948; and *Acta oto-laryngol.*, 1948, Suppl. 72, 76 pp. (*3*) Theory of the acoustical action of the cochlea, *J. Acoust. Soc. Amer.*, 1950, 22, 778-784.

Index